SYMBOLS USED IN THIS TEXT

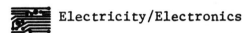 Industrial and Construction Trades

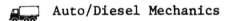 Electricity/Electronics

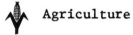 Agriculture

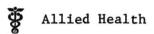 Auto/Diesel Mechanics

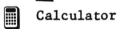 Allied Health

Calculator

ELEMENTARY TECHNICAL MATHEMATICS

FOURTH EDITION

DALE EWEN

Parkland Community College
Champaign, Illinois

C. ROBERT NELSON

Champaign Centennial High School
Champaign, Illinois

Wadsworth Publishing Company
Belmont, California
A Division of Wadsworth, Inc.

Mathematics Editor: Anne Scanlan-Rohrer
Production: Greg Hubit Bookworks
Editorial Assistant: Ruth Singer
Print Buyer: Ruth Cole
Technical Illustrator: Scientific Illustrators

Printed in the United States of America

1 2 3 4 5 6 7 8 9 10—91 90 89 88 87

ISBN 0-534-07626-2

Library of Congress Cataloging in Publication Data

Ewen, Dale, 1941-
    Elementary technical mathematics.

    Includes index.
    1. Mathematics—1961-   . I. Nelson, C. Robert
II. Title.
QA39.2.E9396 1987   512'.1   86-15897
ISBN 0-534-07626-2

PREFACE

Elementary Technical Mathematics is a textbook for students in a technical or trade program. As in the previous editions, this text is designed for the student who has a minimal mathematics background but wants to learn a technical skill. To become a technician, he or she must first become proficient in the basic mathematical skills. We have geared the format to the reading level of most technical students and to their dependence upon visual images to stimulate the problem-solving process. Arithmetic, measurement, and basic algebraic skills are developed with a large number of various applications integrated throughout while emphasizing that these skills are essential for student success in technical as well as supportive courses.

The basic calculator operations are included in the appendix for use as needed by the instructor while the more advanced operations are developed and integrated throughout the text. The authors' basic premise regarding the student use of the calculator is that the student should be proficient in the use of the basic mathematical skills before being introduced to the calculator. We suggest using a calculator after Chapter 3. Sections in which the use of a calculator is especially helpful are identified with a symbol of a calculator in the margin. In all other sections, either the mathematics is basic and students need to become proficient in its skills, or the numbers are such that the mathematics can and should be done mentally.

We are grateful for the courtesy of The L. S. Starrett Company for the photographs of their instruments in Chapter 5. The authors especially thank the many instructors who used the earlier editions and who offered suggestions.

In particular, we thank the following reviewers: Ray Collings, Tri-County Technical College; David C. Ford, Cayuga Community College; Larry Lance, Columbus Technical Institute; Edwin McCravy, Midlands Technical College; John Patterson, Stark Technical College; Robert E. Preller, Illinois Central College; Joan Shack, Hudson Valley Community College; Jennie Thompson, Leeward Community College; Theodora Tsakiris, Maricopa Technical Community College.

We thank our colleagues at Parkland College and the many Parkland College students for their valued assistance. If anyone wishes to correspond with us regarding suggestions, criticisms, questions, or errors, please feel free to contact Dale Ewen directly at Parkland College, 2400 W. Bradley, Champaign, Illinois 61821, or through Wadsworth.

We acknowledge and thank the excellent editorial production staff of Greg Hubit Bookworks and also George Morris for his usual outstanding artwork. Finally, we are especially grateful to our families for their patience, understanding, and encouragement, and to Joyce Ewen for her usual excellent proofing assistance.

Dale Ewen

C. Robert Nelson

CONTENTS

1

BASIC CONCEPTS

1.1   Addition and Subtraction

The *positive integers* are the whole numbers 1, 2, 3, 4, 5, 6, etc. They can also be written as +1, +2, +3, etc., but usually the *positive* (+) sign is omitted. You must be able to add, subtract, multiply, and divide positive integers. The exercises in this chapter will help you review these fundamental operations.

The value of any digit in a number is determined by its place in the particular number. Each place represents a certain power of ten. For example, 2354 means 2 thousands plus 3 hundreds plus 5 tens plus 4 ones. We shall use the following notation for powers of 10:

$10^0 = 1$

$10^1 = 10$

$10^2 = 10 \times 10 = 100$ (the second power of 10)

$10^3 = 10 \times 10 \times 10 = 1000$ (the third power of 10)

$10^4 = 10 \times 10 \times 10 \times 10 = 10,000$ (the fourth power of ten)

etc.

In the number 236,895,174 each digit has been multiplied by some power of 10, as shown below.

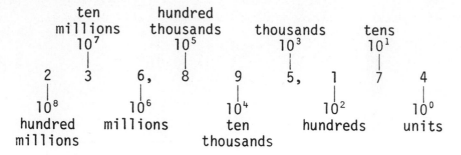

The "+" (plus) symbol is the sign for addition, as in the expression 5 + 7. The result of adding the numbers, in this case 12, is called the *sum*. Integers are added in columns with digits representing like powers of ten in the same vertical line. (Vertical means up and down.)

Example 1. Add 238 + 15 + 9 + 3564.
Think of columns 10^3, 10^2, 10^1, and 10^0, as shown.

10^3	10^2	10^1	10^0
	2	3	8
		1	5
			9
3	5	6	4
3	8	2	6

Subtraction is the inverse operation of addition. Therefore subtraction can be thought of in terms of addition. The "−" (minus) sign is the symbol for subtraction. The quantity 5 − 3 can be thought of as "what number added to 3 gives 5?" The result of subtraction is called the *difference*.

To check a subtraction, add the difference to the second number. If the sum is equal to the first number, the subtraction has been done correctly.

Example 2. Subtract: 2843   first number
                      1928   second number
                       915   difference
                      2843   This sum equals the first
                             number; hence, 915 is the
                             correct difference.

In order to communicate about problems in electricity, technicians have developed a "language" of their own. It is a picture language using symbols and diagrams. The symbols used most often appear in Table 5 of the Appendix.

The circuit diagram is the most common and useful way to show a circuit. Note how each component (part) of the picture (Fig. 1.1a) is represented by its symbol in the circuit diagram (Fig. 1.1b) in its same relative position.

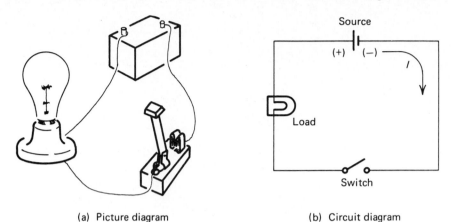

(a) Picture diagram   (b) Circuit diagram

Figure 1.1

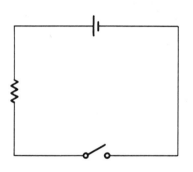

Figure 1.2

The light bulb may be represented as a resistance. Then the circuit diagram in Fig. 1.1b would appear as in Fig. 1.2, where

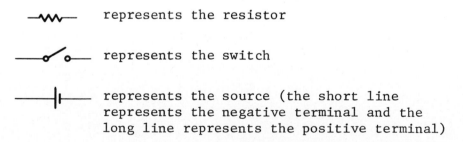

represents the resistor

represents the switch

represents the source (the short line represents the negative terminal and the long line represents the positive terminal)

There are two basic types of electrical circuits: series and parallel. A fuse in a house is wired in series with its outlets. The outlets themselves are wired in parallel.

An electrical circuit with only one path for the current, I, to flow is called a *series* circuit (Fig. 1.3). An electrical circuit with more than one path for the current to flow is called a *parallel* circuit (Fig. 1.4).

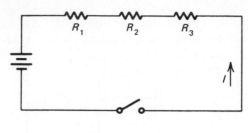

Figure 1.3

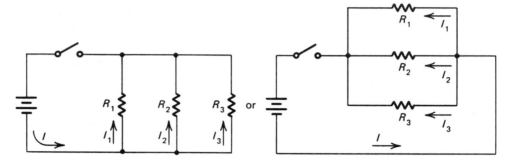

Figure 1.4

Example 3. In a series circuit the total resistance is
equal to the sum of all the resistances in the circuit.
Find the total resistance in the series circuit in
Fig. 1.5.

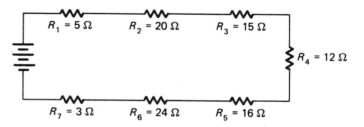

Figure 1.5

The total resistance is   5 Ω
                 20 Ω
                 15 Ω
                 12 Ω
                 16 Ω
                 24 Ω
                  3 Ω
                95 Ω

Example 4. The number of studs needed in an exterior
wall can be estimated by finding the number of linear
feet (ft) of the wall (outside perimeter). How many
studs are needed for the exterior walls in the building
in Fig. 1.6?

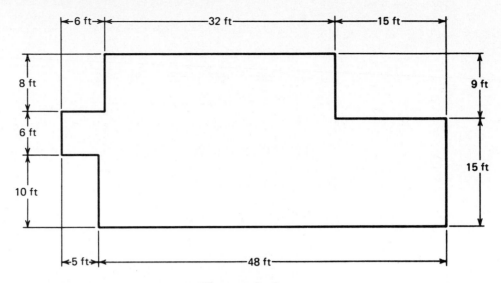

Figure 1.6

The outside perimeter of the building is the sum of the lengths of the sides of the building:

```
  48 ft
  15 ft
  15 ft
   9 ft
  32 ft
   8 ft
   6 ft
   6 ft
   5 ft
  10 ft
 154 ft
```

Therefore, 154 studs are needed in the outside wall.

EXERCISES 1.1   *Add:*

1. 832 + 9 + 56 + 2358

2. 64 + 7 + 97 + 543

3. 16 + 2400 + 47 + 3

4. 324 + 973 + 66 + 9430

5. 245
    1053

6. 619
    297
    3142

7. 384
    291
    147
    632

8. 981
    472
    385
    673
    942

9.   78
    107
     45
    217
      9
    123

10. 654
    376
    190
    521
     79
    100

11. 1082 + 781 + 7 + 55

12. 197 + 1072 + 10,877 + 15,532 + 768,098

13. 8 + 154 + 876 + 39 + 80 + 5

14. 4500 + 80 + 12 + 300 + 34

15. 88,000 + 9400 + 16 + 180

16. 160,000 + 19,000 + 4,160,000 + 506,000

Subtract and check:

17. 283	18. 603	19. 7561	20. 4000
152	438	2397	702

21. 243,178	22. 2843	23. 4095	24. 17,656
98,205	1298	2854	11,178

25. 98,405 − 72,397   26. 417,286 − 287,156

27. 111,779 − 93,743   28. 983,679 − 447,699

29. 3600	30. 48,007	31. 4000	32. 60,000
1481	27,909	1180	9,876

Find the total resistance in each series circuit:

33.

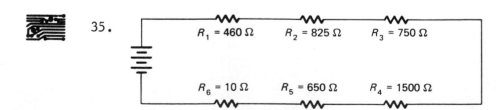

$R_1 = 48\ \Omega$ $R_2 = 30\ \Omega$

$R_4 = 5\ \Omega$ $R_3 = 115\ \Omega$

34.

$R_1 = 55\ \Omega$ $R_2 = 190\ \Omega$

$R_3 = 8\ \Omega$

35.

$R_1 = 460\ \Omega$ $R_2 = 825\ \Omega$ $R_3 = 750\ \Omega$

$R_6 = 10\ \Omega$ $R_5 = 650\ \Omega$ $R_4 = 1500\ \Omega$

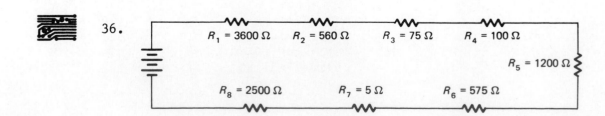
36.

$R_1 = 3600\ \Omega$ $R_2 = 560\ \Omega$ $R_3 = 75\ \Omega$ $R_4 = 100\ \Omega$

$R_5 = 1200\ \Omega$

$R_8 = 2500\ \Omega$ $R_7 = 5\ \Omega$ $R_6 = 575\ \Omega$

37. How many studs are needed for the exterior walls in the building below? (See Example 4.)

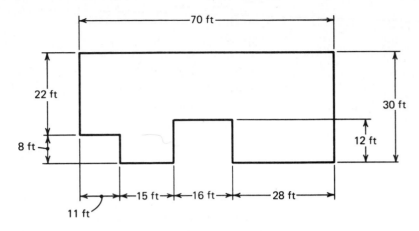

38. Cut a pipe 24 ft long into four pieces, the first 4 ft long, the second 5 ft long, and the third 7 ft long. What is the length of the remaining piece? (Assume no waste from cutting.)

39. The following are the daily cash and credit receipts at Ralph's Service for the week. What are the total cash, total credit, and total receipts for the week?

	Cash	Credit
Sunday	$253	$592
Monday	426	730
Tuesday	216	829
Wednesday	278	637
Thursday	392	634
Friday	415	731
Saturday	237	738

40. During the year a farmer had the following income and expenses. Find the total income, total expenses, and his profit.

	Income
Corn sales	$10,423
Soybean sales	27,771
Wheat sales	6,310
Hog sales	102,507
Beef sales	1,705
Custom work	861
State gas tax refund	120
Federal gas tax refund	96

Expenses

Machinery depreciation	$15,101
Machinery repairs	8,067
Gas and oil	5,494
Fertilizer	4,572
Other agricultural chemicals	2,306
Feed	28,858
Building depreciation	1,337
Labor	12,604
Real estate taxes	5,214
Machine hire	387
Veterinary and medicine	4,476
Utilities	760
Interest	12,548

---

 41. A patient is given 100 milligrams (mg) of Demerol at 8:30 a.m., 75 mg at 1:00 p.m., and 50 mg at 5:30 p.m. Find the total amount given.

 42. A patient on a liquid diet consumes 250 cubic centimetres (cm^3)* of soup, 120 cm^3 of gelatin, and 75 cm^3 of tea. Find the total amount of fluids for this meal.

 43. Total the following input and output (I-O) entries for a patient. Input: 300 cm^3, 550 cm^3, 150 cm^3, 75 cm^3, 150 cm^3, 450 cm^3, 250 cm^3. Output: 325 cm^3, 150 cm^3, 525 cm^3, 250 cm^3, 175 cm^3.

 44. A patient is urged by his doctor to drink 3000 cm^3 of fluids each day. Today he had 480 cm^3 for breakfast, 300 cm^3 at midmorning, 650 cm^3 for lunch, 350 cm^3 at midafternoon, and 850 cm^3 for dinner. How much must he drink before bedtime?

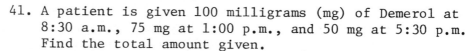

1.2 Multiplication and Division

Repeated addition of the same number can be shortened by multiplication. The "×" (times) and the "•" (dot) are used to indicate multiplication. When adding five 2-by-4s, each 7 ft long, we have 7 ft + 7 ft + 7 ft + 7 ft + 7 ft = 35 ft of 2-by-4s. In multiplication this would be 5 × 7 ft = 35 ft. The 5 and 7 are called *factors* and 35 is called the *product*. Computing areas, volumes, forces, and distances are examples which require skills in multiplication.

---

*Although cm^3 is the "official" metric abbreviation, and will be used throughout this book, some readers may be more familiar with the abbreviation "cc," which is still used in some medical and allied health areas.

Example 1. Multiply   358
                       18
                      ‾‾‾‾
                      2864
                       358
                      ‾‾‾‾
                      6444

Division is the inverse operation of multiplication. The following are all the symbols for division: "÷," "⌐‾‾," "/," and "—." The quantity 15 ÷ 5 can also be thought of as "what times 5 gives 15?" The answer to this question is 3, which is 15 divided by 5. The result, 3, is called the *quotient*. The number you divide by, 5, is called the *divisor*.

Example 2. Divide: 14 ←quotient
                      6⌐84‾
        divisor ↱   6
                     ‾‾‾
                      24
                      24
                     ‾‾‾
                       0 ←remainder

Example 3. Divide: 16 ←quotient
                      7⌐115‾
        divisor ↱   7
                     ‾‾‾
                      45
                      42
                     ‾‾‾
                       3 ←remainder

The *remainder* (when not 0) is usually written one of two ways. One way is with an "r" preceding it, as in Example 4(a). The other is with the remainder written over the divisor with a fraction bar between them, as in Example 4(b). (Fractions are discussed in Chapter 2.)

Example 4(a).   22 r 6   Example 4(b). $22\frac{6}{24}$
                 24⌐534‾   24⌐534‾
                   48   48
                  ‾‾‾‾   ‾‾‾‾
                    54   54
                    48   48
                   ‾‾‾   ‾‾‾
                     6   6

Example 5. Find the total piston displacement of an eight-cylinder engine. Each piston displaces a volume of 41 cubic inches (in^3).

Total displacement = $8 \times 41\ in^3 = 328\ in^3$
                     or 328 cid (cubic inch displacement)

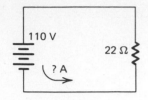

Figure 1.7

<u>Example 6</u>. Ohm's law states that in a simple electrical circuit the current (measured in amps, A) equals the voltage (measured in volts, V) divided by the resistance (measured in ohms, Ω). Find the current in the circuit of Fig. 1.7.

The current equals $\frac{110}{22}$ = 5 A.

<u>Example 7</u>. A corn planter costs $12,600. It has a 10-year life and a salvage value of $1,500. What is the annual depreciation? (Use the straight-line depreciation method.)

The straight-line depreciation method means that the difference between the cost and the salvage value is divided evenly over the life of the item. In this case, the difference between the cost and the salvage value is:

$$\begin{array}{rl} \$12,600 & \text{Cost} \\ -\ \ 1,500 & \text{Salvage} \\ \hline \$11,100 & \text{Difference} \end{array}$$

This difference divided by 10, the life of the item, is $1110. This is the annual depreciation.

EXERCISES 1.2   *Multiply:*

1. 256
   7
	2. 156
	  14
		3. 567
		  48

4. 8374
  203
	5. 3186
	  257
		6. 62,187
		  234

7. 71,263 × 255   8. 98,447 × 536   9. 37,050 × 716

10. 1520 × 320   11. 6800 × 5200   12. 30,010 × 4080

Divide: (Use the remainder form with r.)

13. 4⟌7236   14. 7⟌22,864   15. 8⟌4608

16. 5⟌308,736   17. 6754 ÷ 6   18. 56,772 ÷ 8

19. 4668 ÷ 12   20. 15,648 ÷ 36   21. 26⟌51,482

22. 108⟌376,544   23. 67,560 ÷ 80

24. $\dfrac{188,000}{120}$

25. A car uses gas at a rate of 22 miles per gallon (mi/gal) and has a 24-gal tank. How far can it travel on one tank of gas?

26. A car uses gas at a rate of 8 kilometres per litre (km/L) and has a 75-L tank. How far can it travel on one tank of gas?

27. Find the total piston displacement of an eight-cylinder engine. Each piston displaces 425 cm^3.

 28. An eight-cylinder engine has a total displacement of 400 in^3. Find the displacement in each piston.

 29. A four-cylinder engine has a total displacement of 1300 cm^3. Find the displacement in each piston.

 30. A car travels 1278 mi and uses 71 gal of gasoline. Calculate its mileage in miles per gallon.

 31. A car travels 2352 km and uses 294 L of gasoline. Calculate its "mileage" in kilometres per litre.

 32. A garage buys heater hose in 20-metre (m) rolls. Repairing the average car requires 3 m of hose. How many cars can be repaired with one roll?

 33. Inventory shows the following lengths of 3-in. steel pipe:

$$
\begin{array}{rl}
5 & \text{pieces } 18 \text{ ft long} \\
42 & \text{pieces } 15 \text{ ft long} \\
158 & \text{pieces } 12 \text{ ft long} \\
105 & \text{pieces } 10 \text{ ft long} \\
79 & \text{pieces } 8 \text{ ft long} \\
87 & \text{pieces } 6 \text{ ft long}
\end{array}
$$

What is the total linear feet of pipe in inventory?

 34. An order of lumber contains 36 boards 12 ft long, 28 boards 10 ft long, 36 boards 8 ft long, and 12 boards 16 ft long. How many boards are contained in the order? How many linear feet of lumber are contained in the order?

 35. A steel I beam weighs 74 lb per linear foot. The length is 21 ft. How much does the beam weigh?

 36. A shipment contains a total of 5232 linear feet of steel pipe. Each pipe is 12 ft long. How many pieces should be expected?

 37. How should a window 75 in. wide be placed so that it is centered on a 17-ft-5-in. wall?

 38. A farmer expects to yield 165 bu/acre from 260 acres of corn. If the corn is stored, how many bushels of storage is needed?

 39. A farmer harvests 6864 bushels (bu) of soybeans from 156 acres. What is his yield per acre?

 40. At $3125 per acre, what is the value of a 1275-acre Illinois farm?

 41. A railroad freight car can hold 1825 bu of soybeans. How many freight cars are needed to haul the expected 400,000 bu at a local grain elevator?

42. A railroad freight car can hold 2035 bu of corn. How many freight cars are needed to haul the expected 12,000,000 bu at a local large grain elevator?

43. On a given day eight steers weighed: 856 lb, 754 lb, 1044 lb, 928 lb, 888 lb, 734 lb, 953 lb, and 891 lb. (a) What is the average weight? (b) In 36 days, 4320 lb of feed is consumed. What is the average feed consumption per day per steer?

44. What is the weight (in tons) of a stack of hay bales 6 bales wide, 110 bales long, and 15 bales high? The average weight of each bale is 80 lb (1 ton = 2000 lb.)

45. From a 34 acre field, 92,480 lb of oats are harvested. How many bushels per acre are yielded? (1 bu of oats weighs 32 lb.)

46. A standard bale of cotton weighs approximately 500 lb. How many bales are contained in 15 tons of cotton?

47. A tractor costs $45,000. It has a ten-year life and a salvage value of $3000. What is the annual depreciation? (Use the straight-line depreciation method. See Example 7.)

48. How much pesticide powder would you put in a 400-gal spray tank if 10 gal and 2 lb of pesticide are applied per acre?

Using Ohm's law, find the current in each simple electrical circuit: (See Example 6.)

 49.
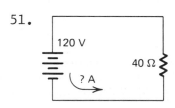

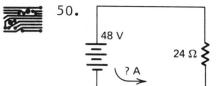 50.

 51.
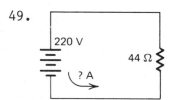

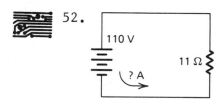 52.

Ohm's law, in another form, states that in a simple circuit the voltage (measured in volts, V) equals the current (measured in amps, A) times the resistance (measured in ohms, Ω). Find the voltage in each simple electrical circuit:

 53.
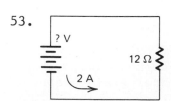

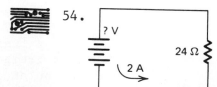 54.

 55.
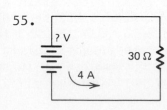

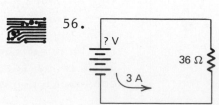 56.

57. A hospital dietitian determines that each patient needs 4 ounces (oz) of orange juice. How many ounces of orange juice must be prepared for 220 patients?

 58. During 24 hours a patient is given three phenobarbital tablets of 25 mg each. How much phenobarbital does the patient receive?

 59. To give 800 mg of quinine sulfate from 200-mg tablets, how many tablets would you use?

 60. A nurse used two 4-grain potassium permanganate tablets in the preparation of a medication. How much potassium permanganate did she use?

 61. The doctor orders 100 mg of Dramamine. The tablets on hand contain 25 mg. How many tablets should the nurse give?

1.3 Parentheses and Order of Operations

Just as we use periods, commas, and other punctuation marks to help make sentences more readable, we use *parentheses* in mathematics to help clarify the meaning of mathematical expressions. Parentheses not only give an expression a particular meaning, they also specify the order to be followed in evaluating and simplifying expressions.

What is the value of 8 − 3 · 2? Is it 10? Is it 2? Or is it some other number? It is very important that each arithmetic operation have only *one* value, and that we each perform the exact same operations on a given computation or problem. As a result, the following order of operations procedure is followed by all.

Order of Operations

1. Always do first what is indicated in parentheses.
2. Next, perform multiplications and divisions in the order they appear as you read from left to right. For example,

$$60 \times 5 \div 4 \div 3 \times 2$$
$$= \quad 300 \div 4 \div 3 \times 2$$
$$= \quad\quad 75 \div 3 \times 2$$
$$= \quad\quad\quad 25 \times 2$$
$$= \quad\quad\quad\quad 50$$

3. Finally, perform additions and subtractions in the order they appear as you read from left to right.

Note. If two parenthetical expressions or a number and a parenthetical expression occur next to one another without any sign between them, multiplication is indicated.

By using the above procedure we find that
$8 - 3 \cdot 2 = 8 - 6 = 2$.

Example 1. Evaluate: $2 + 5(7 + 6)$
$= 2 + 5(13)$
$= 2 + 65$
$= 67$

Example 2. Evaluate: $(9 + 4) \times 16 + 8$
$= \quad 13 \quad \times 16 + 8$
$= \quad\quad\quad 208 \quad + 8$
$= \quad\quad\quad\quad\quad 216$

Example 3. Evaluate: $(6 + 1) \times 3 + (2 + 5)$
$= \quad 7 \quad \times 3 + \quad 7$
$= \quad\quad\quad 21 \quad + \quad 7$
$= \quad\quad\quad\quad\quad 28$

Example 4. Evaluate: $4(16 + 4) + \dfrac{14}{7} - 8$

$= 4(\quad 20 \quad) + \dfrac{14}{7} - 8$

$= \quad\quad 80 \quad + 2 - 8$

$= \quad\quad\quad\quad 74$

EXERCISES 1.3   *Evaluate each expression:*

1. $8 \div 3(4 - 2)$ 2. $(8 + 6)4 + 8$

3. $(8 + 6) - (7 - 3)$ 4. $4 \times (2 \times 6) + (6 + 2) \div 4$

5. $2(9 + 5) - 6 \times (13 + 2) \div 9$

6. $5(8 \times 9) + (13 + 7) \div 4$

7. $27 + 13 \times (7 - 3)(12 + 6) \div 9$

8. $123 - 3(8 + 9) + 17$

9. $16 + 4(7 + 8) - 3$

10. $(18 + 17)(12 + 9) - (7 \times 16)(4 + 2)$

11. $9 - 2(17 - 15) + 18$

12. $(9 + 7)5 + 13$

13. $(39 - 18) - (23 - 18)$

14. $5(3 \times 7) + (8 + 4) \div 3$

15. $3(8 + 6) - 7(13 + 3) \div 14$

16. $6(4 \times 5) + (15 + 9) \div 6$

17. $42 + 12(9 - 3)(12 + 13) \div 30$

18. $228 - 4 \times (7 + 6) - 8(6 - 2)$

19. $38 + 9 \times (8 + 4) - 3(5 - 2)$

20. $(19 + 8)(4 + 3) \div 21 + (8 \times 15) \div (4 \times 3)$

21. $27 - 2 \times (18 - 9) - 3 + 8(43 - 15)$

22. $6 \times 8 \div 2 \times 9 \div 12 + 6$

23. $12 \times 9 \div 18 \times 64 \div 8 + 7$

24. $18 \div 6 \times 24 \div 4 \div 6$

25. $7 + 6(3 + 2) - 7 - 5(4 + 2)$

26. $5 + 3(7 \times 7) - 6 - 2(4 + 7)$

27. $3 + 17(2 \times 2) - 67$

28. $8 - 3(9 - 2) \div 21 - 7$

29. $28 - 4(2 \times 3) + 4 - (16 \times 8) \div (4 \times 4)$

30. $6 + 4(9 + 6) + 8 - 2(7 + 3) - (3 \times 12) \div 9$

31. $24/(6 - 2) + 4 \times 3 - 15/3$

32. $(36 - 6)/(5 + 10) + (16 - 1)/3$

33. $3 \times 15 \div 9 + (13 - 5)/2 \times 4 - 2$

34. $28/2 \times 7 - (6 + 10)/(6 - 2)$

1.4 Area

To measure the length of an object, you must first select a suitable standard unit of length. To measure short lengths, choose a unit such as centimetres or millimetres in the metric system, or inches in the English system. For long distances, choose metres or kilometres in the metric system, or yards or miles in the English system.

To measure the surface area of an object, first select a standard unit of area suitable to the object to be measured. Standard units of area are based on the square and are called square units. For example, a square inch (in^2) is the amount of surface area within a square which measures one inch on a side. A square centimetre (cm^2) is the amount of surface area within a square which is 1 cm on a side. (See Fig. 1.8.) The area of any surface is the number of square units contained on that surface.

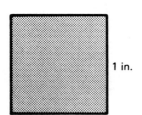

1 in.

1 in.

1 square inch (in^2)

1 cm

1 cm

1 square centimetre (cm^2)

Figure 1.8

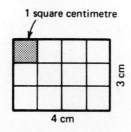

1 square centimetre

3 cm

4 cm

Figure 1.9

<u>Example 1.</u> What is the area of a rectangle measuring 4 cm by 3 cm?

Each square in Fig. 1.9 represents 1 cm^2. By simply counting the number of squares, you find the area of the rectangle is 12 cm^2.

You can also find the area by multiplying the length times the width:

$$\text{Area} = \ell \times w$$
$$= 4 \text{ cm} \times 3 \text{ cm} = 12 \text{ cm}^2$$
$$\text{(length)} \quad \text{(width)}$$

<u>Note.</u> cm $\times$ cm = cm^2

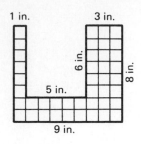

Figure 1.10

Example 2. What is the area of the metal plate represented in Fig. 1.10?

Each square represents 1 square inch. By simply counting the number of squares, you find the area of the metal plate is 42 in^2.

Another way to find the area of the figure would be to find the areas of two rectangles, then find their difference, as follows:

Area of outer rectangle: 9 in. × 8 in. = 72 in^2
Area of inner rectangle: 5 in. × 6 in. = 30 in^2
Area of metal plate: 42 in^2

EXERCISES 1.4

1. How many square yards (yd^2) are contained in a rectangle 12 yd long and 8 yd wide?
2. How many square metres (m^2) are contained in a rectangle 12 m long and 8 m wide?

Find the area of each figure:

3.

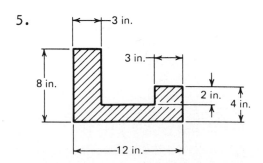

4.

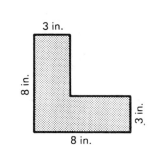

5.

6.

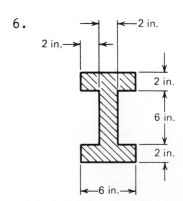

7.

$8 \times 3 = 24$

52

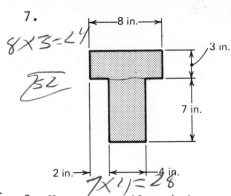

8.

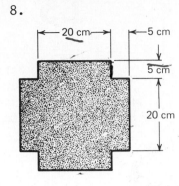

$7 \times 4 = 28$

9. How many tiles 4 in. on a side should be used to cover a portion of a wall 48 in. long by 36 in. high?

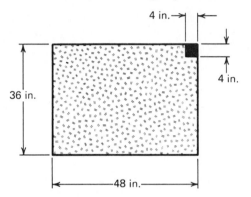

10. How many ceiling tiles 2 ft by 4 ft are needed to tile a ceiling that is 24 ft × 26 ft? Be careful how you arrange the tiles.

11. How many gallons of paint should be purchased to paint 20 motel rooms as shown in the figure? (Do not paint the floor.) It takes 1 gal to paint 400 square feet (ft^2).

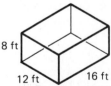

12. How many pieces of 4-ft-by-8-ft dry wall are needed for the 20 motel rooms in Exercise 11? All four walls and the ceiling in each room are to be dry-walled. Assume that the dry wall cut out for windows and doors cannot be salvaged and used again.

13. The replacement cost for construction of houses is $38/ft^2$. Determine how much house insurance should be carried on each of the one-story houses below.

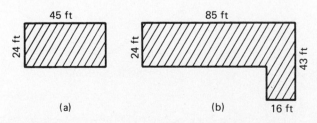

(a)   (b)

1.5 Volume

In area measurement the standard units are based on the square and called square units. For *volume* measurement, the standard units are based on the cube and called cubic units. For example, a cubic foot (ft³) is the amount of space contained in a cube having each edge 1 ft (or 12 in.) in length, as in Fig. 1.11.

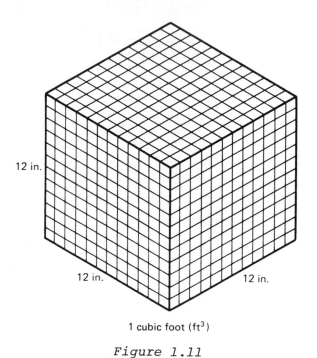

12 in.

12 in.   12 in.

1 cubic foot (ft³)

Figure 1.11

Fig. 1.12 shows the relative sizes of a cubic inch and a cubic centimetre. The cubic inch (in³) is the amount of space contained in a cube that measures 1 in. on each edge. The cubic centimetre (cm³) is the amount of space contained in a cube that measures 1 cm on each edge. The volume of any solid is the number of cubic units contained in the solid.

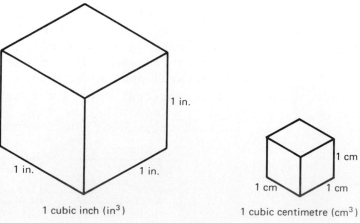

1 in.

1 in.   1 in.

1 cubic inch (in³)

1 cm

1 cm   1 cm

1 cubic centimetre (cm³)

Figure 1.12

Fig. 1.13 shows that the cubic decimetre (litre) is made up of 10 layers, each containing 100 cm^3, for a total of 1000 cm^3.

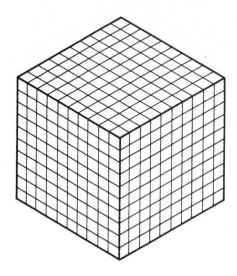

1 litre = 1000 cm^3

Figure 1.13

<u>Example 1</u>. Find the volume of a rectangular box 8 cm long, 4 cm wide, and 6 cm high.

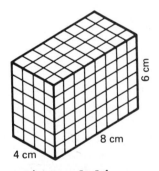

Figure 1.14

Suppose you placed one-centimetre cubes in the box, as in Fig. 1.14. On the bottom layer there would be 8 × 4, or 32, one-cm cubes. In all there are six such layers, or 6 × 32 = 192 one-cm cubes. Therefore, the volume is 192 cm^3.

You can also find the volume by multiplying the length times the width times the height:

$$V = \ell \times w \times h$$
$$= 8 \text{ cm} \times 4 \text{ cm} \times 6 \text{ cm}$$
$$= 192 \text{ cm}^3$$

<u>Note</u>. cm × cm × cm = cm^3.

Example 2. How many cubic inches are in a cubic foot?

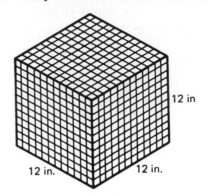

Figure 1.15

The bottom layer of Fig. 1.15 contains 12 × 12, or 144, one-inch cubes. There are 12 such layers, or 12 × 144 = 1728 one-inch cubes. Therefore, 1 ft^3 = 1728 in^3.

EXERCISES 1.5   *Find the volume of each figure:*

1.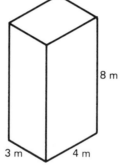

2.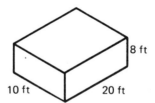

3.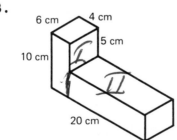

4.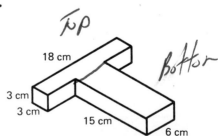

5.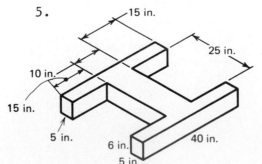

6.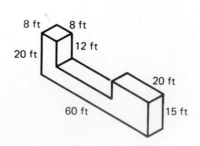

7. Find the volume of a box 10 cm by 12 cm by 5 cm.
8. Find the volume of a room 9 ft by 12 ft by 8 ft.
9. Find the weight of a cement floor that is 15 ft by 12 ft by 2 ft (1 ft^3 of cement weighs 193 lb).
10. A trailer 5 ft by 6 ft by 5 ft is filled with coal. Given that 1 ft^3 of coal weighs 40 lb and 1 ton = 2000 lb, how many tons of coal is the trailer carrying?

1.6 Short Division and Divisibility

Sometimes division can be done mentally. At other times short division is a valuable aid. Short division is a condensed form of long division and usually used when the divisor is 9 or less. The following examples show both methods.

Example 1. Divide 23,284 by 3.

Long form:
$$\begin{array}{r} 7761 \text{ r } 1 \\ 3\overline{\smash)23284} \\ \underline{21} \\ 22 \\ \underline{21} \\ 18 \\ \underline{18} \\ 04 \\ \underline{3} \\ 1 \end{array}$$

Short form:
$$\begin{array}{r} 2101 \\ 3\overline{\smash)23284} \\ \overline{7761 \text{ r } 1} \end{array}$$

With this method, think: 3 divides 23 *seven* times with remainder 2. Place 2 above 3. 3 divides 22 *seven* times with remainder 1. Place 1 above 2. 3 divides 18 *six* times with remainder 0. Place 0 above 8. 3 divides 4 *one* time with remainder 1.

Example 2. Divide 621,523 by 9.

Long form:
$$\begin{array}{r} 69058 \text{ r } 1 \\ 9\overline{\smash)621523} \\ \underline{54} \\ 81 \\ \underline{81} \\ 05 \\ \underline{0} \\ 52 \\ \underline{45} \\ 73 \\ \underline{72} \\ 1 \end{array}$$

Short form:
$$\begin{array}{r} 80571 \\ 9\overline{\smash)621523} \\ \overline{69058 \text{ r } 1} \end{array}$$

With this method, think: 9 divides 62 *six* times with remainder 8. Place 8 above 2. 9 divides 81 *nine* times with remainder 0. Place 0 above 1. 9 divides 5 *zero* times with remainder 5. Place 5 above 5. 9 divides 52 *five* times with remainder 7. Place 7 above 2. 9 divides 73 *eight* times with remainder 1. Place 1 above 3.

A number is *divisible* by a second number if, when you divide the first number by the second number, you get a zero remainder. That is, the second number *divides* the first number.

Example 3. 12 is divisible by 3, because 3 divides 12.

$$3\overline{)12}$$
$$4 \text{ r } 0$$

Example 4. 124 is not divisible by 7, because 7 does not divide 124.

$$7\overline{)\overset{55}{124}}$$
$$17 \text{ r } 5$$

EXERCISES 1.6   *Divide using short division:*

1. 825 ÷ 8   2. 314 ÷ 7   3. 256 ÷ 9

4. 7136 ÷ 5   5. 7136 ÷ 4   6. 3400 ÷ 4

7. 18,420 ÷ 2   8. 19,600 ÷ 6   9. 15,243 ÷ 3

10. 15,000 ÷ 8   11. 318,409 ÷ 3   12. 405,000 ÷ 9

Which numbers are divisible by 3?

13. 15   14. 28   15. 96

16. 172   17. 78   18. 675

Which numbers are divisible by 8?

19. 60   20. 94   21. 6582   22. 18,240   23. 9000

Which numbers are divisible by 9?

24. 216   25. 268   26. 596   27. 414   28. 1851

1.7   Prime Factorization

There are many ways of classifying the positive integers. They can be classified as even or odd, as divisible by 3 or not divisible by 3, as larger than 10 or smaller than 10, and so on. One of the most important classifications involves the concept of a *prime number*. A prime number is an integer greater than 1 that has no divisors except itself and 1. The first ten prime numbers are 2, 3, 5, 7, 11, 13, 17, 19, 23, and 29.

In multiplying two or more positive integers, the positive integers are called the factors of the product. Thus 2 and 5 are factors of 10, since 2 × 5 = 10. The numbers 2, 3, and 5 are factors of 30, since 2 × 3 × 5 = 30.

If the factors are prime numbers, they are called *prime factors*. The process of finding the prime factors of an integer is called *prime factorization*. The prime factorization of a given number is a product of factors. Each of the factors is prime, and their product equals the given number. One of the most useful applications of prime factorization is in finding the least common denominator (LCD) when adding and subtracting fractions. This application is found in Sec. 2.2.

<u>Example 1</u>. Factor 28 into prime factors.

(a) 28 = 4 · 7   (b) 28 = 7 · 4
       = 2 · 2 · 7   = 7 · 2 · 2

(c) 28 = 2 · 14
       = 2 · 7 · 2

In each case you have three prime factors of 28; one factor is 7, the other two are 2's. The factors may be written in any order, but we usually list all the factors in order from smallest to largest. It would not be correct in the examples above to leave 7 · 4, 4 · 7, or 2 · 14 as the factors of 28, since 4 and 14 are not prime numbers.

Short division is a helpful way to find prime factors. Find a prime number that divides the given number. Divide, using short division. Then, find a second prime number that divides the result. Divide, using short division. Keep repeating this process until the final quotient is also prime. The   ne factors will be the product of the divisors and th rinal quotient of the repeated short divisions.

<u>Example 2</u>. Find the prime factorization of 12.

2 | 12
2 | 6   The prime factorization of 12
      3   is 2 · 2 · 3.

<u>Example 3</u>. Find the prime factorization of 56.

2 | 56
2 | 28   The prime factorization of 56
2 | 14   is 2 · 2 · 2 · 7.
      7

<u>Example 4</u>. Find the prime factorization of 17.

17 is already in factored form, because 17 is a prime number. When asked for factors of a prime number, just write "prime" as your answer.

To eliminate some of the guesswork involved in finding prime factors, divisibility tests can be used. Divisibility tests determine whether or not a particular positive integer divides another integer without carrying out the division. The following divisibility tests for certain positive integers are most helpful.

Divisibility by 2. If a number ends with an even digit, then the number is divisible by 2. Note: zero is even.

Example 5. Does 2 divide 4258?

Yes, since 8, the last digit of the number, is even, 4258 is divisible by 2.

Check:
$$2\overline{\smash{)}\,\overset{0010}{4258}}$$
$$2129 \text{ r } 0$$

Example 6. Does 2 divide 215,517?

Since 7, the last digit, is odd, 215,517 is not divisible by 2.

Check:
$$2\overline{\smash{)}\,\overset{011111}{215517}}$$
$$107758 \text{ r } 1$$

Divisibility by 3. If the sum of the digits of a number is divisible by 3, then the number itself is divisible by 3.

Example 7. Does 3 divide 531?

Since the sum of the digits 5 + 3 + 1 = 9, and 9 is divisible by 3, then 531 is divisible by 3.

Check:
$$3\overline{\smash{)}\,\overset{220}{531}}$$
$$177 \text{ r } 0$$

Example 8. Does 3 divide 551?

Since the sum of the digits 5 + 5 + 1 = 11, and 11 is not divisible by 3, then 551 is not divisible by 3.

Check:
$$3\overline{\smash{)}\,\overset{212}{551}}$$
$$183 \text{ r } 2$$

Divisibility by 5. If a number has 0 or 5 as its last digit, then the number is divisible by 5.

Example 9. Does 5 divide 2372?

Since the last digit of 2372 is neither 0 nor 5, then 5 does not divide 2372.

Check:
$$5\overline{\smash{)}\,\overset{322}{2372}}$$
$$474 \text{ r } 2$$

Example 10. Does 5 divide 3210?

Since the last digit of 3210 is 0, then 3210 is divisible by 5.

Check:

$$5\overline{\smash{\big)}3210} \\ 642\ r\ 0$$

(with 210 written above 3210)

Example 11. Find the prime factorization of 204.

$$2\,\lfloor 204 \quad \text{(Last digit is even.)}$$
$$2\,\lfloor 102 \quad \text{(Last digit is even.)}$$
$$3\,\lfloor 51 \quad \text{(Sum of digits is divisible by 3.)}$$
$$17$$

The prime factorization of 204 is 2 · 2 · 3 · 17.

Example 12. Find the prime factorization of 630.

$$2\,\lfloor 630 \quad \text{(Last digit is even.)}$$
$$3\,\lfloor 315 \quad \text{(Sum of digits is divisible by 3.)}$$
$$3\,\lfloor 105 \quad \text{(Sum of digits is divisible by 3.)}$$
$$5\,\lfloor 35 \quad \text{(Last digit is 5.)}$$
$$7$$

The prime factorization of 630 is 2 · 3 · 3 · 5 · 7.

Note: As a general rule of thumb:

1. Keep dividing by 2 until the quotient is not even.
2. Keep dividing by 3 until the quotient's sum of digits is not divisible by 3.
3. Keep dividing by 5 until the quotient does not end in 0 or 5.

That is, if you divide out all the factors of 2, 3, and 5, the remaining factors, if any, will be much smaller and easier to work with, and perhaps prime.

Other divisibility tests are given in Sec. 2.1.

EXERCISES 1.7   *Classify each number as prime or not prime:*

| 1. 53 | 2. 57 | 3. 93 | 4. 121 | 5. 16 |
| 6. 123 | 7. 39 | 8. 87 | | |

Test for divisibility by 2:

| 9. 458 | 10. 12,746 | 11. 315,817 |
| 12. 877,778 | 13. 1367 | 14. 1205 |

Test for divisibility by 3 and check your results by short division:

15. 387   16. 1254   17. 453,128

18. 178,213   19. 218,745   20. 15,690

Test for divisibility by 5 and check your results by short division:

21. 20,505   22. 14,670   23. 3666

24. 256,665   25. 63,227   26. 14,601

Test the divisibility of each first number by the second number:

27. 56; 2   28. 42; 3   29. 218; 3

30. 375; 5   31. 528; 5   32. 2184; 3

33. 198; 3   34. 2236; 3   35. 1,820,670; 2

36. 2,817,638; 2   37. 7,215,720; 5   38. 5,275,343; 3

Find the prime factorization of each number: (Use divisibility tests where helpful.)

39. 20   40. 18   41. 66   42. 30

43. 36   44. 25   45. 27   46. 59

47. 51   48. 56   49. 42   50. 45

51. 120   52. 72   53. 171   54. 360

55. 105   56. 78   57. 252   58. 444

59. 540   60. 2012

1.8 Addition of Signed Numbers

As a technician you will use negative numbers in many ways. In an experiment using low temperatures, for example, you would record 10° below zero as -10°. Or, consider sea level as zero altitude. If a submarine dives 75 m, you could consider its depth as -75 m. See Fig. 1.16.

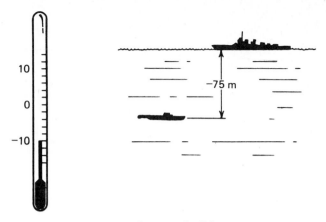

Figure 1.16

These measurements indicate a need for numbers other than positive integers, which are the only numbers that we have used up to now. They can be represented by points evenly spaced on a line above a point representing zero, as shown in Fig. 1.17.

Using the same spacing, mark off points below zero. We get points that correspond to the negative integers. We represent negative integers by using the names of the positive integers preceded by a negative (–) sign. For example, –3 is read "negative 3," –5 is read "negative 5." Each positive integer corresponds to a negative integer. For example, 3 and –3 are corresponding integers. Note that the distances from 3 to 0 and from 0 to –3 are equal.

Figure 1.17

The *absolute value* of a number can be thought of as the distance between the number and zero on the number line. Therefore, the absolute value of a number is never negative. We write the absolute value of a number x as $|x|$; it is read "the absolute value of x." Thus, $|x| \geqslant 0$ ("$\geqslant$" means "is greater than or equal to"). For example, $|+6| = 6$, $|4| = 4$, and $|0| = 0$. However, if a number is less than 0 (negative), its absolute value is the corresponding positive number. For example, $|-6| = 6$ and $|-7| = 7$.

The use of *signed numbers* is one of the most important operations that we will study. Signed numbers are used in work with exponents and certain dials. Operations with signed numbers are also essential for success in the basic algebra which follows later.

Since you will need to work problems involving signed numbers, you must know the basic operations.

Adding Two Numbers with Like Signs

To add two positive numbers, add their absolute values. A positive sign may or may not be used before the result. It is usually omitted.

To add two negative numbers, add their absolute values and place a negative sign before the result.

<u>Example 1</u>. Add:

(a) $\begin{array}{r} +2 \\ +3 \\ \hline +5 \end{array}$ (b) $\begin{array}{r} -4 \\ -6 \\ \hline -10 \end{array}$

(c) $(+4) + (+5) = +9$ (d) $(-8) + (-3) = -11$

Adding Two Numbers with Different Signs

To add a negative number and a positive number, find the difference of their absolute values. The sign of the number having the larger absolute value is placed before the result.

<u>Example 2</u>. Add:

(a) $\begin{array}{r} +4 \\ -7 \\ \hline -3 \end{array}$ (b) $\begin{array}{r} -3 \\ +8 \\ \hline +5 \end{array}$ (c) $\begin{array}{r} -9 \\ +2 \\ \hline -7 \end{array}$ (d) $\begin{array}{r} +8 \\ -5 \\ \hline +3 \end{array}$

(e) $(+6) + (-1) = +5$ (f) $(-8) + (+6) = -2$

(g) $(-2) + (+5) = +3$ (h) $(+3) + (-11) = -8$

> To add three or more signed numbers:
>
> <u>Step 1</u>: Add the positive numbers.
>
> <u>Step 2</u>: Add the negative numbers.
>
> <u>Step 3</u>: Add the sums from Steps 1 and 2 according to the rules for addition of signed numbers.

<u>Example 3</u>. Add: (-8) + (+22) + (-7) + (-19) + (+3)

Step 1:	*Step 2:*	*Step 3:*
+22	-8	+25
+ 3	-7	-34
+25	-19	-9
	-34	

Therefore, (-8) + (+22) + (-7) + (-19) + (+3) = -9.

<u>Example 4</u>. Add: (+4) + (-7) + (-2) + (+6) + (-3) + (-5)

Step 1:	*Step 2:*	*Step 3:*
+4	-7	+10
+6	-2	-17
+10	-3	-7
	-5	
	-17	

Therefore, (+4) + (-7) + (-2) + (+6) + (-3) + (-5) = -7.

EXERCISES 1.8   *Find the absolute value of each number:*

1. 3   2. -4   3. -6   4. 0   5. +4

6. +8   7. 17   8. -37   9. -15   10. 49

Add:

11. +4   12. -5   13. +9   14. -10   15. +5
    +6   -9   -2   +4   -7

16. -4   17. -3   18. +4   19. 12   20. -12
    +6   -9   -9   -6   6

21. 3   22. -2   23. 0   24. -5   25. -7
    9   0   4   -10   -7

26. 18   27. +15   28. -11   29. 1   30. -22
    -18   -20   -3   -13   +14

31. (-4) + (-5)   32. (+2) + (-11)   33. (-3) + (+7)

34. (+4) + (+6)   35. (-5) + (+2)   36. (-7) + (-6)

37. (+7) + (-8)   38. (8) + (-3)   39. (-10) + (6)

40. (+4) + (-11)   41. (-8) + (2)   42. (+3) + (+7)

43. $(-2) + (0)$ 44. $(0) + (+3)$ 45. $(9) + (-5)$

46. $(+9) + (-9)$ 47. $(16) + (-7)$ 48. $(-19) + (-12)$

49. $(-6) + (+9)$ 50. $(+20) + (-30)$

51. $(-1) + (-3) + (+8)$ 52. $(+5) + (-3) + (+4)$

53. $(+1) + (+7) + (-1)$ 54. $(-5) + (-9) + (-4)$

55. $(-9) + (+6) + (-4)$ 56. $(+8) + (+7) + (-2)$

57. $(+8) + (-8) + (+7) + (-2)$ 58. $(-6) + (+5) + (-8) + (+4)$

59. $(-4) + (-7) + (-7) + (-2)$ 60. $(-3) + (-9) + (+5) + (+6)$

61. $(-1) + (-2) + (+9) + (-8)$ 62. $(+6) + (+5) + (-7) + (-3)$

63. $(-6) + (+2) + (+7) + (-3)$ 64. $(+8) + (-1) + (+9) + (+6)$

65. $(-5) + (+1) + (+3) + (-2) + (-2)$

66. $(+5) + (+2) + (-3) + (-9) + (-9)$

67. $(-5) + (+6) + (-9) + (-4) + (-7)$

68. $(-9) + (+7) + (-6) + (+5) + (-8)$

69. $(+1) + (-4) + (-2) + (+2) + (-9)$

70. $(-1) + (-2) + (-6) + (-3) + (-5)$

71. $(-2) + 8 + (-4) + 6 + (-1)$

72. $14 + (-5) + (-1) + 6 + (-3)$

73. $5 + 6 + (-2) + 9 + (-7)$

74. $(-5) + 4 + (-1) + 6 + (-7)$

75. $(-3) + 8 + (-4) + (-7) + 10$

76. $16 + (-7) + (-5) + 20 + (-5)$

77. $3 + (-6) + 7 + 4 + (-4)$

78. $(-8) + 6 + 9 + (-5) + (-4)$

79. $(-5) + 4 + (-7) + 2 + (-8)$

80. $7 + 9 + (-6) + (-4) + 9 + (-2)$

1.9   Subtraction of Signed Numbers

Subtracting Two Signed Numbers

To subtract signed numbers, change the sign of the
number being subtracted and add according to the
rules for addition of signed numbers.

Example 1. Subtract:

(a) Subtract: $+2$ ↔ Add: $+2$ To subtract, change the
$\underline{+5}$ $\underline{-5}$ sign of the number being
-3 -3 subtracted, $+5$, and add.

(b) Subtract: -7 ↔ Add: -7 To subtract, change the
$\underline{-6}$ $\underline{+6}$ sign of the number being
-1 -1 subtracted, -6, and add.

(c) Subtract: $+6$ ↔ Add: $+6$
$\underline{-4}$ $\underline{+4}$
$+10$ $+10$

(d) Subtract: -8 ↔ Add: -8
$\underline{+3}$ $\underline{-3}$
-11 -11

(e) $(+4) - (+6) = (+4) + (-6)$ To subtract, change the
$\qquad = -2$ sign of the number being
subtracted, $+6$, and add.

(f) $(-8) - (-10) = (-8) + (+10) = +2$

(g) $(+9) - (-6) = (+9) + (+6) = +15$

(h) $(-4) - (+7) = (-4) + (-7) = -11$

Subtracting More Than Two Signed Numbers

When more than two signed numbers are involved in subtraction, change the sign of <u>each</u> number being subtracted and add the resulting signed numbers.

Example 2. Subtract: $(-4) - (-6) - (+2) - (-5) - (+7)$

$= (-4) + (+6) + (-2) + (+5) + (-7)$

Step 1: $+6$ *Step 2:* -4 *Step 3:* $+11$
$\underline{+5}$ -2 $\underline{-13}$
$+11$ $\underline{-7}$ -2
-13

Therefore, $(-4) - (-6) - (+2) - (-5) - (+7) = -2$.

When combinations of additions and subtractions of signed numbers occur in the same problem, change *only* the sign of each number being subtracted. Then add the resulting signed numbers.

Example 3. Perform the indicated operations.

$$(+4) - (-5) + (-6) - (+8) - (-2) + (+5) - (+1)$$
$$= (+4) + (+5) + (-6) + (-8) + (+2) + (+5) + (-1)$$

Step 1: +4   Step 2: −6   Step 3: +16

Step 1: +4	Step 2: −6	Step 3: +16
+5	−8	−15
+2	−1	+1
+5	−15	
+16		

Therefore, $(+4) - (-5) + (-6) - (+8) - (-2) + (+5) - (+1)$

$$= +1.$$

Example 4. Perform the indicated operations.

$$(-12) + (-3) - (-5) - (+6) - (-1) + (+4) - (+3) - (-8)$$
$$= (-12) + (-3) + (+5) + (-6) + (+1) + (+4) + (-3) + (+8)$$

Step 1: +5	Step 2: −12	Step 3: +18
+1	−3	−24
+4	−6	−6
+8	−3	
+18	−24	

Therefore, $(-12) + (-3) - (-5) - (+6) - (-1) + (+4)$

$$- (+3) - (-8) = -6.$$

EXERCISES 1.9   *Subtract:*

1. +4	2. −5	3. +9	4. −10	5. +5
+6	−9	−2	−4	+7
6. −4	7. −3	8. +4	9. 12	10. −12
+6	−9	−9	+6	6
11. 3	12. −2	13. 0	14. −5	15. −7
9	0	4	−10	−7
16. 18	17. +15	18. −11	19. 1	20. −22
−18	−20	−3	−13	+14

Perform the indicated operations:

21. $(-4) - (-5)$ 22. $(+2) - (-11)$ 23. $(-3) - (+7)$

24. $(+4) - (+6)$ 25. $(-5) - (+2)$ 26. $(-7) - (-6)$

27. $(+7) - (-8)$ 28. $(8) - (-3)$ 29. $(-10) - (6)$

30. $(+4) - -11$ 31. $(-8) - (+2)$ 32. $(+3) - (+7)$

33. $(-2) - (0)$ 34. $(0) - (+3)$ 35. $(9) - (+5)$

36. $(+9) - (+9)$ 37. $(16) - (+7)$ 38. $(-19) - (+12)$

39. $(-6) - (+9)$ 40. $(+20) - (-30)$

41. $(+6) - (-3) - (+1)$ 42. $(+3) - (-7) - (+6)$

43. $(-3) - (-7) - (+8)$ 44. $(+3) - (4) - (-9)$

45. $(+3) - (-6) - (+9) - (-8)$

46. $(+10) - (-4) - (6) - (-9)$

47. $(+5) - (-5) + (-8)$ 48. $(+1) + (-7) - (-7)$

49. $(-3) + (-5) - (+0) - (+7)$ ~15

50. $(+4) - (-3) + (+6) - (+8)$ +1 5

51. $(+4) - (-11) - (+12) + (-6)$ ~3

52. $(8) - (-6) - (+18) - (4)$

53. $(-7) - (+6) + (-3) - (-2) - (+9)$

54. $(-3) + (-4) + (+7) - (-2) - (+6)$

55. $-9 + 8 - 5 + 6 - 4$ 56. $-12 + 2 + 30 - 6$

57. $-8 + 12 - 7 - 4 + 6$ 58. $7 + 4 - 8 - 9 + 3$

59. $16 - 18 + 4 - 7 - 2 + 9$

60. $3 - 7 + 5 - 6 - 7 + 2$

61. $8 + 10 - 20 + 4 - 5 - 6 + 1$

62. $5 - 6 - 7 + 2 - 8 + 10$

63. $9 - 7 + 4 + 3 - 8 - 6 - 6 + 1$

64. $-4 + 6 - 7 - 5 + 6 - 7 - 1$

1.10 Multiplication and Division of Signed Numbers

> To multiply two signed numbers:
>
> 1. If the two numbers have the same signs, multiply their absolute values. This product is always positive.
> 2. If the two numbers have different signs, multiply their absolute values and place a negative sign before the product.

Example 1. Multiply:

(a) +2   (b) −4   (c) −2   (d) −6
    +3   −7   +4   +5
    ——   ———   ——   ———
    +6   +28   −8   −30

(e) (+3)(+4) = +12   (f) (−6)(−9) = +54

(g) (−5)(+7) = −35   (h) (+4)(−9) = −36

To multiply more than two signed numbers:

1. If the number of negative factors is even (divisible by 2), multiply the absolute values of the numbers. This product is positive.
2. If the number of negative factors is odd, multiply the absolute values of the numbers and place a negative sign before the product.

Example 2. Multiply: (−11)(+3)(−6) = +198

The number of negative factors is even; therefore, the product is positive.

Example 3. Multiply: (−5)(−4)(+2)(−7) = −280

The number of negative factors is odd; therefore, the product is negative.

Since multiplication and division are related operations, the same rules for signed numbers apply to both operations.

To divide two signed numbers:

1. If the two numbers have the same signs, divide their absolute values. This quotient is always positive.
2. If the two numbers have different signs, divide their absolute values and place a negative sign before the quotient.

Example 4. Divide:

(a) $\frac{+12}{+2} = +6$ (b) $\frac{-18}{-6} = +3$ (c) $\frac{+20}{-4} = -5$ (d) $\frac{-24}{+6} = -4$

(e) $(+30) \div (+5) = +6$ (f) $(-42) \div (-2) = +21$

(g) $(+16) \div (-4) = -4$ (h) $(-45) \div (+9) = -5$

EXERCISES 1.10   *Multiply:*

1.	+4	2.	−5	3.	+9	4.	−10	5.	+5
	+6		−9		−2		+4		−7

6.	−4	7.	−3	8.	+4	9.	12	10.	−12
	+6		−9		−9		−6		6

11.	3	12.	−2	13.	0	14.	−5	15.	−7
	9		0		4		−10		−7

16.	18	17.	+15	18.	−11	19.	1	20.	−22
	−18		−20		−3		−13		+14

21. $(+3)(-2)$ 22. $(+5)(+7)$ 23. $(-6)(-8)$

24. $(-9)(+2)$ 25. $(-7)(-3)$ 26. $(-4)(+4)$

27. $(-8)(+2)$ 28. $(+5)(-3)$ 29. $(-6)(-9)$

30. $(+4)(+7)$ 31. $(-6)(+1)$ 32. $(+4)(-2)$

33. $(-8)(-3)$ 34. $(-2)(+6)$ 35. $(-3)(-9)$

36. $(8)(-4)$ 37. $(-9)(7)$ 38. $(-8)(0)$

39. $(9)(-1)$ 40. $(-4)(-10)$ 41. $(+3)(-2)(+1)$

42. $(-5)(-9)(+2)$ 43. $(-3)(+3)(-4)$ 44. $(-2)(-8)(-3)$

45. $(+5)(+2)(+3)$ 46. $(-4)(+3)(0)(+3)$

47. $(-3)(-2)(4)(-7)$ 48. $(-3)(-1)(+1)(+2)$

49. $(-9)(-2)(+3)(+1)(-3)$ 50. $(-6)(-2)(-4)(-1)(-2)(+2)$

Divide:

51. $\dfrac{+10}{+2}$ 52. $\dfrac{-8}{-4}$ 53. $\dfrac{+27}{-3}$ 54. $\dfrac{-48}{+6}$ 55. $\dfrac{-32}{-4}$

56. $\dfrac{+39}{-13}$ 57. $\dfrac{+14}{+7}$ 58. $\dfrac{-45}{+15}$ 59. $\dfrac{-54}{-6}$ 60. $\dfrac{+72}{-9}$

61. $\dfrac{-100}{+25}$ 62. $\dfrac{+84}{+12}$ 63. $\dfrac{-75}{-25}$ 64. $\dfrac{+36}{-6}$ 65. $\dfrac{+85}{+5}$

66. $\dfrac{-270}{+9}$ 67. $\dfrac{+480}{+12}$ 68. $\dfrac{-350}{+70}$ 69. $\dfrac{-900}{-60}$ 70. $\dfrac{+4800}{-240}$

71. $(-49) \div (-7)$ 72. $(+9) \div (-3)$

73. $(+80) \div (+20)$ 74. $(-60) \div (+12)$

75. $(+45) \div (-15)$ 76. $(+120) \div (-6)$

77. $(-110) \div (-11)$ 78. $(+84) \div (+6)$

79. $(-96) \div (-12)$ 80. $(-800) \div (+25)$

1.11 Formulas

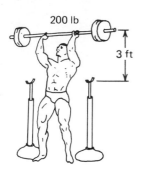

Figure 1.18

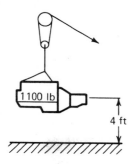

Figure 1.19

A *formula* is a statement of a rule using letters to represent certain unknown quantities. In physics one of the basic rules states that *work* is equal to *force* times *distance*. If a man (Fig. 1.18) lifts a 200-lb weight a distance of 3 ft, we say the work done is 200 ft × 3 lb = 600 foot-pounds (ft-lb). The work, W, is equal to the force, f, times the distance, d, or $W = f × d$.

A man pushes against a car weighing 2700 lb but does not move it. The work done is 2700 lb × 0 ft = 0 ft-lb. An automotive technician (Fig. 1.19) moves a diesel engine weighing 1100 lb from the floor to a workbench 4 ft high. The work done in moving the engine is 1100 lb × 4 ft = 4400 ft-lb.

To summarize, if you know the amount of force and the distance the force is moved, the work can be found by simply multiplying the force and distance. By formula, $W = f × d$ is often written: $W = f \cdot d$, or simply $W = fd$. Whenever no symbol appears between a number and a letter, or between two letters in a formula, the operation to be performed is multiplication.

<u>Example</u>. If $W = fd$, and $f = 10$, $d = 16$, what is S?

$$W = fd$$

$$W = (10)(16) = 160$$

Therefore, $W = 160$

There are many other formulas used in science and technology. Some examples are given here:

(a) $d = vt$ (b) $W = IEt$ (c) $f = ma$

(d) $P = IE$ (e) $I = \dfrac{E}{R}$ (f) $P = \dfrac{V^2}{R}$

EXERCISES 1.11

Use the formula $W = fd$, where f represents a force, and d represents the distance that the force is moved. Find the work done, W:

1. $f = 30$, $d = 20$ 2. $f = 17$, $d = 9$

3. $f = 1125$, $d = 10$ 4. $f = 203$, $d = 27$

5. $f = 176$, $d = 326$ 6. $f = 2400$, $d = 120$

Use the formulas above to work the following:

1600 · 24 = 38,400

7. If m = 1600 and a = 24, find f.

8. If V = 120 and R = 24, find P.   600

9. If E = 120 and R = 15, find I.   8

10. If v = 372 and t = 18, find d.   6696

11. If I = 29 and E = 173, find P.   5017

12. If I = 11, E = 95, and t = 46, find W.   48,070

The area of a rectangle whose length and width are known is given by length × width (provided both dimensions are in the same unit). The formula is given by $A = \ell w$. Find A if:

13. ℓ = 8 m, w = 7 m

14. ℓ = 24 in., w = 15 in.

15. ℓ = 12 yd, w = 15 ft

16. The volume of a rectangular solid is given by $V = \ell w h$, where ℓ is the length, w is the width, and h is the height of the solid. Find V if ℓ = 25 cm, w = 15 cm, and h = 12 cm.

17. Given $F = \frac{9}{5}C + 32$, and C = 55, find F. $\frac{9}{5} \times 55 + 32 = 131$

18. Given $Q = CV$, C = 12 and V = 2500, find Q.   30000

19. Given $I = \frac{E}{Z}$, E = 240, and Z = 15, find I.   16

20. Given $P = I^2R$, I = 4 and R = 2000, find P.   32000

Chapter 1 Review

1. Add: 435 + 2600 + 18 + 5184 + 6

2. Subtract: 60,000   3. Multiply: 7060 × 1300
            4,803

4. Divide: 68,040 ÷ 300   5. Evaluate: 12 − 3(5 − 2)

6. Evaluate: (6 + 4)8 ÷ 2 + 3

7. Evaluate: 18 ÷ 2 × 5 ÷ 3 − 6 + 4 × 7

8. Evaluate 18/(5 − 3) + (6 − 2) × 8 − 10

9. Find the area of the figure.

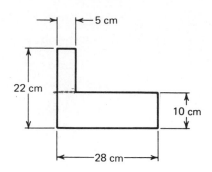

10. Find the volume of the figure.

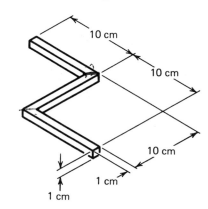

11. Divide 436 ÷ 8 using short division.
12. Is 460 divisible by 3?
13. Find the prime factorization of 54.
14. Find the prime factorization of 330.

Find the absolute value of:

15. +5   16. −16   17. 13

Add:

18. −4   19. −6   20. (+5) + (−8)
    +7   −2
21. (−9) + (+2) + (−6)

Subtract:

22. 3   23. −7   24. (+9) − (+10)
    −6   −4
25. (−6) − (+4) − (+8)

Find the value of:

26. (−2) − (+7) + (+4) + (−5) − (+10)
27. 5 − 6 + 4 − 9 + 4 + 3 − 12 − 8

Multiply:

28. $(-6)(+4)$ 29. $(+4)(+9)$ 30. $(-9)(-8)$

31. $(-2)(-7)(+1)(+3)(-2)$

Divide:

32. $\dfrac{-18}{-3}$ 33. $(+30) \div (-5)$ 34. $\dfrac{+45}{+9}$

35. Given the formula $P = \dfrac{Fs}{t}$, $F = 600$, $s = 50$, and $t = 10$, find P.

FRACTIONS AND THE ENGLISH SYSTEM

2.1 Introduction to Fractions

The English system of measurement is basically a system whose units are expressed as common fractions and mixed numbers. The metric system of measurement is a system whose units are expressed as decimal fractions and powers of ten. As we convert from the English system to the metric system, more computations will be done with decimals, which are easier—especially with a calculator. Fewer computations will be done with fractions, which are more difficult and are not easily done with a calculator. During the transition period we will need to feel comfortable with both systems.

The English system is emphasized in this chapter, while the metric system is developed in Chapter 4.

A *common fraction* may be defined as the ratio or quotient of two integers, say a and b, in the form $\frac{a}{b}$ (where $b \neq 0$). Examples are $\frac{1}{2}$, $\frac{7}{11}$, $-\frac{3}{8}$, and $\frac{37}{22}$. The integer below the line is called the *denominator*. It gives the denomination (size) of equal parts into which the fraction unit is divided. The integer above the line is called the *numerator*. It numerates (counts) the number of times the denominator is used. Look at one inch on a ruler, and then look at one inch enlarged, as shown in Figs. 2.1 and 2.2.

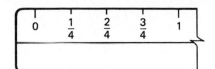

Figure 2.1

$\frac{1}{4}$ in. means 1 of 4 equal parts of an inch.

$\frac{2}{4}$ in. means 2 of 4 equal parts of an inch.

$\frac{3}{4}$ in. means 3 of 4 equal parts of an inch.

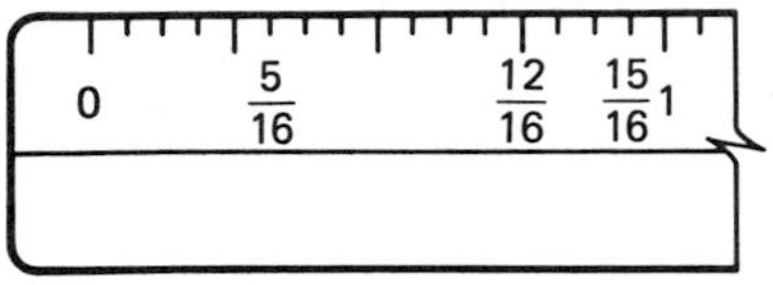

Figure 2.2

$\frac{5}{16}$ in. means 5 of 16 equal parts of an inch.

$\frac{12}{16}$ in. means 12 of 16 equal parts of an inch.

$\frac{15}{16}$ in. means 15 of 16 equal parts of an inch.

Two fractions $\frac{a}{b}$ and $\frac{c}{d}$ are equal or equivalent if $ad = bc$. That is, $\frac{a}{b} = \frac{c}{d}$ if $ad = bc$ ($b \neq 0$ and $d \neq 0$). For example, $\frac{2}{4}$ and $\frac{8}{16}$ are names for the same fraction because $(2)(16) = (4)(8)$. There are many other names for this same fraction, such as $\frac{1}{2}$, $\frac{9}{18}$, $\frac{10}{20}$, $\frac{-5}{-10}$, $\frac{-3}{-6}$, and so on.

$\frac{2}{4} = \frac{1}{2}$ because $2(2) = 4(1)$, and $\frac{2}{4} = \frac{-5}{-10}$ because $2(-10) = 4(-5)$.

We have two rules for finding equal (or *equivalent*) fractions:

Equivalent Fractions

1. The numerator and denominator of any fraction may be <u>multiplied</u> by the same number (except zero) without changing the value of the given fraction. For example, $\frac{4}{9} = \frac{4 \cdot 5}{9 \cdot 5} = \frac{20}{45}$.

2. The numerator and denominator of any fraction may be <u>divided</u> by the same number (except zero) without changing the value of the given fraction. For example, $\frac{6}{10} = \frac{6 \div 2}{10 \div 2} = \frac{3}{5}$.

Equivalent fractions are used to reduce a fraction to lowest terms and to change a fraction to higher terms when adding and subtracting fractions with unequal denominators.

To *simplify* a fraction means to find an equivalent fraction whose numerator and denominator are *relatively prime*—that is, whose numerator and denominator have no common divisor. This is also called "reducing a fraction to lowest terms."

Fractions with rather large numerators and denominators are usually more difficult to simplify. Factoring each numerator and denominator by using the divisibility tests from Sec. 1.7 is recommended. There are other divisibility tests that are especially useful in simplifying fractions. They are presented below.

<u>Divisibility by 4</u>. Consider the last two digits of the number. If the number formed by these two digits is divisible by 4, then the given number is divisible by 4.

<u>Example 1</u>. Does 4 divide 5228?

Since the number formed by the last two digits is 28 and 28 is divisible by 4, then 5228 is divisible by 4.

Check:
$$4\overline{\smash{\big)}\,5228}$$
$$1307 \text{ r } 0$$

<u>Example 2</u>. Does 4 divide 3238?

Since the number formed by the last two digits is 38 and 38 is not divisible by 4, then 3238 is not divisible by 4.

Check:
$$4\overline{\smash{\big)}\,3238}$$
$$809 \text{ r } 2$$

<u>Divisibility by 9</u>. If the sum of the digits of a number is divisible by 9, the number is divisible by 9.

<u>Example 3</u>. Does 9 divide 321,372?

Since the sum of the digits $3 + 2 + 1 + 3 + 7 + 2 = 18$, and 18 is divisible by 9, then 321,372 is divisible by 9.

Check:
$$9\overline{\smash{\big)}\,321372}$$
$$35708 \text{ r } 0$$

<u>Example 4</u>.Does 9 divide 423,179?

Since the sum of the digits $4 + 2 + 3 + 1 + 7 + 9 = 26$, and 26 is not divisible by 9, then 423,179 is not divisible by 9.

Check:
$$9\overline{\smash{\big)}\,423179}$$
$$47019 \text{ r } 8$$

OTHER HELPFUL DIVISIBILITY TESTS

<u>For 10</u>. If a number ends with 0, the number is divisible by 10.

<u>For 6</u>. If a number's last digit is even and the sum of the digits of that number is divisible by 3, then the number is divisible by 6.

<u>For 8</u>. Consider the last three digits of the number. If the number formed by these three digits is divisible by 8, then the number is divisible by 8.

The divisibility test for 7 is too long to be useful.

<u>Example 5</u>. Does 6 divide 39,534?

The last digit, 4, is even. The sum of the digits, $3 + 9 + 5 + 3 + 4 = 24$, is divisible by 3. So 39,534 is divisible by 6.

$$\textit{Check:} \qquad 6 \overline{\smash{\big)}\, 39534} \atop \quad 6589 \text{ r } 0$$

<u>Example 6</u>. Does 8 divide 325,216?

The number formed by the last three digits, 216, is divisible by 8;

$$8 \overline{\smash{\big)}\, 216} \atop \quad 27 \text{ r } 0$$

therefore, 325,216 is divisible by 8.

$$\textit{Check:} \qquad 8 \overline{\smash{\big)}\, 325216} \atop \quad 40652 \text{ r } 0$$

<u>Example 7</u>. Simplify $\dfrac{35}{50}$.

$$\frac{35}{50} = \frac{\cancel{5} \cdot 7}{\cancel{5} \cdot 10} = \frac{7}{10}$$

Note the use of divisibility test for 5. A last digit of 0 or 5 indicates the number is divisible by 5.

<u>Example 8</u>. Simplify $\dfrac{117}{963}$.

$$\frac{117}{963} = \frac{\cancel{9} \cdot 13}{\cancel{9} \cdot 107} = \frac{13}{107}$$

Note the use of divisibility test for 9. The sum of the digits of the numerator, 9, and the denominator, 18, are each divisible by 9.

<u>Example 9</u>. Simplify $\frac{84}{300}$.

$$\frac{84}{300} = \frac{\cancel{3} \cdot \cancel{4} \cdot 7}{\cancel{3} \cdot \cancel{4} \cdot 25} = \frac{7}{25}$$

There are some special rules for simplifying fractions that we should note here:

1. Any number (except zero) divided by itself is equal to one.

   For example, $\frac{3}{3} = 1$; $\frac{-5}{-5} = 1$; $\frac{171}{171} = 1$.

2. Any number divided by one is equal to itself.

   For example, $\frac{5}{1} = 5$; $\frac{-9}{1} = -9$; $\frac{25}{1} = 25$.

3. Zero divided by any number (except zero) is equal to zero.

   For example, $\frac{0}{6} = 0$; $\frac{0}{13} = 0$; $\frac{0}{-8} = 0$.

Any number *divided by zero* is not meaningful and is called *undefined*. For example, $\frac{4}{0}$ is undefined.

A *proper fraction* is a fraction whose numerator is less than its denominator. Examples of proper fractions are $\frac{2}{3}$, $\frac{5}{14}$, and $\frac{3}{8}$. An *improper fraction* is a fraction whose numerator is greater than or equal to its denominator. Examples of improper fractions are $\frac{7}{5}$, $\frac{11}{11}$, and $\frac{9}{4}$.

A *mixed number* is an integer plus a proper fraction. Examples of mixed numbers are $1\frac{3}{4}$, $14\frac{1}{9}$, and $5\frac{2}{15}$.

Changing An Improper Fraction to a Mixed Number

To change an improper fraction to a mixed number, divide the numerator by the denominator. The quotient is the whole number part. The remainder over the divisor is the proper fraction part of the mixed number.

Example 10. Change $\frac{17}{3}$ to a mixed number.

$$\frac{17}{3} = 17 \div 3 = 3\underline{\big|17} = 5\frac{2}{3}$$
$$5 \text{ r } 2$$

Example 11. Change $\frac{78}{7}$ to a mixed number.

$$\frac{78}{7} = 78 \div 7 = 7\underline{\big|78} = 11\frac{1}{7}$$
$$11 \text{ r } 1$$

If the improper fraction is not reduced to lowest terms, you may find it easier to reduce it before changing it to a mixed number. Of course you may reduce the proper fraction after the division if you prefer.

Example 12. Change $\frac{324}{48}$ to a mixed number and simplify.

Method 1: Reduce the improper fraction to lowest terms first.

$$\frac{324}{48} = \frac{\cancel{12} \cdot 27}{\cancel{12} \cdot 4} = \frac{27}{4} = 4\underline{\big|27} = 6\frac{3}{4}$$
$$6 \text{ r } 3$$

Method 2: Change the improper fraction to a mixed number first.

$$\frac{324}{48} = 48\overline{\big)324}^{6 \text{ r } 36} = 6\frac{36}{48} = 6\frac{\cancel{12} \cdot 3}{\cancel{12} \cdot 4} = 6\frac{3}{4}$$
$$\underline{288}$$
$$36$$

One way to change a mixed number to an improper fraction is to multiply the integer by the denominator of the fraction and then add the numerator of the fraction. Then place this sum over the original denominator.

Example 13. Change $2\frac{1}{3}$ to an improper fraction.

$$2\frac{1}{3} = \frac{(2 \times 3) + 1}{3} = \frac{7}{3}$$

Example 14. Change $5\frac{3}{8}$ to an improper fraction.

$$5\frac{3}{8} = \frac{(5 \times 8) + 3}{8} = \frac{43}{8}$$

Example 15. Change $10\frac{5}{9}$ to an improper fraction.

$$10\frac{5}{9} = \frac{(10 \times 9) + 5}{9} = \frac{95}{9}$$

EXERCISES 2.1   *Simplify:*

1. $\dfrac{12}{28}$ 2. $\dfrac{9}{12}$ 3. $\dfrac{36}{42}$ 4. $\dfrac{12}{18}$ 5. $\dfrac{9}{48}$

6. $\dfrac{8}{10}$ 7. $\dfrac{13}{39}$ 8. $\dfrac{24}{36}$ 9. $\dfrac{48}{60}$ 10. $\dfrac{72}{96}$

11. $\dfrac{9}{9}$ 12. $\dfrac{15}{1}$ 13. $\dfrac{0}{8}$ 14. $\dfrac{6}{6}$ 15. $\dfrac{9}{0}$

16. $\dfrac{6}{8}$ 17. $\dfrac{14}{16}$ 18. $\dfrac{7}{28}$ 19. $\dfrac{27}{36}$ 20. $\dfrac{15}{18}$

21. $\dfrac{12}{16}$ 22. $\dfrac{9}{18}$ 23. $\dfrac{20}{25}$ 24. $\dfrac{12}{36}$ 25. $\dfrac{12}{40}$

26. $\dfrac{54}{72}$ 27. $\dfrac{32}{48}$ 28. $\dfrac{60}{72}$ 29. $\dfrac{112}{128}$ 30. $\dfrac{330}{360}$

31. $\dfrac{72}{84}$ 32. $\dfrac{9}{144}$ 33. $\dfrac{126}{210}$ 34. $\dfrac{270}{480}$ 35. $\dfrac{57}{111}$

36. $\dfrac{30}{162}$ 37. $\dfrac{58}{87}$ 38. $\dfrac{198}{462}$ 39. $\dfrac{112}{144}$ 40. $\dfrac{525}{1155}$

Change each fraction to a mixed number in simplest form:

41. $\dfrac{78}{5}$ 42. $\dfrac{11}{4}$ 43. $\dfrac{28}{3}$ 44. $\dfrac{21}{3}$ 45. $\dfrac{45}{36}$

46. $\dfrac{67}{16}$ 47. $\dfrac{57}{6}$ 48. $\dfrac{84}{9}$ 49. $\dfrac{104}{5}$ 50. $\dfrac{47}{3}$

51. $\dfrac{221}{9}$ 52. $8\dfrac{11}{5}$ 53. $5\dfrac{15}{12}$ 54. $2\dfrac{70}{16}$

Change each mixed number to an improper fraction:

55. $3\dfrac{5}{6}$ 56. $6\dfrac{3}{4}$ 57. $2\dfrac{1}{8}$ 58. $5\dfrac{2}{3}$ 59. $1\dfrac{7}{16}$

60. $4\dfrac{1}{2}$ 61. $6\dfrac{7}{8}$ 62. $5\dfrac{2}{3}$ 63. $10\dfrac{3}{5}$ 64. $12\dfrac{5}{6}$

2.2 Addition and Subtraction of Fractions

Technicians must be able to compute fractions accurately, since accuracy is very important in their work and mistakes on the job can be quite costly. Also, many shop drawing dimensions are given in fractions.

Rule for Adding Fractions

$$\frac{a}{c} + \frac{b}{c} = \frac{a + b}{c} \quad (c \neq 0)$$

That is, to add two or more fractions with the same, or common, denominator, first add their numerators. Then place this sum over the common denominator.

Example 1. Add: $\frac{1}{8} + \frac{3}{8} = \frac{1 + 3}{8} = \frac{4}{8} = \frac{1}{2}$

Example 2. Add: $\frac{2}{16} + \frac{5}{16} = \frac{2 + 5}{16} = \frac{7}{16}$

Example 3. Add: $\frac{2}{31} + \frac{7}{31} + \frac{15}{31} = \frac{2 + 7 + 15}{31} = \frac{24}{31}$

To add fractions with different denominators, we first need to find a common denominator. You found in reducing fractions to lowest terms in Sec. 2.1 that you could *divide* both numerator and denominator by the same nonzero number without changing the value of the fraction. Similarly, you can *multiply* both numerator and denominator by the same nonzero number without changing the value of the fraction.

How do you determine what number to use for multiplying? It is customary to find the *least common denominator (LCD)* for fractions with unlike denominators. The LCD is the smallest number that has all the denominators as divisors. The following example shows how to find the LCD.

Example 4. Find the LCD of the following fractions.

$$\frac{1}{6}, \ \frac{1}{8}, \ \frac{1}{18}$$

Step 1: Factor each denominator into prime factors. (Prime factorization may be reviewed in Sec. 1.7.)

$$6 = 2 \cdot 3$$
$$8 = 2 \cdot 2 \cdot 2$$
$$18 = 2 \cdot 3 \cdot 3$$

Step 2: Write each prime factor that number of times it appears *most* in any *one* denominator in Step 1. The LCD is the product of these prime factors.

Here, 2 appears once as a factor of 6, three times as a factor of 8, and once as a factor of 18. So 2 appears at most *three* times in any one denominator. Therefore, you have $2 \cdot 2 \cdot 2$ as factors of the LCD. The factor 3 appears at most twice in any one denominator, so you have $3 \cdot 3$ as factors of the LCD. Now 2 and 3 are the only factors of the three given denominators. The LCD for $\frac{1}{6}$, $\frac{1}{8}$, and $\frac{1}{18}$ must be $2 \cdot 2 \cdot 2 \cdot 3 \cdot 3 = 72$. Note that 72 has divisors 6, 8, and 18.

This procedure could be listed in a table as follows:

| | | Number of times that prime factor appears: | |
Prime factor	Denominator	in given denominator	*most* in any one denominator
2	6	once	
	8	three times	three times
	18	once	
3	6	once	
	8	none	
	18	twice	twice

That is, the LCD contains the factor 2 three times, and the factor 3 two times. Thus, LCD = $2 \cdot 2 \cdot 2 \cdot 3 \cdot 3 = 72$.

Example 5. Find the LCD of $\frac{3}{4}$, $\frac{3}{8}$, $\frac{3}{16}$.

$$4 = 2 \cdot 2$$
$$8 = 2 \cdot 2 \cdot 2$$
$$16 = 2 \cdot 2 \cdot 2 \cdot 2$$

2 appears at most *four* times in any one denominator, so the LCD is $2 \cdot 2 \cdot 2 \cdot 2 = 16$. Note that 16 has divisors 4, 8, and 16.

Example 6. Find the LCD of $\frac{2}{5}$, $\frac{4}{15}$, $\frac{3}{20}$.

$$5 = 5$$
$$15 = 3 \cdot 5 \qquad \text{LCD is } 2 \cdot 2 \cdot 3 \cdot 5 = 60.$$
$$20 = 2 \cdot 2 \cdot 5$$

Of course, if you can find the LCD by inspection, you need not go through the method shown in the examples.

Example 7. Find the LCD of $\frac{3}{8}$ and $\frac{5}{16}$.

By inspection, the LCD is 16, because 16 has divisors 8 and 16.

After finding the LCD of the fractions you wish to add, change each of the original fractions to a fraction of equal value, with the LCD as its denominator.

Example 8. Add: $\frac{3}{8} + \frac{5}{16}$

First, find the LCD of $\frac{3}{8}$ and $\frac{5}{16}$. The LCD is 16. Now you must change $\frac{3}{8}$ to a fraction of equal value with a denominator of 16.

Write: $\frac{3}{8} = \frac{?}{16}$.   Think: $8 \times ? = 16$.

Since $8 \times 2 = 16$, we multiply both the numerator and the denominator by 2. We get 6 for the numerator and 16 for the denominator.

$$\frac{3}{8} \times \frac{2}{2} = \frac{6}{16}$$

Now, using the rule for adding fractions:

$$\frac{3}{8} + \frac{5}{16} = \frac{6}{16} + \frac{5}{16} = \frac{6 + 5}{16} = \frac{11}{16}$$

Now try adding some fractions for which the LCD is more difficult to find.

Example 9. Add: $\frac{1}{4} + \frac{1}{6} + \frac{1}{16} + \frac{1}{12}$

First find the LCD.

$$4 = 2 \cdot 2$$
$$6 = 2 \cdot 3$$
$$16 = 2 \cdot 2 \cdot 2 \cdot 2$$
$$12 = 2 \cdot 2 \cdot 3$$

Note that 2 is used as a factor at most four times in any one denominator and 3 as a factor at most once. Thus the LCD = $2 \cdot 2 \cdot 2 \cdot 2 \cdot 3 = 48$.

Second, change each fraction to a fraction of equal value having 48 as its denominator.

$$\frac{1}{4} = \frac{?}{48} \qquad \frac{1 \times 12}{4 \times 12} = \frac{12}{48}$$

$$\frac{1}{6} = \frac{?}{48} \qquad \frac{1 \times 8}{6 \times 8} = \frac{8}{48}$$

$$\frac{1}{16} = \frac{?}{48} \qquad \frac{1 \times 3}{16 \times 3} = \frac{3}{48}$$

$$\frac{1}{12} = \frac{?}{48} \qquad \frac{1 \times 4}{12 \times 4} = \frac{4}{48}$$

Therefore, $\frac{1}{4} + \frac{1}{6} + \frac{1}{16} + \frac{1}{12} = \frac{12}{48} + \frac{8}{48} + \frac{3}{48} + \frac{4}{48}$

$= \frac{12 + 8 + 3 + 4}{48} = \frac{27}{48}$. Writing the result in lowest

terms we have $\frac{27}{48} = \frac{9 \cdot \cancel{3}}{16 \cdot \cancel{3}} = \frac{9}{16}$.

Rule for Subtracting Fractions

$$\frac{a}{c} - \frac{b}{c} = \frac{a - b}{c} \qquad (c \neq 0)$$

To subtract two or more fractions with a common denominator, subtract their numerators and place the difference over the common denominator.

<u>Example 10</u>. Subtract: $\frac{3}{5} - \frac{2}{5} = \frac{3 - 2}{5} = \frac{1}{5}$

<u>Example 11</u>. Subtract: $\frac{5}{8} - \frac{3}{8} = \frac{5 - 3}{8} = \frac{2}{8} = \frac{\cancel{2} \cdot 1}{\cancel{2} \cdot 4} = \frac{1}{4}$

To subtract two fractions that have different denominators, first find the LCD. Then, express each fraction as an equivalent fraction using the LCD, and subtract

<u>Example 12</u>. Subtract: $\frac{5}{6} - \frac{4}{15} = \frac{25}{30} - \frac{8}{30} = \frac{25 - 8}{30} = \frac{17}{30}$

To add mixed numbers, find the LCD of the proper fractions. Add the whole numbers, then add the fractions. Finally, add these two results.

Example 13. Add $2\frac{1}{2}$ and $3\frac{3}{5}$.

$$2\frac{1}{2} = 2\frac{5}{10}$$
$$3\frac{3}{5} = 3\frac{6}{10}$$
$$5\frac{11}{10} = 5 + \frac{11}{10} = 5 + 1 + \frac{1}{10} = 6\frac{1}{10}$$

To subtract mixed numbers after finding the LCD of the proper fractions, subtract the whole numbers. Then subtract the proper fractions.

Example 14. Subtract $8\frac{2}{3}$ from $13\frac{3}{4}$.

$$13\frac{3}{4} = 13\frac{9}{12}$$
$$8\frac{2}{3} = 8\frac{8}{12}$$
$$5\frac{1}{12}$$

If the larger mixed number does not also have the larger proper fraction, borrow 1 from the whole number. Then add it to the proper fraction before subtracting.

Example 15. Subtract $2\frac{3}{5}$ from $4\frac{1}{2}$.

$$4\frac{1}{2} = 4\frac{5}{10} = 3\frac{15}{10}$$
$$2\frac{3}{5} = 2\frac{6}{10} = 2\frac{6}{10}$$
$$1\frac{9}{10}$$

Example 16. Subtract $2\frac{3}{7}$ from $8\frac{1}{4}$.

$$8\frac{1}{4} = 8\frac{7}{28} = 7\frac{35}{28}$$
$$2\frac{3}{7} = 2\frac{12}{28} = 2\frac{12}{28}$$
$$5\frac{23}{28}$$

<u>Example 17</u>. Subtract: $12 - 4\frac{3}{8}$.

$$12 \quad = 11\frac{8}{8}$$
$$\underline{4\frac{3}{8} \quad = \quad 4\frac{3}{8}}$$
$$7\frac{5}{8}$$

EXERCISES 2.2   *Perform the indicated operations and simplify:*

1. $\frac{1}{16} + \frac{7}{16} + \frac{3}{16}$ 2. $\frac{5}{8} + \frac{3}{8} + \frac{7}{8}$

3. $\frac{5}{32} + \frac{7}{32} + \frac{3}{32}$ 4. $\frac{9}{64} + \frac{17}{64} + \frac{3}{64} + \frac{5}{64}$

Find the LCD of each set of fractions:

5. $\frac{1}{2}, \frac{1}{8}, \frac{1}{16}$ 6. $\frac{1}{3}, \frac{2}{5}, \frac{3}{7}$ 7. $\frac{1}{6}, \frac{3}{10}, \frac{1}{14}$

8. $\frac{1}{9}, \frac{1}{15}, \frac{5}{21}$ 9. $\frac{1}{3}, \frac{1}{16}, \frac{7}{8}$ 10. $\frac{1}{5}, \frac{3}{14}, \frac{4}{35}$

11. $\frac{2}{7}, \frac{3}{10}, \frac{2}{15}$ 12. $\frac{5}{11}, \frac{3}{22}, \frac{7}{10}$ 13. $\frac{9}{13}, \frac{5}{6}, \frac{7}{26}$

14. $\frac{3}{16}, \frac{7}{12}, \frac{5}{24}$

Perform the indicated operations and simplify:

15. $\frac{2}{3} + \frac{1}{6}$ 16. $\frac{1}{2} + \frac{3}{8}$ 17. $\frac{1}{16} + \frac{3}{32}$ 18. $\frac{5}{6} + \frac{1}{18}$

19. $\frac{2}{7} + \frac{3}{28}$ 20. $\frac{1}{9} + \frac{2}{45}$ 21. $\frac{3}{8} + \frac{5}{64}$ 22. $\frac{3}{10} + \frac{7}{100}$

23. $\frac{1}{5} + \frac{3}{20}$ 24. $\frac{3}{4} + \frac{3}{16}$ 25. $\frac{4}{5} + \frac{1}{2}$ 26. $\frac{2}{3} + \frac{4}{9}$

27. $\frac{3}{5} + \frac{7}{10}$ 28. $\frac{1}{8} + \frac{2}{3}$ 29. $\frac{7}{8} + \frac{5}{16}$ 30. $\frac{11}{16} + \frac{17}{32}$

31. $\frac{1}{2} + \frac{15}{32}$ 32. $\frac{3}{4} + \frac{7}{8}$ 33. $\frac{2}{23} + \frac{3}{69}$ 34. $\frac{11}{35} + \frac{3}{7}$

35. $\frac{1}{3} + \frac{1}{6} + \frac{3}{16} + \frac{1}{12}$ 36. $\frac{3}{16} + \frac{1}{8} + \frac{1}{3} + \frac{1}{4}$ 37. $\frac{1}{20} + \frac{1}{30} + \frac{1}{40}$

38. $\frac{1}{14} + \frac{1}{15} + \frac{1}{6}$ 39. $\frac{3}{10} + \frac{1}{14} + \frac{4}{15}$ 40. $\frac{5}{36} + \frac{11}{72} + \frac{5}{6}$

41. $\frac{4}{9} + \frac{13}{27} + \frac{2}{3}$ 42. $\frac{3}{7} + \frac{4}{21} + \frac{3}{14}$ 43. $\frac{31}{121} + \frac{4}{33} + \frac{5}{6}$

44. $\frac{23}{27} + \frac{5}{18} + \frac{5}{36}$ 45. $\frac{5}{16} - \frac{3}{16}$ 46. $\frac{5}{7} - \frac{2}{7}$

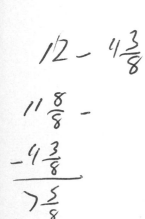

47. $\dfrac{7}{8} - \dfrac{3}{4}$ 48. $\dfrac{9}{64} - \dfrac{2}{128}$ 49. $\dfrac{4}{5} - \dfrac{3}{10}$ 50. $\dfrac{7}{16} - \dfrac{1}{3}$

51. $\dfrac{9}{14} - \dfrac{3}{42}$ 52. $\dfrac{8}{9} - \dfrac{5}{24}$ 53. $\dfrac{9}{16} - \dfrac{13}{32} - \dfrac{1}{8}$ 54. $\dfrac{7}{8} - \dfrac{2}{9} - \dfrac{1}{12}$

55. $2\dfrac{1}{2} + 4\dfrac{3}{4}$ 56. $3\dfrac{5}{8} + 5\dfrac{3}{4}$ 57. $6\dfrac{7}{8} + 3\dfrac{3}{16}$ 58. $7\dfrac{15}{16} - 4\dfrac{9}{16}$

59. $3 - \dfrac{3}{8}$ 60. $8 - 5\dfrac{3}{4}$ 61. $8\dfrac{3}{16} - 3\dfrac{7}{16}$ 62. $5\dfrac{3}{8} + 2\dfrac{3}{4}$

63. $7\dfrac{3}{16} - 4\dfrac{7}{8}$ 64. $8\dfrac{1}{4} - 4\dfrac{7}{16}$ 65. $3\dfrac{4}{5} + 9\dfrac{8}{9}$ 66. $4\dfrac{5}{12} + 6\dfrac{17}{20}$

67. $5\dfrac{6}{7} - 4\dfrac{11}{12}$ 68. $7\dfrac{2}{9} - 4\dfrac{5}{6}$ 69. $3\dfrac{9}{16} + 4\dfrac{7}{12} + 3\dfrac{1}{6}$

70. $5\dfrac{2}{5} + 3\dfrac{7}{10} + 4\dfrac{7}{15}$ 71. $16\dfrac{5}{8} - 4\dfrac{7}{12} - 2\dfrac{1}{2}$ 72. $12\dfrac{9}{16} - 3\dfrac{1}{6} + 2\dfrac{1}{4}$

2.3 Applications Involving Addition and Subtraction of Fractions

An electrical circuit with more than one path for the current to flow is called a *parallel circuit*.

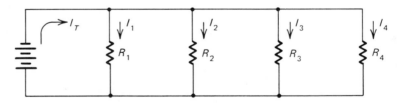

Figure 2.3

The current in a parallel circuit as in Fig. 2.3 is divided among the branches in the circuit. How it is divided depends on the resistance in each branch. Since the current is divided among the branches, the total current (I_T) of the circuit is the same as the current from the source. This equals the sum of the currents through the individual branches of the circuit. That is, $I_T = I_1 + I_2 + I_3 + \ldots$.

Example 1. Find the total current in the parallel circuit in Fig. 2.4.

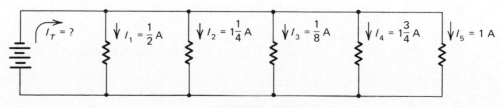

Figure 2.4

$$I_T = I_1 + I_2 + I_3 + I_4 + I_5$$

$$\frac{1}{2}\,A = \frac{4}{8}\,A$$

$$1\frac{1}{4}\,A = 1\frac{2}{8}\,A$$

$$\frac{1}{8}\,A = \frac{1}{8}\,A$$

$$1\frac{3}{4}\,A = 1\frac{6}{8}\,A$$

$$\underline{1\ \ A = 1\ \ A}$$

$$3\frac{13}{8}\,A = 4\frac{5}{8}\,A$$

<u>Example 2</u>. Find the missing dimension in Fig. 2.5.

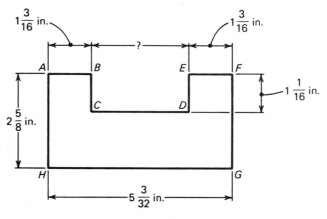

Figure 2.5

To find the length of the missing dimension, subtract the sum of side *AB* and side *EF* from side *HG*. That is, add:

$$1\frac{3}{16}\text{ in.}$$

$$\underline{1\frac{3}{16}\text{ in.}}$$

$$2\frac{6}{16}\text{ in. or } 2\frac{3}{8}\text{ in.}$$

Then subtract: $5\dfrac{3}{32}$ in. $= 5\dfrac{3}{32}$ in. $= 4\dfrac{35}{32}$ in.

$$\underline{2\frac{3}{8}\text{ in.} = 2\frac{12}{32}\text{ in.} = 2\frac{12}{32}\text{ in.}}$$

$$2\frac{23}{32}\text{ in.}$$

Therefore, the missing dimension is $2\dfrac{23}{32}$ in.

Example 3. Find the perimeter (distance around) of Fig. 2.5.

The perimeter is the sum of the lengths of all the sides of the figure.

AB:	$1\frac{3}{16}$ in.	$= 1\frac{6}{32}$ in.
BC:	$1\frac{1}{16}$ in.	$= 1\frac{2}{32}$ in.
CD:	$2\frac{23}{32}$ in.	$= 2\frac{23}{32}$ in.
DE:	$1\frac{1}{16}$ in.	$= 1\frac{2}{32}$ in.
EF:	$1\frac{3}{16}$ in.	$= 1\frac{6}{32}$ in.
FG:	$2\frac{5}{8}$ in.	$= 2\frac{20}{32}$ in.
GH:	$5\frac{3}{32}$ in.	$= 5\frac{3}{32}$ in.
HA:	$2\frac{5}{8}$ in.	$= 2\frac{20}{32}$ in.

$$15\frac{82}{32} \text{ in.} = 15 + 2\frac{18}{32} \text{ in.}$$

$$= 17\frac{9}{16} \text{ in.}$$

EXERCISES 2.3   1. Find the perimeter of the triangular field below.

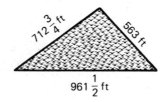

Find (a) *the length of the missing dimension and* (b) *the perimeter of each figure:*

2.

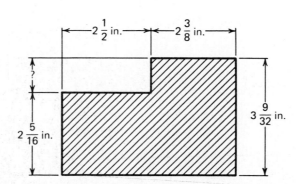

3.

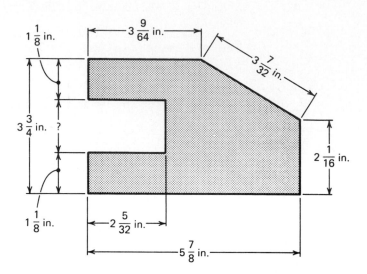

4.

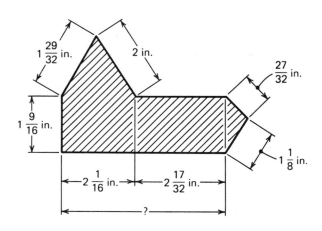

5.

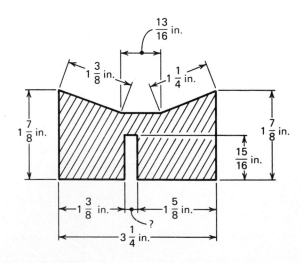

6. The perimeter (sum of the sides) of a triangle is $59\frac{9}{32}$ in. One side is $19\frac{5}{8}$ in., and a second side is $17\frac{13}{16}$ in. How long is the remaining side?

Find the total current in each parallel circuit:

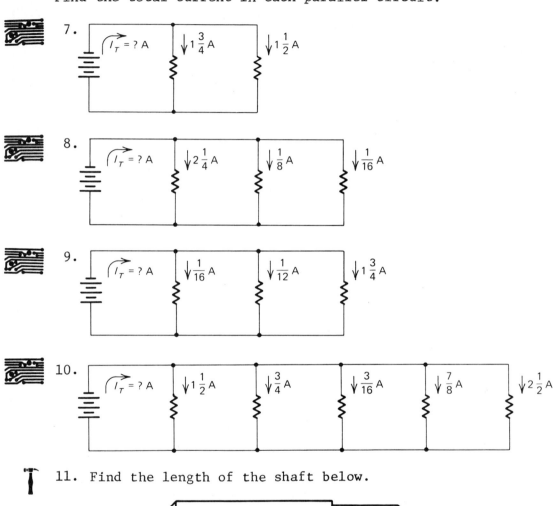

7. $I_T = ?$ A $1\frac{3}{4}$ A $1\frac{1}{2}$ A

8. $I_T = ?$ A $2\frac{1}{4}$ A $\frac{1}{8}$ A $\frac{1}{16}$ A

9. $I_T = ?$ A $\frac{1}{16}$ A $\frac{1}{12}$ A $1\frac{3}{4}$ A

10. $I_T = ?$ A $1\frac{1}{2}$ A $\frac{3}{4}$ A $\frac{3}{16}$ A $\frac{7}{8}$ A $2\frac{1}{2}$ A

11. Find the length of the shaft below.

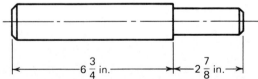

$6\frac{3}{4}$ in. $2\frac{7}{8}$ in.

12. Find the distance between the centers of the two end-holes of the plate below.

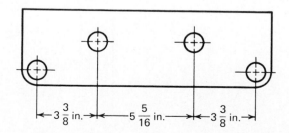

$3\frac{3}{8}$ in. $5\frac{5}{16}$ in. $3\frac{3}{8}$ in.

13. Find (a) the length of the tool and (b) the diameter of the part indicated below by the question mark.

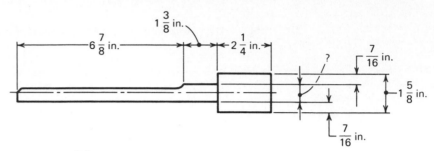

14. A rod $13\frac{13}{16}$ in. has been cut as shown below. Assume the waste in each cut is $\frac{1}{16}$ in. What is the length of the remaining piece?

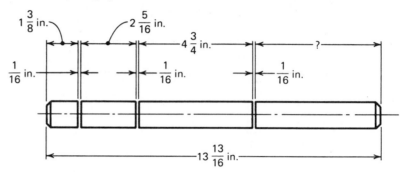

15. Find (a) the length and (b) the diameter of the shaft in the figure below.

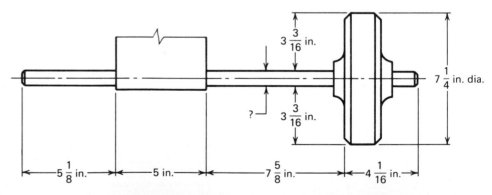

16. Find the missing dimension in the taper below.

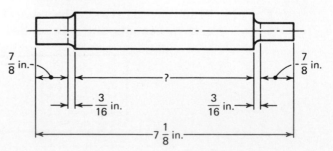

17. How much must the diameter of a $\frac{7}{8}$-in. shaft be reduced so that its diameter will be $\frac{51}{64}$ in.?

18. What is the difference in thickness between a $\frac{7}{16}$-in. steel plate and a $\frac{5}{8}$-in. steel plate?

19. A planer takes a $\frac{3}{32}$-in. cut on a plate which is $1\frac{7}{8}$ in. thick. What is the thickness of the plate after one cut? What is the thickness of the plate after three cuts?

20. A home is built on a $65\frac{3}{4}$ ft-wide lot. The house is $5\frac{5}{12}$ ft from one side of the lot and is $43\frac{5}{6}$ ft wide. What is the distance of the house from the other side of the lot?

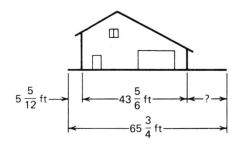

21. What width and length steel strip is needed in order to drill three holes of $3\frac{5}{16}$ in.? Allow $\frac{7}{32}$ in. between, and on each side of, the holes.

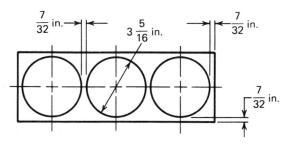

22. Find length x below.

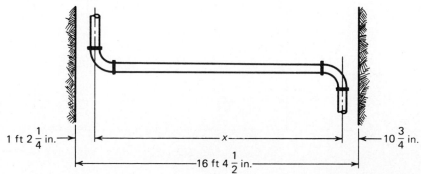

23. What is the length of the line *AB* in the figure?

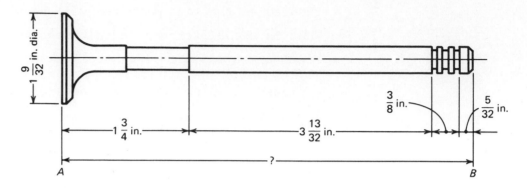

 24. Find the total length of the shaft shown below.

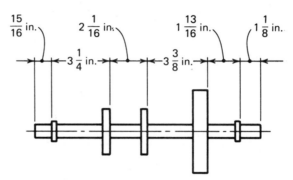

 25. A mechanic needs the following lengths of $\frac{3}{8}$ - in.

copper tubing: $15\frac{3}{8}$ in., $7\frac{3}{4}$ in., $11\frac{1}{2}$ in., $7\frac{7}{32}$ in., and $10\frac{5}{16}$ in. What is the total length of tubing needed?

26. An end view and side view of a shaft are shown below.
(a) Find the diameter of the largest part of the shaft.
(b) Find the length of *A* of the shaft.

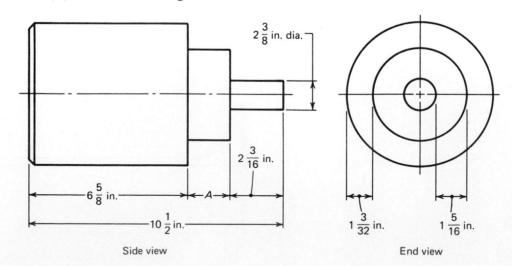

Side view   End view

2.4 Multiplication and Division of Fractions

Rule for Multiplying Fractions

$$\frac{a}{b} \times \frac{c}{d} = \frac{a \cdot c}{b \cdot d} \qquad (b \neq 0,\ d \neq 0)$$

To multiply fractions, multiply the numerators and multiply the denominators.

<u>Example 1</u>. Multiply: $\dfrac{5}{9} \times \dfrac{3}{10} = \dfrac{5 \cdot 3}{9 \cdot 10} = \dfrac{15}{90} = \dfrac{\cancel{15} \cdot 1}{\cancel{15} \cdot 6} = \dfrac{1}{6}$

To simplify the work, consider the following:

$$\frac{\overset{1}{\cancel{5}}}{\underset{3}{\cancel{9}}} \times \frac{\overset{1}{\cancel{3}}}{\underset{2}{\cancel{10}}} = \frac{1}{3 \cdot 2} = \frac{1}{6}$$

This method divides the numerator by 15, or (5 × 3), and the denominator by 15, or (5 × 3). It does not change the value of the fraction.

<u>Example 2</u>. Multiply: $\dfrac{\overset{2}{\cancel{18}}}{25} \times \dfrac{7}{\underset{3}{\cancel{27}}} = \dfrac{2 \cdot 7}{25 \cdot 3} = \dfrac{14}{75}$

As a short cut, divide a numerator by 9 and a denominator by 9. Then multiply.

<u>Example 3</u>. Multiply: $8 \times 3\dfrac{3}{4} = \dfrac{\overset{2}{\cancel{8}}}{1} \times \dfrac{15}{\underset{1}{\cancel{4}}} = 30$

Whenever you multiply several fractions, you may divide *any* numerator and *any* denominator by the same number.

<u>Example 4</u>. Multiply: $\dfrac{9}{16} \times \dfrac{5}{22} \times \dfrac{4}{7} \times 3\dfrac{2}{3}$

$$= \frac{\overset{3}{\cancel{9}}}{\underset{4}{\cancel{16}}} \times \frac{5}{\underset{2}{\cancel{22}}} \times \frac{\overset{1}{\cancel{4}}}{7} \times \frac{\overset{1}{\cancel{11}}}{\underset{1}{\cancel{3}}} = \frac{15}{56}$$

Rule for Dividing Fractions

$$\frac{a}{b} \div \frac{c}{d} = \frac{a}{b} \times \frac{d}{c} = \frac{a \cdot d}{b \cdot c} \qquad (b \neq 0,\ c \neq 0,\ d \neq 0)$$

To divide a fraction by a fraction, invert the fraction (interchange numerator and denominator) that follows the division sign ($\div$). Then multiply the resulting fractions.

Example 5. Divide: $\dfrac{5}{6} \div \dfrac{2}{3} = \dfrac{5}{\overset{}{\underset{2}{6}}} \times \dfrac{\overset{1}{3}}{2} = \dfrac{5 \cdot 1}{2 \cdot 2} = \dfrac{5}{4}$ or $1\dfrac{1}{4}$

Example 6. Divide: $7 \div \dfrac{2}{5} = \dfrac{7}{1} \times \dfrac{5}{2} = \dfrac{35}{2}$ or $17\dfrac{1}{2}$

Example 7. Divide: $\dfrac{3}{5} \div 4 = \dfrac{3}{5} \times \dfrac{1}{4} = \dfrac{3}{20}$

Example 8. Divide: $\dfrac{9}{10} \div 5\dfrac{2}{5} = \dfrac{9}{10} \div \dfrac{27}{5} = \dfrac{\overset{1}{9}}{\underset{2}{10}} \times \dfrac{\overset{1}{5}}{\underset{3}{27}} = \dfrac{1}{6}$

When both multiplication and division of fractions occur, invert the first fraction that follows a division sign ($\div$). Then proceed according to the rules for multiplying fractions.

Example 9. $\dfrac{2}{5} \times \dfrac{1}{3} \div \dfrac{3}{4}$

$= \dfrac{2}{5} \times \dfrac{1}{3} \times \dfrac{4}{3} = \dfrac{2 \cdot 1 \cdot 4}{5 \cdot 3 \cdot 3} = \dfrac{8}{45}$

Example 10. $7\dfrac{1}{3} \div 4 \times \dfrac{2}{1} = \dfrac{\overset{11}{22}}{3} \times \dfrac{1}{\underset{\underset{1}{2}}{4}} \times \dfrac{\overset{1}{2}}{1} = \dfrac{11 \cdot 1 \cdot 1}{3 \cdot 1 \cdot 1}$

$= \dfrac{11}{3}$ or $3\dfrac{2}{3}$

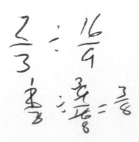

EXERCISES 2.4   *Perform the indicated operations and simplify:*

1. $\frac{1}{3} \times 7$

2. $8 \times \frac{1}{2}$

3. $\frac{3}{4} \times 12$

4. $3\frac{1}{2} \times \frac{2}{5}$

5. $1\frac{3}{4} \times \frac{5}{16}$

6. $\frac{1}{3} \times \frac{1}{3} \times \frac{1}{3}$

7. $\frac{16}{21} \times \frac{7}{8}$

8. $\frac{7}{12} \times \frac{45}{56}$

9. $\frac{2}{7} \times 35$

10. $\frac{9}{16} \times \frac{2}{3} \times 1\frac{6}{15}$

11. $\frac{5}{8} \times \frac{7}{10} \times \frac{2}{7}$

12. $\frac{9}{16} \times \frac{5}{9} \times \frac{4}{25}$

13. $2\frac{1}{3} \times \frac{5}{8} \times \frac{6}{7}$

14. $\frac{5}{28} \times \frac{3}{5} \times \frac{2}{3} \times \frac{2}{9}$

15. $\frac{6}{11} \times \frac{26}{35} \times 1\frac{9}{13} \times \frac{7}{12}$

16. $\frac{3}{8} \div \frac{1}{4}$

17. $\frac{3}{5} \div \frac{10}{12}$

18. $\frac{10}{12} \div \frac{3}{5}$

19. $4\frac{1}{2} \div \frac{1}{4}$

20. $18\frac{2}{3} \div 6$

21. $15 \div \frac{3}{8}$

22. $\frac{77}{6} \div 6$

23. $\frac{7}{11} \div \frac{3}{5}$

24. $7 \div 3\frac{1}{8}$

25. $\frac{2}{5} \times 3\frac{2}{3} \div \frac{3}{4}$

26. $\frac{7}{8} \times \frac{1}{2} \div \frac{2}{7}$

27. $\frac{16}{5} \times \frac{3}{2} \times \frac{10}{4} \div 5\frac{1}{3}$

28. $6 \times 6 \times \frac{21}{7} \div 48$

29. $\frac{7}{9} \times \frac{3}{8} \div \frac{28}{81}$

30. $2\frac{1}{3} \times \frac{5}{8} \div \frac{10}{4}$

31. $\frac{2}{7} \times \frac{5}{9} \times \frac{3}{10} \div 6$

32. $\frac{9}{4} \times \frac{9}{4} \times \frac{21}{7} \div 81$

33. $\frac{7}{16} \div \frac{3}{8} \times \frac{1}{2}$

34. $\frac{5}{8} \div \frac{25}{64} \times \frac{5}{6}$

Fraction Review. *Perform the indicated operations and simplify:*

35. $\frac{1}{2} + \frac{3}{4}$

36. $1\frac{1}{3} + \frac{3}{5}$

37. $\frac{3}{4} - \frac{2}{3}$

38. $2\frac{3}{4} - \frac{7}{8}$

39. $\frac{1}{2} \times \frac{3}{4}$

40. $\frac{7}{8} \times 2\frac{1}{2}$

41. $2\frac{1}{2} \div 4$

42. $3\frac{3}{4} \div \frac{7}{8}$

43. $\frac{5}{8} + 2\frac{3}{4}$

44. $5 - \frac{5}{16}$

45. $3\frac{1}{2} \times \frac{3}{4}$

46. $2\frac{1}{2} \div \frac{1}{2}$

47. $1\frac{7}{8} + 2\frac{3}{4}$ 48. $3\frac{1}{2} - 2\frac{7}{8}$ 49. $14\frac{1}{2} - 5\frac{7}{16}$

50. $2\frac{5}{8} \times 5$ 51. $7\frac{3}{8} \div \frac{1}{3}$ 52. $4\frac{1}{8} - 2\frac{21}{64}$

53. $7\frac{15}{16} + \frac{7}{8}$ 54. $5\frac{1}{4} - 3\frac{15}{16}$ 55. $3\frac{1}{16} + \frac{3}{32}$

56. $3\frac{1}{8} \times 2\frac{3}{4}$ 57. $3\frac{1}{2} \div 2\frac{1}{4}$ 58. $5\frac{1}{2} \times 3\frac{1}{8}$

59. $\frac{3}{5} \times 2\frac{1}{3} \div 4\frac{2}{3}$ 60. $1\frac{7}{8} \div 2\frac{13}{16} \div 8$

2.5 Applications Involving Multiplication and Division of Fractions

Lumber is usually measured in board feet. One *board foot* is the amount of wood contained in a piece of wood that measures one inch thick and one square foot in area, or its equivalent. (See Fig. 2.6.) The number of board feet in lumber may be found by the formula:

$$\text{bd ft} = \frac{\substack{\text{number of} \\ \text{boards}} \times \substack{\text{thickness} \\ \text{(in in.)}} \times \substack{\text{width} \\ \text{(in in.)}} \times \substack{\text{length} \\ \text{(in ft)}}}{12}$$

The 12 in the denominator comes from the fact that the simplest form of one board foot can be thought of as a board that is 1 in. thick × 12 in. wide × 1 ft long.

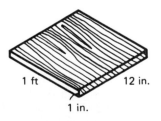

1 ft   12 in.

1 in.

One board foot

Figure 2.6

Lumber is either rough or dressed. *Rough stock* is lumber that is not planed or dressed. *Dressed stock* is lumber that is planed on one or more sides. When measuring lumber, we compute the full size. That is, we compute the measure of the rough stock that is required to make the desired dressed piece. When lumber is dressed or planed, $\frac{1}{16}$ in. is taken off each side when the lumber is less than $1\frac{1}{2}$ in. thick. If the lumber is $1\frac{1}{2}$ in. or more in thickness, $\frac{1}{8}$ in. is taken off each side. Lumber for framing houses usually measures $\frac{1}{2}$ in. less than the name that we call the piece. For example, a "two-by-four," a piece 2 in. by 4 in., actually measures $1\frac{1}{2}$ in. by $3\frac{1}{2}$ in.

Example 1. Find the number of board feet contained in 6 pieces of lumber 2 in. × 8 in. × 16 ft (Fig. 2.7).

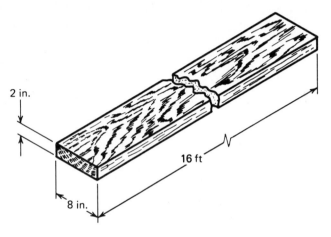

2 in.

16 ft

8 in.

Figure 2.7

$$\text{bd ft} = \frac{\overset{\text{number of}}{\underset{\text{boards}}{}} \times \overset{\text{thickness}}{\underset{\text{(in in.)}}{}} \times \overset{\text{width}}{\underset{\text{(in in.)}}{}} \times \overset{\text{length}}{\underset{\text{(in ft)}}{}}}{12}$$

$$= \frac{6 \times 2 \times 8 \times 16}{12} = 128 \text{ bd ft}$$

Example 2. Energy in the form of electrical power is used and depended on by industry and consumers alike. Power (in watts, W) equals the voltage (in volts, V) times the current (in amperes, or amps, A). A soldering iron draws a current of $7\frac{1}{2}$ A on a 110-V circuit. What is its wattage rating—the power it uses?

$$\text{Power} = (\text{voltage}) \times (\text{current})$$

$$= \quad 110 \quad \times \quad 7\frac{1}{2}$$

$$= \quad 110 \quad \times \quad \frac{15}{2}$$

$$= \quad 825 \text{ W}$$

Power may also be found by computing the product of the square of the current (in amps, A) and the resistance (in ohms, Ω).

<u>Example 3</u>. To give $\frac{1}{5}$ grain of Myleran from $\frac{1}{30}$ - grain tablets, how many tablets would be given?

To find how many tablets would be given, we divide the amount to be given by the amount in each tablet.

$$\frac{1}{5} \div \frac{1}{30} = \frac{1}{\cancel{5}_{1}} \times \frac{\cancel{30}^{6}}{1}$$

$$= 6 \text{ tablets}$$

<u>Example 4</u>. One form of Ohm's law states that the current (in amps, A) in a simple circuit equals the voltage (in volts, V) divided by the resistance (in ohms, Ω). What current is required for a heating element with a resistance of $7\frac{1}{2}$ Ω operating in a 12-V circuit?

$$\text{Current} = (\text{voltage}) \div (\text{resistance})$$

$$= \quad 12 \quad \div \quad 7\frac{1}{2}$$

$$= \quad 12 \quad \div \quad \frac{15}{2}$$

$$= \quad 12 \quad \times \quad \frac{2}{15}$$

$$= \quad \frac{8}{5} \text{ or } 1\frac{3}{5} \text{ A}$$

EXERCISES 2.5

1. A barrel has a capacity of 42 gal. How many gallons does it contain when it is $\frac{3}{4}$ full?

2. Find the area of the rectangle with the length $6\frac{1}{3}$ ft and width $3\frac{3}{4}$ ft. (Area = length × width.)

3. Find the perimeter of a rectangle with length $6\frac{1}{3}$ ft and width $3\frac{3}{4}$ ft.

4. Find the perimeter of a rectangle whose length is $2\frac{1}{4}$ times its width and whose width is 6 ft.

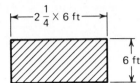

5. A length of $71\frac{1}{2}$ in. is to be divided into 16 equal parts. What is the length of each part?

Find the number of board feet in each quantity of lumber:

6. One piece 2 in. × 6 in. × 8 ft

7. 10 pieces 2 in. × 4 in. × 12 ft

8. 24 pieces 4 in. × 4 in. × 16 ft

9. 175 pieces 1 in. × 8 in. × 14 ft

10. What is the total length of eight pieces of steel each $5\frac{3}{4}$ in. long?

11. The outside diameter of a pipe is $4\frac{9}{32}$ in. The walls are $\frac{7}{32}$ in. thick. Find the inside diameter.

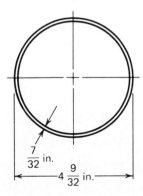

12. Two strips of metal are riveted together in a straight line, with nine rivets equally spaced $2\frac{5}{16}$ in. apart. What is the distance between the first and last rivet?

13. Two metal strips are riveted together in a straight line, with 16 equally spaced rivets. The distance between the first and last rivet is $28\frac{1}{8}$ in. What is the distance between any two consecutive rivets?

14. Determine length x in the figure.

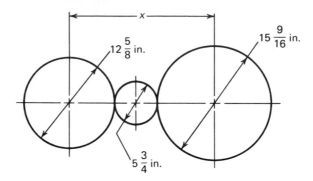

15. From a steel rod 36 in. long, the following pieces are cut:

$$3 \text{ pieces } 2\frac{1}{8} \text{ in.}$$

$$2 \text{ pieces } 5\frac{3}{4} \text{ in.}$$

$$6 \text{ pieces } \frac{7}{8} \text{ in.}$$

$$1 \text{ piece } 3\frac{1}{2} \text{ in.}$$

Assume there is $\frac{1}{16}$ in. waste in each cut. What is the length of the remaining piece?

16. A piece of drill rod 2 ft 6 in. long is to be cut into pins, each $2\frac{1}{2}$ in. long. (a) Assume no loss of material in cutting. How many pins will you get? (b) Assume $\frac{1}{16}$ in. waste in each cut. How many pins will you get?

17. The cutting tool on a lathe moves $\frac{3}{128}$ in. along the piece being turned for each revolution of work. The cutting tool revolves at 45 revolutions per minute. How long will it take to turn a length of $9\frac{9}{64}$ in. stock in one operation?

18. The outer diameter of the gear shown is $7\frac{21}{22}$ in. Find the circumference of the gear. (Hint: Circumference = $\frac{22}{7}$ × diameter.)

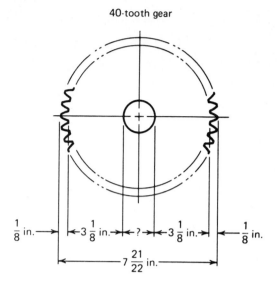

40-tooth gear

 19. Find the diameter of the hole in the gear above.

 20. How many $5\frac{1}{4}$ in. complete lengths of radiator hose can be cut from a 6-ft roll?

 21. Find the load of a circuit that takes $12\frac{1}{4}$ A at 220 V. (See Example 2.)

 22. An electric iron requires $4\frac{1}{4}$ A and has a resistance of $24\frac{1}{2}$ Ω. What voltage does it require to operate? ($V = IR$)

 23. A hand drill draws $3\frac{3}{4}$ A and has a resistance of $14\frac{1}{2}$ Ω. What power does it use? ($P = I^2R$)

24. A wiring job requires the following lengths of BX cable:

$$12 \text{ pieces} \quad 8\frac{1}{2} \text{ ft}$$

$$7 \text{ pieces} \quad 18\frac{1}{2} \text{ ft}$$

$$24 \text{ pieces} \quad 1\frac{3}{4} \text{ ft}$$

$$12 \text{ pieces} \quad 6\frac{1}{2} \text{ ft}$$

$$2 \text{ pieces} \quad 34\frac{1}{4} \text{ ft}$$

How much cable is needed to complete the job?

25. What current is required for a heating element with a resistance of $10\frac{1}{2}$ Ω operating in a 24-V circuit? (See Example 4.)

 26. How many lengths of wire, each $3\frac{3}{4}$ in. long, can be cut from a 25-ft roll?

 27. A total of 19 ceiling outlets are to be equally spaced in a straight line between two points which are $130\frac{1}{2}$ ft apart between centers in a hallway. How far apart will the ceiling outlets be?

 28. If $8\frac{3}{4}$ dozen electrical clips cost $7.35, what did each clip cost?

 29. A steer gains $1\frac{3}{4}$ lb a day. How many pounds will it gain in 36 days?

 30. Tom needs to apply $1\frac{3}{4}$ gal of Amiben per acre of soybeans. How many gallons of Amiben are needed for 120 acres?

 31. An airplane sprayer tank holds 60 gal. If $\frac{3}{4}$ gal of water and $\frac{1}{2}$ lb of pesticide are applied per acre, how much pesticide powder is needed per tankful?

 32. If 1 ft^3 of cotton weighs $22\frac{1}{2}$ lb, how many cubic feet are contained in a bale of cotton weighing 500 lb? In 15 tons of cotton?

 33. A test plot of $\frac{1}{20}$ acre produces 448 lb of shelled corn. Find the yield in bushels per acre. (1 bu of shelled corn weighs 56 lb.)

 34. A farmer wishes to concrete his feed lot, which measures 120 ft by 180 ft. He chooses to have a base of 4 in. of gravel covered with $3\frac{1}{2}$ in. of concrete. (a) How many cubic yards of each material must he purchase? (b) What is his total materials cost? Concrete costs $50/yd^3 delivered and gravel costs $5/ton delivered. (1 yd^3 of gravel weighs approximately 2500 lb.)

 35. The medicine in a bottle contains $\frac{1}{5}$ alcohol. The bottle holds $2\frac{1}{2}$ oz of medicine. How many ounces of alcohol does the bottle contain?

36. A patient receives $\frac{1}{4}$ of a $\frac{1}{4}$-grain morphine sulfate tablet. How much of the drug did the patient receive?

37. To give 50 mg of ascorbic acid from 100-mg tablets, how many tablets should be given?

38. To give 50 mg of ascorbic acid from 250-mg tablets, how many tablets should be used?

39. A patient is given $\frac{1}{4}$ of a 5-grain aspirin. How much aspirin did the patient receive?

40. To give 1 grain of digitalis from $1\frac{1}{2}$-grain tablets, how many tablets should be used?

41. A patient is given $\frac{3}{4}$ grain of codeine from $\frac{3}{8}$-grain tablets. How many tablets are given?

42. If you used $\frac{2}{3}$ of a $7\frac{1}{2}$-grain tablet, how much would you use?

43. A patient is given $\frac{1}{2}$ grain of Valium from $\frac{1}{6}$-grain tablets. How many tablets are given?

44. The doctor orders 45 mg of prednisone and each tablet contains 10 mg. How many tablets would be given to the patient?

45. To give 750 mg of ascorbic acid from 100-mg tablets, how many tablets would you use?

2.6 The English System of Weights and Measures

Centuries ago the thumb, hand, foot, and length from nose to outstretched fingers were used as units of measurement. These methods, of course, did not turn out to be very satisfactory. In the fourteenth century, King Edward II proclaimed the length of the English inch to be the same as three barleycorn grains laid end to end. (See Fig. 2.8.) This proclamation helped some, but it did not eliminate disputes over the length of the English inch.

Each of these methods furnishes estimations of measurements. Actually, *measurement* is the comparison of an *observed* quantity with a *standard unit* quantity. In the estimation above, there is no one standard unit. A standard unit that is constant, accurate, and accepted by all is needed for technical measurements. Today, most nations have bureaus to set national standards for all measures.

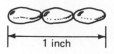

Figure 2.8

The English system makes it necessary for us all to understand fractions and be able to use fractions in everyday life. After we have converted to the metric system, the importance of fractional computations will be greatly reduced.

Take a moment now to look at the table of English weights and measures, Table 1, in the Appendix. You should become familiar with most of this table because, when solving problems, you will have to use it when changing units.

<u>Example 1</u>. Change 5 ft 9 in. to inches.

$$1 \text{ ft} = 12 \text{ in., so } 5 \text{ ft} = 5 \times 12 \text{ in.} = 60 \text{ in.}$$
$$5 \text{ ft } 9 \text{ in.} = 60 \text{ in.} + 9 \text{ in.} = 69 \text{ in.}$$

When converting from one set of units to another, most people know that multiplication or division by some quantity is usually involved. Unfortunately, many just guess whether to multiply or to divide. This gives them only a 50-50 chance of being correct. We are now going to develop a method of converting which may be used even when one is not familiar with the units.

We know that we can multiply any number or quantity by 1 (one) without changing the value of the original quantity. We also know that any fraction whose numerator and denominator are the same is equal to 1. For example, $\frac{5}{5} = 1$, $\frac{16}{16} = 1$, and $\frac{7 \text{ ft}}{7 \text{ ft}} = 1$. Also since 12 in. = 1 ft, $\frac{12 \text{ in.}}{1 \text{ ft}} = 1$, and likewise, $\frac{1 \text{ ft}}{12 \text{ in.}} = 1$ because the numerator equals the denominator. We call this name for 1 a *conversion factor*. The information necessary for forming the conversion factor is found in the tables in the Appendix, in case you do not remember it.

Conversion Factor

The correct conversion factor is equal to <u>one</u> in fraction form; that is, the numerator equals the denominator. The numerator is expressed in the new units and the denominator is expressed in the old units.

<u>Example 2</u>. Change 19 ft to inches.

Since 1 ft = 12 in., the two possible conversion factors are $\frac{1 \text{ ft}}{12 \text{ in.}} = 1 = \frac{12 \text{ in.}}{1 \text{ ft}}$. We want to choose the one whose numerator is expressed in the new units (in.) and wnose denominator is expressed in the old units (ft); that is, $\frac{12 \text{ in.}}{1 \text{ ft}}$. Therefore,

$$19 \; \cancel{\text{ft}} \times \frac{12 \text{ in.}}{1 \; \cancel{\text{ft}}} = 19 \times 12 \text{ in.} = 228 \text{ in.}$$

$\uparrow$——conversion factor

<u>Example 3</u>. Change 8 yd to feet.

$$3 \text{ ft} = 1 \text{ yd, so } 8 \; \cancel{\text{yd}} \times \frac{3 \text{ ft}}{1 \; \cancel{\text{yd}}} = 8 \times 3 \text{ ft} = 24 \text{ ft}$$

<u>Example 4</u>. Change 76 oz to pounds.

$$76 \; \cancel{\text{oz}} \times \frac{1 \text{ lb}}{16 \; \cancel{\text{oz}}} = \frac{76}{16} \text{ lb} = \frac{19}{4} \text{ lb} = 4\frac{3}{4} \text{ lb}$$

Sometimes it is necessary to use more than one conversion factor.

<u>Example 5</u>. Change 6 mi to yards.

In Table 1 of the Appendix, there is no expression equating miles with yards. Thus, it is necessary to use two conversion factors.

$$6 \; \cancel{\text{mi}} \times \frac{5280 \; \cancel{\text{ft}}}{1 \; \cancel{\text{mi}}} \times \frac{1 \text{ yd}}{3 \; \cancel{\text{ft}}} = \frac{6 \times 5280}{3} \text{ yd} = 10,560 \text{ yd}$$

Example 6. How could a technician mixing chemicals express 5000 fluid ounces (fl oz) as gallons?

No conversions between fluid ounces and gallons are given in the tables. Use the conversion factors for fluid ounces to pints (pt); pints to quarts (qt); and quarts to gallons.

$$5000 \ \cancel{fl \ oz} \times \frac{1 \ \cancel{pt}}{16 \ \cancel{fl \ oz}} \times \frac{1 \ \cancel{qt}}{2 \ \cancel{pt}} \times \frac{1 \ gal}{4 \ \cancel{qt}} = \frac{5000}{16 \times 2 \times 4} \ gal$$

$$= 39\frac{1}{16} \ gal$$

EXERCISES 2.6   *Fill in each blank:*

1. 3 ft 7 in. = _____ in.

2. 6 yd 4 ft = _____ ft

3. 5 lb 3 oz = _____ oz

4. 7 yd 3 ft 6 in. = _____ in.

5. 4 qt 1 pt = _____ pt

6. 6 gal 3 qt = _____ pt

7. 2 bu 2 pecks (pk) = _____ pk

8. 5 bu 3 pk = _____ pk

9. 8 ft = _____ in.   10. 5 yd = _____ ft

11. 3 qt = _____ pt   12. 4 mi = _____ ft

13. 96 in. = _____ ft   14. 72 ft = _____ yd

15. 10 pt = _____ qt   16. 54 in. = _____ ft

17. 88 oz = _____ lb   18. 32 fl oz = _____ pt

19. 14 qt = _____ gal   20. 3 bu = _____ pk

21. 56 fl oz = _____ pt   22. 7040 ft = _____ mi

23. 92 ft = _____ yd   24. 9000 lb = _____ tons

25. 2 mi = _____ yd   26. 6000 fl oz = _____ gal

27. 500 fl oz = _____ qt   28. 3 mi = _____ rods

29. 6 bu = _____ pt

30. A plane is flying at 22,000 ft. How many miles high is it?

31. Express in inches the length of a shaft $12\frac{3}{4}$ ft long.

32. A machinist has 15 wrought-iron rods to mill. Each rod weighs 24 oz. What is the total weight of the rods in pounds?

33. The instructions on a carton of chemicals call for mixing 144 fl oz of water, 24 fl oz of chemical No. 1, and 56 fl oz of chemical No. 2. How many quarts are contained in the final mixture?

34. The resistance of 1 ft of No. 32-gauge copper wire is $\frac{4}{25}$ Ω. What is the resistance of 12 yd of this wire?

35. A farmer wishes to wire a shed which is 1 mi from the electricity source in his barn. He uses No. 0-gauge copper wire, which has a resistance of $\frac{1}{10}$ ohm (Ω) per 1000 ft. What is the resistance for the mile of wire?

36. To mix an order of feed, the following quantities of feed are combined: 4200 lb, 600 lb, 5800 lb, 1300 lb, and 2100 lb. How many tons are in the final mixture?

2.7 More Formulas

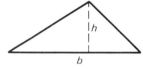

Figure 2.9

The area of a triangle is given by the formula $A = \frac{1}{2}bh$, where b is the length of the base and h is the length of the altitude to the base (Fig. 2.9).

EXERCISES 2.7   *Find the area of each triangle:*

1. $b = 4\frac{7}{8}$ in., $h = 2\frac{1}{2}$ in.

2. $b = 8$ in., $h = 3\frac{3}{4}$ in.

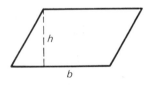

Figure 2.10

The area of a *parallelogram* is given by the formula $A = bh$, where b is the length of the base and h is the length of the altitude to the base (Fig. 2.10).

Find the area of each parallelogram:

3. $b = 6$ in., $h = 2\frac{31}{32}$ in.

4. $b = 8\frac{1}{4}$ in., $h = 3\frac{3}{8}$ in.

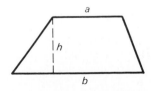

Figure 2.11

The area of a *trapezoid* is given by the formula $A = \left(\frac{a + b}{2}\right)h$, where a and b are the lengths of the parallel sides, and h is the length of the altitude to the base (Fig. 2.11).

Find the area of each trapezoid:

5. $a = 6$ in., $b = 16$ in., $h = 3\frac{5}{8}$ in.

6. $a = 9\frac{1}{8}$ in., $b = 14\frac{5}{8}$ in., $h = 5\frac{3}{16}$ in.

A formula for *distance*, d, when the rate, r, and the time, t, are known is $d = rt$. Find d:

7. $r = 50$ mi/h and $t = 3\frac{1}{2}$ h.

8. $r = 78\frac{1}{2}$ mi/h and $t = 4$ h.

The formula $r = \frac{d}{t}$ gives the average rate, r, when the distance, d, and the time, t, are known. Find r:

9. $d = 39$ mi and $t = \frac{3}{4}$ h.

10. $d = 550$ mi and $t = 2\frac{1}{2}$ h.

11. Find the average rate of a car that travels 644 mi in $12\frac{1}{4}$ h.

12. In the formula $P = \frac{N + 2}{r}$, find P when $N = 27$ and $r = 6\frac{1}{4}$.

13. Given the formula $I = \frac{N - 2}{P}$, find I when $N = 42\frac{1}{2}$ and $P = 8\frac{3}{4}$.

14. To change from Celsius to Fahrenheit we use the formula $F = \frac{9}{5}C + 32$. Find F when $C = 40$.

15. To change from Fahrenheit to Celsius we use the formula $C = \frac{5}{9}(F - 32)$. Find C when $F = 77$.

Piston displacement of a cylinder is the volume inside the cylinder that the piston displaces as it moves from the bottom of its stroke to the top of its stroke (Fig. 2.12). The piston displacement is given by the formula

$$PD = \frac{\pi d^2 h}{4}$$

where d = the diameter of the piston and h = the length of the stroke.

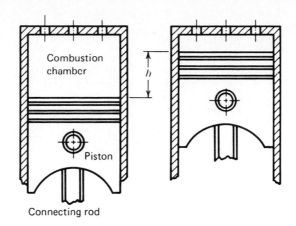

Figure 2.12

 16. Find the piston displacement of a cylinder with a
diameter of $2\frac{1}{2}$ in. and a stroke of $1\frac{3}{4}$ in. (Use $\pi = \frac{22}{7}$.)

 17. Find the piston displacement of a cylinder with a diameter of 4 in. and a stroke of $2\frac{3}{16}$ in. Also find the
total displacement of all four cylinders. (Use $\pi = \frac{22}{7}$.)

In a parallel circuit, the total resistance (R_T) is
given by the formula

$$R_T = \frac{1}{\dfrac{1}{R_1} + \dfrac{1}{R_2} + \dfrac{1}{R_3} + \cdots}$$

Note: The three dots mean that you should use as many
fractions in the denominator as there are resistances in
the circuit.

Find the total resistance in each parallel circuit.

 18.

 19.

 20.

2.8 Signed Fractions

The rules for operations with signed integers in Chapter 1 also apply to fractions; therefore, let us state them once again.

Adding Two Signed Numbers with Like Signs

1. To add two positive numbers, add their absolute values. The result is positive.
2. To add two negative numbers, add their absolute values and place a negative sign before the result.

<u>Example 1.</u> $(-3) + (-2) = -5$

<u>Example 2.</u> $\left(-\frac{1}{4}\right) + \left(-\frac{3}{16}\right) = \left(-\frac{4}{16}\right) + \left(-\frac{3}{16}\right) = -\frac{7}{16}$

Adding Two Signed Numbers with Opposite Signs

3. To add two numbers with opposite signs, find the difference of their absolute values. The sign of the number having the larger absolute value is placed before the result.

<u>Example 3.</u> $5 + (-7) = -2$

<u>Example 4.</u> $(-5) + 7 = 2$

<u>Example 5.</u> $\frac{3}{5} + \left(-\frac{2}{3}\right) = \frac{9}{15} + \left(-\frac{10}{15}\right) = -\frac{1}{15}$

<u>Example 6.</u> $\left(-\frac{4}{9}\right) + \frac{2}{3} = \left(-\frac{4}{9}\right) + \frac{6}{9} = \frac{2}{9}$

Subtracting Two Signed Numbers

4. To subtract one signed number from another, change the sign of the second number and add according to the rules for addition of signed numbers.

Example 7. $5 - (-12) = 5 + (+12) = 17$

Example 8. $\left(-\frac{5}{9}\right) - \left(-\frac{5}{12}\right) = \left(-\frac{20}{36}\right) + \left(+\frac{15}{36}\right) = -\frac{5}{36}$

Multiplying or Dividing Two Signed Numbers

5. To multiply or divide two numbers that have the same sign, multiply or divide their absolute values. The result will be positive.
6. To multiply or divide two numbers that have opposite signs, multiply or divide their absolute values and place a negative sign before the result.

Example 9. $(-3)(-6) = 18$

Example 10. $\left(-\frac{2}{5}\right)\left(-\frac{5}{8}\right) = \frac{1}{4}$

Example 11. $(-21) \div (-3) = 7$

Example 12. $\left(-\frac{3}{7}\right) \div \left(-\frac{9}{14}\right) = \left(-\frac{3}{7}\right) \times \left(-\frac{14}{9}\right) = \frac{2}{3}$

Example 13. $(3)(-5) = -15$

Example 14. $\left(\frac{4}{15}\right)\left(-\frac{5}{2}\right) = -\frac{2}{3}$

Example 15. $(-45) \div 5 = -9$

Example 16. $\left(-\frac{11}{15}\right) \div \frac{2}{3} = \left(-\frac{11}{15}\right) \times \frac{3}{2} = -\frac{11}{10}$ or $-1\frac{1}{10}$

Although these same rules apply for operations with fractions, one more rule about fractions that will help you is:

Equivalent Signed Fractions

$$\frac{a}{-b} = \frac{-a}{b} = -\frac{a}{b}$$

That is, a negative fraction may be written in three different but equivalent forms. However, the form $-\frac{a}{b}$ is the customary form.

For example, $\frac{3}{-4} = \frac{-3}{4} = -\frac{3}{4}$.

Note: $\frac{-a}{-b} = \frac{a}{b}$ using the rules for dividing signed numbers.

Example 17. $\frac{3}{-4} + \frac{-2}{3} = \left(-\frac{3}{4}\right) + \left(-\frac{2}{3}\right) = \left(-\frac{9}{12}\right) + \left(-\frac{8}{12}\right) = -\frac{17}{12}$

Example 18. $\frac{3}{-4} + \frac{2}{3} = \left(-\frac{3}{4}\right) + \frac{2}{3} = \left(-\frac{9}{12}\right) + \left(\frac{8}{12}\right) = -\frac{1}{12}$

Example 19. $\frac{3}{-4} - \frac{2}{3} = \left(-\frac{3}{4}\right) + \left(-\frac{2}{3}\right) = \left(-\frac{9}{12}\right) + \left(-\frac{8}{12}\right) = -\frac{17}{12}$

Example 20. $\left(-\frac{1}{2}\right)\left(\frac{-3}{5}\right) = \left(-\frac{1}{2}\right)\left(-\frac{3}{5}\right) = \frac{3}{10}$

Example 21. $\left(\frac{-1}{2}\right)\left(\frac{3}{5}\right) = \left(-\frac{1}{2}\right)\left(\frac{3}{5}\right) = -\frac{3}{10}$

Example 22. $\left(\frac{-2}{3}\right) \div 3 = \left(-\frac{2}{3}\right) \div 3 = \left(-\frac{2}{3}\right)\left(\frac{1}{3}\right) = -\frac{2}{9}$

Example 23. $\left(\frac{-3}{7}\right) \div \frac{-5}{6} = \left(-\frac{3}{7}\right) \div \left(-\frac{5}{6}\right) = \left(-\frac{3}{7}\right)\left(-\frac{6}{5}\right) = \frac{18}{35}$

EXERCISES 2.8   *Perform the indicated operations and simplify:*

1. $\frac{1}{8} + \left(-\frac{5}{16}\right)$ 2. $\left(-\frac{2}{3}\right) + \left(\frac{-2}{7}\right)$ 3. $\frac{1}{2} + \left(\frac{7}{-16}\right)$

4. $\frac{2}{3} + \left(-\frac{7}{9}\right)$ 5. $\left(-5\frac{3}{4}\right) + \left(-6\frac{2}{5}\right)$ 6. $\left(\frac{-1}{4}\right) - \left(\frac{1}{-5}\right)$

7. $\left(\frac{-2}{9}\right) - \left(\frac{1}{2}\right)$ 8. $\left(3\frac{1}{6}\right) - \left(\frac{4}{15}\right)$ 9. $\frac{5}{8} - \left(\frac{-5}{16}\right)$

10. $\left(\frac{-1}{3}\right) - \left(-3\frac{1}{2}\right)$ 11. $\frac{1}{4} \times \left(\frac{1}{-5}\right)$ 12. $\frac{-1}{9} \times \frac{1}{7}$

13. $\left(\frac{-2}{3}\right)\left(-\frac{1}{2}\right)$ 14. $\left(-2\frac{1}{3}\right)\left(-1\frac{4}{5}\right)$ 15. $\frac{21}{8} \times 1\frac{7}{9}$

16. $\frac{4}{5} \div \left(-\frac{8}{9}\right)$ 17. $\left(-1\frac{1}{4}\right) \div \frac{3}{5}$ 18. $\left(-\frac{7}{9}\right) \div \left(-\frac{8}{3}\right)$

19. $32 \div \left(\frac{-2}{3}\right)$ 20. $2\frac{3}{4} \div \left(-3\frac{1}{6}\right)$ 21. $\left(\frac{-4}{5}\right) + \left(-1\frac{1}{2}\right)$

22. $2\frac{3}{4} - \left(-3\frac{1}{4}\right)$ 23. $\left(\frac{-5}{-8}\right) \times \left(-5\frac{1}{3}\right)$ 24. $\left(-1\frac{3}{5}\right) \div \left(-3\frac{1}{5}\right)$

25. $\left(\frac{-6}{8}\right) - (-4)$ 26. $\left(\frac{-3}{2}\right) + \left(\frac{-8}{3}\right)$

27. $\left(\frac{-2}{3}\right) + \left(-\frac{5}{6}\right) + \frac{1}{4} + \frac{1}{8}$ 28. $\left(\frac{-3}{4}\right) + \left(\frac{2}{-3}\right) - \left(\frac{-1}{-2}\right) - \left(\frac{-5}{6}\right)$

29. $\left(\frac{-2}{5}\right)\left(\frac{3}{-4}\right)\left(\frac{-15}{-18}\right)$ 30. $\left(-2\frac{3}{4}\right) \div \left(1\frac{3}{5}\right) \times \left(\frac{-2}{5}\right)$

31. $\left(\frac{-2}{3}\right) + \left(-\frac{1}{2}\right)\left(\frac{5}{-6}\right)$ 32. $\left(\frac{-4}{5}\right) \div \left(-1\frac{1}{2}\right) - \left(\frac{2}{-5}\right)$

Chapter 2 Review

Simplify:

1. $\frac{36}{56}$ 2. $\frac{180}{216}$

Change each to a mixed number in simplest form:

3. $\frac{25}{6}$ 4. $3\frac{18}{5}$

Change each mixed number to an improper fraction:

5. $2\frac{5}{8}$ 6. $3\frac{7}{16}$

Perform the indicated operations and simplify:

7. $\dfrac{3}{8} + \dfrac{7}{8} + \dfrac{6}{8}$ 8. $\dfrac{1}{4} + \dfrac{5}{12} + \dfrac{5}{6}$ 9. $\dfrac{29}{36} - \dfrac{7}{30}$

10. $5\dfrac{3}{14} + 9\dfrac{5}{12}$ 11. $6\dfrac{3}{8} - 4\dfrac{7}{12}$ 12. $18 - 6\dfrac{2}{5}$

13. $16\dfrac{2}{3} + 1\dfrac{1}{4} - 12\dfrac{11}{12}$ 14. $\dfrac{5}{6} \times \dfrac{3}{10}$

15. $3\dfrac{6}{7} \times 4\dfrac{2}{3}$ 16. $\dfrac{3}{8} \div 6$ 17. $\dfrac{2}{3} \div 1\dfrac{7}{9}$

18. $1\dfrac{4}{5} \div 1\dfrac{9}{16} \times 11\dfrac{2}{3}$

19. Find the missing dimensions in the figure.

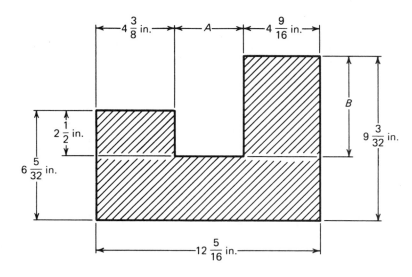

Fill in each blank:

20. 6 lb 9 oz = _____ oz   21. 168 ft = _____ in.

22. 72 ft = _____ yd   23. 36 mi = _____ yd

24. Given $P = 2\ell + 2w$, $\ell = 3\dfrac{5}{6}$ in. and $w = 1\dfrac{3}{4}$ in., find P.

Perform the indicated operations and simplify:

25. $\left(-\dfrac{6}{7}\right) - \left(\dfrac{5}{-6}\right)$ 26. $\dfrac{-3}{16} \div \left(-2\dfrac{1}{4}\right)$

27. $\dfrac{-5}{8} + \left(-\dfrac{5}{6}\right) - \left(+1\dfrac{2}{3}\right)$ 28. $\left(-\dfrac{7}{16}\right) \times 1\dfrac{11}{14}$

3

DECIMAL FRACTIONS

3.1 Introduction to Decimals

As we change to the metric system of measurement—which is a decimal system—decimal calculations and measuring instruments calibrated in decimals will become the basic tools for measurement. Also, the common use and importance of the electronic calculator, which does decimal calculations with ease, make necessary a basic understanding of the principles of decimals.

A *decimal fraction* is a fraction whose denominator is a power of 10. You do not write the denominator when writing a decimal fraction. You write only the numerator and use the decimal point to indicate the ones place. Thus, you may think of decimal fractions as a shorthand way of writing fractions with denominators that are powers of 10.

Recall the place values of the digits of a whole number from Chapter 1. Each digit to the left of the decimal point represents a multiple of a power of 10. Each digit to the right of the decimal point represents a multiple of a power of $\frac{1}{10}$. Study the following table of place values for decimals.

180

Place Values for Decimals

Number	Words	Product form	Exponential form
1,000,000	One million	$10 \times 10 \times 10 \times 10 \times 10 \times 10$	10^6
100,000	One hundred thousand	$10 \times 10 \times 10 \times 10 \times 10$	10^5
10,000	Ten thousand	$10 \times 10 \times 10 \times 10$	10^4
1,000	One thousand	$10 \times 10 \times 10$	10^3
100	One hundred	10×10	10^2
10	Ten	10	10^1
1	One	1	10^0
0.1	One tenth	$\dfrac{1}{10}$	$\left(\dfrac{1}{10}\right)^1$ or 10^{-1}
0.01	One hundredth	$\dfrac{1}{10} \times \dfrac{1}{10}$	$\left(\dfrac{1}{10}\right)^2$ or 10^{-2}
0.001	One thousandth	$\dfrac{1}{10} \times \dfrac{1}{10} \times \dfrac{1}{10}$	$\left(\dfrac{1}{10}\right)^3$ or 10^{-3}
0.0001	One ten-thousandth	$\dfrac{1}{10} \times \dfrac{1}{10} \times \dfrac{1}{10} \times \dfrac{1}{10}$	$\left(\dfrac{1}{10}\right)^4$ or 10^{-4}
0.00001	One hundred-thousandth	$\dfrac{1}{10} \times \dfrac{1}{10} \times \dfrac{1}{10} \times \dfrac{1}{10} \times \dfrac{1}{10}$	$\left(\dfrac{1}{10}\right)^5$ or 10^{-5}
0.000001	One millionth	$\dfrac{1}{10} \times \dfrac{1}{10} \times \dfrac{1}{10} \times \dfrac{1}{10} \times \dfrac{1}{10} \times \dfrac{1}{10}$	$\left(\dfrac{1}{10}\right)^6$ or 10^{-6}

Note that $10^0 = 1$. (See Sec. 3.8.)

<u>Example 1</u>. In the decimal number 123.456 find the place value of each digit and the number it represents.

Digit	Place value	Number represented
1	Hundreds	1×10^2
2	Tens	2×10^1
3	Ones or units	3×10^0 or 3×1
4	Tenths	$4 \times \dfrac{1}{10}$ or 4×10^{-1}
5	Hundredths	$5 \times \left(\dfrac{1}{10}\right)^2$ or 5×10^{-2}
6	Thousandths	$6 \times \left(\dfrac{1}{10}\right)^3$ or 6×10^{-3}

Recall that place values to the left of the decimal point are powers of 10 and place values to the right of the decimal point are powers of $\frac{1}{10}$.

Example 2. Write each decimal in words: 0.05; 0.0006; 24.41; 234.001207.

Decimal	Word form
0.05	Five hundredths
0.0006	Six ten-thousandths
24.41	Twenty-four *and* forty-one hundredths
234.001207	Two hundred thirty-four *and* one thousand two hundred seven millionths

Note that the decimal point is read "and."

Example 3. Write each number as a decimal and as a common fraction.

Number	Decimal	Common fraction
One hundred four and sixteen hundredths	104.16	$104\frac{16}{100}$
Fifty and four thousandths	50.004	$50\frac{4}{1000}$
Five hundred eleven hundred-thousandths	0.00511*	$\frac{511}{100,000}$

*We will follow the common practice of writing a zero before the decimal point in a decimal less than one.

Often common fractions are easier to use if they are expressed as decimal equivalents. Every common fraction can be expressed as a repeating decimal. A *repeating decimal* is a decimal in which a digit or a group of digits repeats again and again.

A bar over a digit or group of digits means that this digit or group of digits is repeated without ending.

Example 4. Each of the following is a repeating decimal.

$$0.33\overline{3} \qquad 72.644\overline{4} \qquad 0.2121\overline{21} \qquad 6.00120\overline{120}$$

Note that decimals such as 3.00, 0.75, 0.6, and 0.47962 are all repeating decimals since zeros can be supplied indefinitely at the right of the last digit without changing the value of the number. These same four decimals can be written as follows:

$$3.00\overline{0} \qquad 0.75\overline{0} \qquad 0.6\overline{0} \qquad 0.47962\overline{0}$$

These are frequently called *terminating decimals*.

To change a common fraction to a decimal, divide the numerator of the fraction by the denominator.

Example 5. Change $\frac{3}{4}$ to a decimal.

$$
\begin{array}{r}
0.75 \\
4\overline{)3.00} \\
2\ 8 \\
\hline
20 \\
20 \\
\hline
\end{array}
$$
$\qquad \frac{3}{4} = 0.75$ (a terminating decimal)

Example 6. Change $\frac{8}{15}$ to a decimal.

$$
\begin{array}{r}
0.533\overline{3} \\
15\overline{)8.0000} \\
7\ 5 \\
\hline
50 \\
45 \\
\hline
50 \\
45 \\
\hline
50 \\
45 \\
\hline
5
\end{array}
$$
$\qquad \frac{8}{15} = 0.533\overline{3}$ (a repeating decimal)

The result could be written $0.53\overline{3}$ or $0.5\overline{3}$. It is not necessary to continue the division after the digit 3 in the quotient has begun to repeat.

Since a decimal fraction can be written as a common fraction with a denominator that is a power of 10, it is easy to change a decimal fraction to a common fraction. Simply use the digits that appear to the right of the decimal point (disregarding beginning zeros) as the numerator. Use the place value of the last digit as the denominator. Any digits to the left of the decimal point will be the whole number part of the resulting mixed number.

$\underline{\text{Example 7}}$. Change each decimal to a common fraction or a mixed number.

Decimal	Common fraction or mixed number
(a) 0.3	$\dfrac{3}{10}$
(b) 0.17	$\dfrac{17}{100}$
(c) 0.25	$\dfrac{25}{100} = \dfrac{1}{4}$
(d) 0.125	$\dfrac{125}{1000} = \dfrac{1}{8}$
(e) 0.86	$\dfrac{86}{100} = \dfrac{43}{50}$
(f) 8.1	$8\dfrac{1}{10}$
(g) 13.64	$13\dfrac{64}{100} = 13\dfrac{16}{25}$
(h) 5.034	$5\dfrac{34}{1000} = 5\dfrac{17}{500}$

EXERCISES 3.1   *Write each decimal in words:*

1. 0.004
2. 0.021
3. 0.0005
4. 7.1
5. 1.00421
6. 14.014023
7. 62.384
8. 1042.007
9. 6.092
10. 8.1461
11. 246.124
12. 12.60048

Write each number in both decimal and common fraction or mixed number forms.

13. Five and two hundredths
14. One hundred twenty-three and six thousandths
15. Seventy-one and twenty-one ten-thousandths
16. Sixty-five thousandths
17. Sixty-five thousand
18. Three and four ten-thousandths
19. Forty-three and one hundred one ten-thousandths
20. Five hundred sixty-three millionths

Change each common fraction to a decimal:

21. $\dfrac{3}{8}$ 22. $\dfrac{16}{25}$ 23. $\dfrac{11}{15}$ 24. $\dfrac{2}{5}$

25. $\dfrac{17}{50}$ 26. $\dfrac{11}{9}$ 27. $\dfrac{14}{11}$ 28. $\dfrac{128}{25}$

29. $\dfrac{128}{7}$ 30. $\dfrac{603}{24}$ 31. $\dfrac{308}{9}$ 32. $\dfrac{230}{6}$

Change each decimal to a common fraction or a mixed number:

33. 0.7   34. 0.6   35. 0.11   36. 0.75

37. 0.8425   38. 3.14   39. 14.613   40. 614.43

41. 10.76   42. 148.255

43. The purchasing department of a large industry finds that, in the catalog of replacement parts, dimensions are listed in decimal fractions in inches. The original equipment specifications were given in common fractions. The decimal parts of the numbers most frequently given are 0.0625, 0.3125, 0.125, 0.875, and 0.03125. What are their common fraction equivalents?

3.2 Addition and Subtraction of Decimal Fractions

Many times, in on-the-job situations, it is more convenient to add, subtract, multiply, and divide measurements that are in decimal form rather than in fractional form. Except for the placement of the decimal point, the four arithmetic operations are the same for decimal fractions as they are for whole numbers.

    <u>Example 1</u>. Add 13.2, 8.42, and 120.1.

    (a) Using decimal fractions:

$$
\begin{array}{r}
13.2 \\
8.42 \\
\underline{120.1} \\
141.72
\end{array}
$$

    (b) Using common fractions:

$$
\begin{aligned}
13\tfrac{2}{10} &= 13\tfrac{20}{100} \\
8\tfrac{42}{100} &= 8\tfrac{42}{100} \\
120\tfrac{1}{10} &= 120\tfrac{10}{100} \\
&\quad\ \ \overline{141\tfrac{72}{100}} = 141.72
\end{aligned}
$$

To add or subtract decimal fractions:

Step 1: Write the decimals so that the digits having the same place value are in vertical columns. (Make certain that the decimal points are lined up vertically.)

Step 2: Add or subtract as with whole numbers.

Step 3: Place the decimal point between the ones digit and the tenths digit of the sum or the difference. (Be certain the decimal point is in the same vertical line as the other decimal points.)

Example 2. Subtract 1.28 from 17.9.

```
17.90     Note. Zeros can be supplied after the last
 1.28     digit at the right of the decimal point
16.62     without changing the value of a number.
```

Example 3. Add 24.1, 26, and 37.02.

```
24.10     Note. A decimal point can be placed at the
26.00     right of any whole number and zeros supplied
37.02     without changing the value of the number.
87.12
```

Example 4. Perform the indicated operations:
51.6 - 2.45 + 7.3 - 14.92

```
    51.60
  -  2.45
    49.15     Difference
  +  7.30
    56.45     Sum
  - 14.92
    41.53     Final difference
```

Example 5. Find the missing dimension in Fig. 3.1.

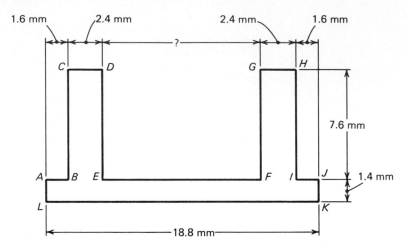

Figure 3.1

The missing dimension *EF* equals the sum of the lengths *AB*, *CD*, *GH*, and *IJ* subtracted from the length *LK*. That is, add

AB:	1.6	mm
CD:	2.4	mm
GH:	2.4	mm
IJ:	1.6	mm
	8.0	mm

Then subtract

LK:	18.8	mm
	8.0	mm
	10.8	mm

That is, length *EF* =   10.8 mm.

Example 6. As we saw on page 53, the total current in a parallel circuit is equal to the sum of the currents in each of the branches in the circuit. Find the total current in the parallel circuit of Fig. 3.2.

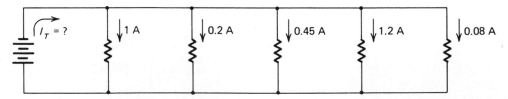

Figure 3.2

1	A
0.2	A
0.45	A
1.2	A
0.08	A
2.93	A

EXERCISES 3.2   *Find each sum:*

1. 137.64
    7.14
    0.008
    6.1

2. 63
    4.7
    19.45
    120.015

3. 147.49 + 7.31 + 0.004 + 8.4

4. 47 + 6.3 + 20.71 + 170.027

5. 127.11 + 48.143 + 63.908

Subtract:

6. 72.4 from 159

7. 3.12 from 4.7

8. 49.41 from 64.718

9. 16.412 from 140

10. 80.18 from 180.14

Perform the indicated operations:

11. 18.4 - 13.72 + 4

12. -4.14 - 8.7 - 16.5

13. 0.37 - 4.5 + 0.008

14. 51.7 - 1.11 + 4.6 - 84.1

15. -1.511 + 6.1743 - 14.714

16. 0.0056 + 0.023 + 0.00456 - 0.9005

 Use a calculator to do the rest of the exercises. See Section A.3 in the Appendix if you need to.

17. **Find the missing dimensions in the figure below.**
18. **Find the perimeter of the figure below.**

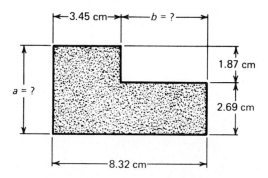

19. **Find the length of the shaft on page 92.**

20. The perimeter of the hexagon below is 6.573 in. Find the length of side *x*.

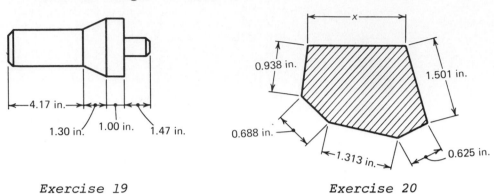

0.938 in.

1.501 in.

0.688 in.

1.313 in.

0.625 in.

Exercise 19   Exercise 20

21. Find the perimeter of the figure in Example 5.
22. The weather bureau reported the rainfall for the months of June and July was 5.31 in. and 3.50 in., respectively. (a) How much less rain fell in July than in June? (b) What was the total rainfall for the two months?

23. Find the total current in the parallel circuit below.

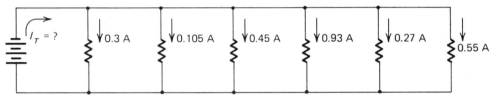

I_T = ?   0.3 A   0.105 A   0.45 A   0.93 A   0.27 A   0.55 A

As we saw on page 4, in a series circuit the total resistance is equal to the sum of the resistances in the circuit. Find the total resistance in each series circuit.

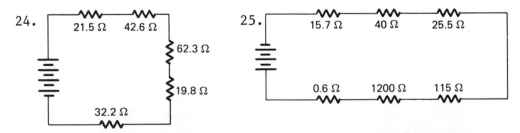

24.   21.5 Ω   42.6 Ω   62.3 Ω   19.8 Ω   32.2 Ω

25.   15.7 Ω   40 Ω   25.5 Ω   0.6 Ω   1200 Ω   115 Ω

26. In a series circuit the voltage of the source equals the sum of the separate voltage (drops) in the circuit. Find the voltage of the source in the circuit below.

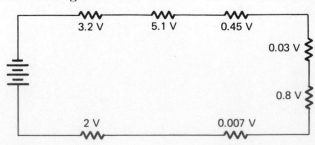

3.2 V   5.1 V   0.45 V   0.03 V   0.8 V   2 V   0.007 V

27. Find the difference in the diameters of the ends of the taper shown below.

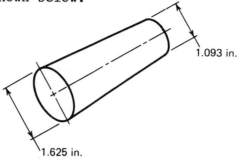

1.093 in.

1.625 in.

28. Find the missing dimension in each figure below.

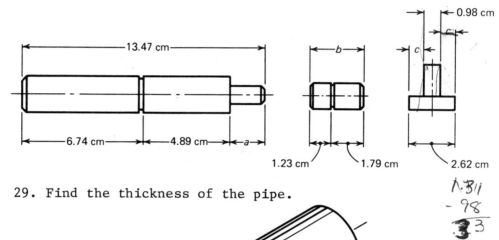

13.47 cm

6.74 cm   4.89 cm   a

0.98 cm

c

c

b

1.23 cm   1.79 cm   2.62 cm

29. Find the thickness of the pipe.

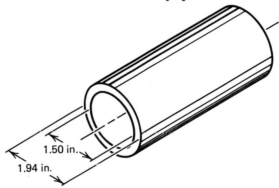

1.50 in.

1.94 in.

1.3̸1
- 98
3̸3

30. Find the length, ℓ, of the socket below. Also, find the length of A.

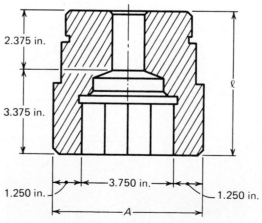

2.375 in.

3.375 in.

ℓ

3.750 in.

1.250 in.   1.250 in.

A

31. In order to seat a valve properly in an automobile
    engine, the factory part which measures 1.732 in. must
    be ground 0.005 in. Find the size of the valve after
    it is ground.

 32. The standard diameter of a certain piston is 6.733 in.
    The oversize piston has a diameter that is 0.029 in.
    greater. Find the diameter of the oversize piston.

 33. The standard width of a new piston ring is 0.1675 in.
    The used ring measures 0.1643 in. How much has it worn?

3.3 Rounding Numbers

There are times when an estimate of a number or a measurement is desirable. When a truck driver is to make a
delivery from one side of a city to another, he can only
estimate the time it will take to make the trip. An automobile technician must estimate the cost of a repair job
and the number of mechanics to assign to that job. On such
occasions estimates are usually *rounded*.

Earlier you found that $\frac{1}{3} = 0.33\overline{3}$. You can see that there
is no exact decimal value to use in a calculation. You have
to round $0.33\overline{3}$ to a certain number of decimal places,
depending on the accuracy needed in a given situation.

There are many rounding procedures that are in common
use today. Some are complicated and some are simple. We
will use one of the simplest methods, which we first
outline in the next examples and then state in the form
of a rule.

Example 1. Round 25,348 to the nearest thousand.

Note that 25,348 is more than 25,000 and less than
26,000.

26,000
25,900
25,800
25,700
25,600
25,500
25,400
25,348   As you can see, 25,348 is closer
25,300   to 25,000 than to 26,000. There-
25,200   fore, 25,348 rounded to the
25,100   nearest thousand is 25,000.
25,000

Example 2. Round 2.5271 to the nearest hundredth.

Note that 2.5271 is more than 2.5200 and less than 2.5300.

<u>2.5300</u> = 2.53
2.5290
2.5280
<u>2.5271</u>   2.5271 is nearer to 2.53 than to 2.52.
2.5270   Therefore, 2.5271 rounded to the
2.5260   nearest hundredth is 2.53.
2.5250
2.5240
2.5230
2.5220
2.5210
<u>2.5200</u> = 2.52

<u>Note</u>: If a number is halfway between two numbers, round up to the larger number.

Rounding Numbers

To round a number to a particular place value that is the tens place or greater:

1. If the digit in the next place to the right is less than 5, replace, with zeros, that digit and all other following digits to the left of the decimal point. Drop the decimal point, and drop all other digits to the right of the decimal point.
2. If the digit in the next place to the right is 5 or greater, add 1 to the digit in the place to which you are rounding. Replace with zeros all other following digits to the left of the decimal point. Drop the decimal point, and drop all other digits to the right of the decimal point.

To round a number to a particular place value that is the ones place or less:

1. If the digit in the next place to the right is less than 5, drop that digit and all other following digits.
2. If the digit in the next place to the right is 5 or greater, add 1 to the digit in the place to which you are rounding. Drop all other following digits.

Example 3. Round each number in the left column to the place indicated in each of the other columns.

Number	Hundred	Ten	Unit	Tenth	Hundredth	Thousandth
158.6147	200	160	159	158.6	158.61	158.615
4,562.7155	4,600	4,560	4,563	4,562.7	4,562.72	4,562.716
7.12579	0	10	7	7.1	7.13	7.126
63,576.15	63,600	63,580	63,576	63,576.2	63,576.15	—

EXERCISES 3.3   *Round each number to: (a) the nearest hundred, and (b) the nearest ten:*

1. 1652   2. 1760   3. 3125.4

4. 73.82   5. 18,675

Round each number to (a) the nearest tenth, and (b) the nearest thousandth:

6. 3.7654   7. 3.1416   8. 0.161616

9. 0.05731   10. 0.9836

Round each number in the left column to the place indicated in each of the other columns:

	Number	Hundred	Ten	Unit	Tenth	Hundredth	Thousandth
11.	659.1725	65917.25					
12.	1,451.5254						
13.	17,159.1666						
14.	8.171717						
15.	1,543,679						
16.	41,892.1565						
17.	10,649.83						

3.4 Multiplication and Division of Decimal Fractions

To multiply two decimal fractions:

1. Multiply the numbers as you would whole numbers.
2. Count the total number of digits to the right of the decimal points in the two numbers being multiplied. Then place the decimal in the product so there is that same number of digits to the right of the decimal point.

Example 1. Multiply 42.6 by 1.73.

```
   42.6      Note that 42.6 has one digit to the right
   1.73      of the decimal point and 1.73 has two
  12 78      digits to the right of the decimal point.
 298 2       The product should have three digits to
 426         the right of the decimal point.
 73.698
```

Example 2. Multiply: 30.6 × 4200

```
      30.6
      4200
    61200
   1224
   128520.0
```

To divide two decimal fractions:

Step 1: Use the same form as in dividing two whole numbers.

Step 2: Multiply both the divisor and the dividend (numerator and denominator) by a power of ten that makes the divisor a whole number.

Step 3: Divide as you did with whole numbers, and place the decimal point in the quotient directly above the decimal point in the dividend.

Example 3. Divide 24.32 by 6.4.

$$\frac{24.32}{6.4} \times \frac{10}{10} = \frac{243.2}{64}$$

```
        3.8
  64 | 243.2
       192
       512
       512
```

Or,

```
          3.8
 6.4. | 24.3.2
  →    19 2
        5 1 2
        5 1 2
```

Example 4. Divide 75.1 by 1.62 and round to the nearest hundredth.

To round to the nearest hundredth, you must carry the division out to the thousandths place and then round to hundredths. We show two methods:

Method 1: $$\frac{75.1}{1.62} \times \frac{100}{100} = \frac{7510}{162}$$

```
            46.358
   162 | 7510.000
         648
        1030
         972
         580
         486
         940
         810
        1300
        1296
           4
```

Method 2:
```
              46.358
  1.62. | 75.10.000
   →     64 8
        10 30
         9 72
         58 0
         48 6
          9 40
          8 10
          1 300
          1 296
              4
```

In both methods you need to add zeros after the decimal point and carry the division out to the thousandths place. Then round to the nearest hundredth. This gives 46.36 as the result.

Example 5. If a gasoline station leases for $1155 per month, how much gasoline must be sold each month to make the cost of the lease equal to 3.5¢ ($0.035) per gallon?

Divide the cost of the lease per gallon into the cost of the lease per month.

$$
0.035. \overline{)1155.000.} \\
$$

```
             33 000.
0.035. 1155.000.
  →    105    →
       ───
       105
       ───
       105
```

That is, 33,000 gal of gasoline must be sold each month.

Example 6. The inductive reactance (in ohms, Ω) in an ac circuit equals the product of 2π times the frequency (in hertz, Hz—cycles/second) times the inductance (in henries, H). Find the inductive reactance in the ac circuit in Fig. 3.3. (Use π = 3.14 if your calculator does not have a π button.)

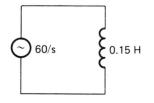

Figure 3.3

The inductive reactance is

```
 2π    × frequency × inductance
2 × π ×     60     ×    0.15    = 56.5 Ω
```

Example 7. The effect of both resistance and inductance in a circuit is called impedance. Ohm's law for an ac circuit states that the current (in amps, A) is equal to the voltage (in volts, V) divided by the impedance (in ohms, Ω). Find the current in a 110-V ac circuit which has an impedance of 65 Ω.

```
current = voltage ÷ impedance

      =    110   ÷    65

      = 1.69 A
```

Example 8. A sprayer tank holds 350 gal. Suppose 20 gal of water and 1.25 gal of pesticide are applied to each acre. (a) How many acres can be treated on one tankful? (b) How much pesticide is needed per tankful?

(a) To find the number of acres treated on one tankful, divide the number of gallons of water *and* pesticide into the number of gallons of a full tank.

```
              16.4      or approximately
21.25. 350.00.0         16 acres/tankful
  →    212 5   →
       ─────
       137 50
       127 50
       ──────
        10 00 0
         8 50 0
         ──────
         1 50 0
```

(b) To find the amount of pesticide needed per tankful, multiply the number of gallons of pesticide applied per acre times the number of acres treated on one tankful.

$$
\begin{array}{r}
1.25 \\
\underline{16} \\
7\ 50 \\
\underline{12\ 5\ \ } \\
20.00
\end{array}
$$
or approximately 20 gal/tankful

EXERCISES 3.4   *Multiply:*

1. 3.7
   0.15

2. 14.1
   1.7

3. 25.03
   0.42

4. 4.162
   3.14

5. 480
   3.14

6. 6.08
   420

7. 3050
   5.04

8. 4000
   6.75

9. 5800
   1600

10. 90,000
    0.00705

Divide:

11. 36 ÷ 1.2

12. 5.1 ÷ 1.7

13. 0.6 ÷ 0.04

14. 14.356 ÷ 0.74

15. 16.2932 ÷ 0.11

16. 328.314 ÷ 2.1

17. 145.58 ÷ 25.1

18. 19.62 ÷ 9

Divide and round to the nearest hundredth:

19. 17,500 ÷ 70.5

20. 7900 ÷ 1.52

21. 75,000 ÷ 20.4

22. 1850 ÷ 0.75

 Use a calculator to do the rest of the exercises.

23. Find the perimeter of the outside square.

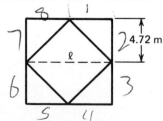

24. Find the length of the center line, ℓ, in the figure in Exercise 23.

25. Find the perimeter of the following octagon, which has eight equal sides.

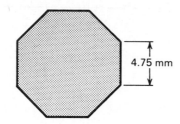

4.75 mm

26. A 78-ft cable is to be cut into 3.25 ft lengths. Into how many such lengths can the cable be cut?

27. A steel rod 32.63 in. long is to be cut into 8 pieces. Each piece is 3.56 in. long. Each cut wastes 0.15 in. of rod in shavings. How many inches of the rod are left?

28. How high would a pile of 32 metal sheets be if each sheet is 0.045 in. thick?

29. How many metal sheets are in a stack that measures 18 in. high if each sheet is 0.0060 in. thick?

30. A building measures 45 ft 3 in. by 64 ft 6 in. inside. How many square feet of possible floor space does it contain?

31. The cost of excavation is $0.75/yd^3. What is the cost of excavating a basement 87 feet long, 42 ft wide, and 8 ft deep?

32. Each cut on a lathe is 0.018 in. deep. How many cuts would be needed to turn down 2.640-in. stock to 2.388 in.?

33. Find the total length of the crankshaft shown below.

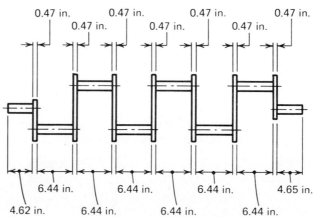

0.47 in.   0.47 in.   0.47 in.   0.47 in.

0.47 in.   0.47 in.   0.47 in.

6.44 in.   6.44 in.   6.44 in.   4.65 in.

4.62 in.   6.44 in.   6.44 in.   6.44 in.

34. A shop foreman may spend $335 to use as overtime to complete a job. Overtime pay is $16.75 per hour. How many hours of overtime may he use?

35. Find the total piston displacement of an eight-cylinder engine if each piston displaces a volume of 56.25 in^3.

36. Find the total piston displacement of a six-cylinder engine if each piston displaces 0.9 litres (L).

37. A four-cylinder engine has a total displacement of 8.4 L. Find the displacement in each piston.

 38. An eight-cylinder engine has a total displacement of 318 in^3. Find the displacement of each piston.

 39. The diameter of a new piston is 4.675 in. The average wear per 10,000 mi is 0.007 in. *uniformly over the piston.* (a) What will be the average wear after 80,000 mi? (b) What will be the diameter after 100,000 mi?

 40. A certain job requires 500 man-hours to complete. How many days will it take for five men working eight hours per day to complete the job?

 41. How many gallons of Amiben are needed for 150 acres of soybeans if 1.6 gal/acre are applied?

 42. Suppose 10 gal of water and 1.7 lb of pesticide are to be applied per acre. (a) How much pesticide would you put in a 300-gal spray tank? (b) How many acres are applied per tankful?

 43. A cattle feeder buys some feeder cattle which average 550 lb at $55/hundredweight (per hundred pounds, or 55¢/lb). His cost of adding 500 lb of weight gain while the cattle are in his feedlot is $39/hundredweight. The price he receives when he sells them as slaughter cattle is $60/hundredweight. What is his net profit per head?

 44. Malathion insecticide is to be applied at a rate of 2 pt/100 gal. How many pints are needed for a tank which holds 20 gal? 60 gal? 150 gal? 350 gal?

Find the inductive reactance in each ac circuit. (See Example 6.)

 45.

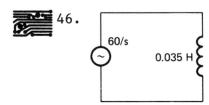

46.

Power (in watts, W) equals voltage times current. Find the power in each circuit.

 47.

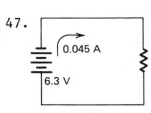

48.

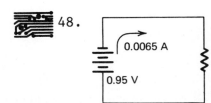

49. Find the current in a 220-V ac circuit that has an impedance of 35.5 Ω. (See Example 7.)

50. A flashlight bulb is connected to a 1.5-V dry cell. If it draws 0.25 A, what is its resistance? (Resistance equals voltage divided by current.)

51. A lamp that requires 0.84 A of current is connected to a 115-V source. What is the lamp's resistance?

52. A coffee pot operates on 24 V. If it draws 1.6 A, what is its resistance?

53. A heating element operates on a 115-V line. If it has a resistance of 18 Ω, what current does it draw? (Current equals voltage divided by resistance.)

54. One gives three tablets of glyceryl trinitrate from 0.150-grain tablets. How many grains are given?

55. One gives two tablets of ephedrine which contain 0.75 grains each. How many grains are given?

56. An order reads 0.5 mg of digitalis and each tablet contains 0.1 mg. How many tablets should be given?

57. An order reads 1.25 mg of digoxin and the tablets on hand are 0.25 mg. How many tablets should be given?

3.5 Scientific Notation

Scientific notation is a method that is especially useful for writing very large or very small numbers. To write a number in scientific notation, write it as a product of a number between 1 and 10 and a power of 10.

Example 1. Write 226 in scientific notation.

$$226 = 2.26 \times 10^2$$

Remember that 10^2 is a short way of writing $10 \times 10 = 100$. Since multiplying 2.26 by 100 gives 226, you have simply moved the decimal point two places to the left.

Example 2. Write 52,800 in scientific notation.

$$52,800 = 5.28 \times 10,000 = 5.28 \times (10 \times 10 \times 10 \times 10)$$
$$= 5.28 \times 10^4$$

Writing a Decimal Number in Scientific Notation

1. Reading from left to right, place a decimal point after the first nonzero digit.
2. Place a caret ($\wedge$) at the position of the original decimal point.
3. If the decimal point is to the <u>left</u> of the caret, the exponent of 10 is the same as the number of places from the caret to the decimal point.

$$26,638 = 2.6638. \times 10^{④} = 2.6638 \times 10^4$$
④

4. If the decimal point is to the <u>right</u> of the caret, the exponent of 10 is the same as the negative of the number of places from the caret to the decimal point.

$$0.00986 = 0.009.86 \times 10^{-③} = 9.86 \times 10^{-3}$$
③

5. If the decimal point is already after the first nonzero digit, the exponent of 10 is zero.

$$2.15 = 2.15 \times 10^0$$

Example 3. Write 2,738 in scientific notation.

$$2,738 = 2.738 \times 10^{③} = 2.738 \times 10^3$$
③

Example 4. Write 0.0000006842 in scientific notation.

$$0.0000006842 = 0\ 0000006.842 \times 10^{-⑦} = 6.842 \times 10^{-7}$$
⑦

> Writing a Number in Scientific Notation
> in Decimal Form
>
> To change from scientific notation to a decimal, you
> must be able to multiply by a power of 10.
>
> 1. To multiply by a positive power of 10, move the
>   decimal point <u>to the right</u> the same number of
>   places as indicated by the exponent of 10. Write
>   zeros when needed.
> 2. To multiply by a negative power of 10, move the
>   decimal point <u>to the left</u> the same number of
>   places as the absolute value of the exponent of
>   10. Write zeros when needed.

<u>Example 5.</u> Write 2.67×10^2 as a decimal.

$$2.67 \times 10^2 = 267$$

Note that you move the decimal point two places to the
right since the exponent of 10 is +2.

<u>Example 6.</u> Write 8.67×10^4 as a decimal.

$$8.67 \times 10^4 = 86,700$$

Note that you move the decimal point four places to the
right since the exponent of 10 is +4. It is necessary to
write two zeros.

<u>Example 7.</u> Write 5.13×10^{-4} as a decimal.

$$5.13 \times 10^{-4} = 0.000513$$

Note that you move the decimal point four places to the
left since the exponent of 10 is −4. It is necessary to
write three zeros.

You may find it useful to note that a number in scientific notation with

      (a) a positive exponent is *greater than* 10 and

      (b) a negative exponent is *less than* 1.

That is, a number in scientific notation with a positive
exponent represents a relatively large number. And, a
number in scientific notation with a negative exponent
represents a relatively small number.

Scientific notation is used to compare two numbers expressed as decimals. First, write both numbers in scientific notation. The number having the greater power of 10 is the larger. If the powers of 10 are equal, compare the parts of the numbers that are between 1 and 10.

Example 8. Which is greater, 0.000876 or 0.0004721?

$$0.000876 = 8.76 \times 10^{-4} \qquad 0.0004721 = 4.721 \times 10^{-4}$$

Since the exponents are the same, compare 8.76 and 4.721. Since 8.76 is greater than 4.721, 0.000876 is greater than 0.0004721.

Example 9. Which is greater, 0.0062 or 0.0382?

$$0.0062 = 6.2 \times 10^{-3} \qquad 0.0382 = 3.82 \times 10^{-2}$$

Since −2 is greater than −3, 0.0382 is greater than 0.0062.

EXERCISES 3.5   *Write each number in scientific notation:*

1. 236 2.36×10^2
2. 798 7.98×10^2
3. 862.4 8.624×10^2
4. 0.00613 6.13×10^{-3}
5. 9.218×10^0
6. 482,300 4.823×10^5
7. 0.00118 1.18×10^{-3}
8. 218,380,000 $\times 10^8$
9. 0.000188 1.88×10^{-4}
10. 720,000 7.20×10^5
11. 823,000,000,000,000,000 8.23×10^{17}
12. 0.000000000000000315 3.15×10^{-16}

Write each number in decimal form:

13. 2.68×10^0 2.68
14. 7.68×10^2 768.
15. 1.36×10^{-4} .000136
16. 1.45×10^3 1450
17. 8.67×10^{-1} .867
18. 4.99×10^1 49.9
19. 8.88×10^8 888,000,000
20. 9.14×10^{-3} .00914
21. 3.96×10^1 39.6
22. 8.73×10^{-2} .0873
23. 1.69×10^4 16900
24. 3.37×10^3 3370

Find the larger number:

25. 0.0037; 0.0048
26. 0.029; 0.0083
27. 0.000042; 0.00091
28. 148,000; 96,988
29. 0.00037; 0.000094
30. 0.8216; 0.792
31. 0.0613; 0.00812
32. 0.0000613; 0.01200
33. 296,888; 29,687
34. 0.4918; 0.00999

Find the smaller number:

35. 0.008; 0.0009
36. 295,682; 295,681
37. 1.003; 1.0009
38. 21.8; 30.2

39. 0.00000000998; 0.01   40. 0.10108; 0.10102

41. 0.000314; 0.000271   42. 0.00812; 0.0318

43. 25,618; 18,215   44. 0.001; 0.0009

3.6 Powers of Ten

Scientific notation is especially helpful for multiplying and dividing very large and very small numbers. To perform these operations you must first know some rules for exponents. You should also know that many of the electronic calculators can perform multiplication, division, and powers of numbers entered in scientific notation. Calculators may also give the results, when very large or very small, in scientific notation.

Let's first study some rules of exponents.

Multiplying Powers of 10

To multiply two powers of 10, add the exponents as follows:

$$10^a \times 10^b = 10^{a+b}$$

Example 1. Multiply $(10^2)(10^3)$

Method 1: $(10^2)(10^3) = (10 \cdot 10)(10 \cdot 10 \cdot 10)$

$$= 10^5$$

Method 2: $(10^2)(10^3) = 10^{2+3} = 10^5$

Example 2. Multiply each of the following powers of 10:

(a) $(10^9)(10^{12}) = 10^{9+12} = 10^{21}$

(b) $(10^{-12})(10^{-7}) = 10^{(-12)+(-7)} = 10^{-19}$

(c) $(10^{-9})(10^6) = 10^{(-9)+6} = 10^{-3}$

(d) $(10^{10})(10^{-6}) = 10^{10+(-6)} = 10^4$

(e) $10^5 \cdot 10^{-8} \cdot 10^4 \cdot 10^{-3} = 10^{5+(-8)+4+(-3)} = 10^{-2}$

Dividing Powers of 10

To divide two powers of 10, subtract the exponents as follows:

$$10^a \div 10^b = 10^{a-b}$$

Example 3. Divide $\dfrac{10^6}{10^2}$

Method 1: $\dfrac{10^6}{10^2} = \dfrac{\cancel{10} \cdot \cancel{10} \cdot 10 \cdot 10 \cdot 10 \cdot 10}{\cancel{10} \cdot \cancel{10}} = 10^4$

Method 2: $\dfrac{10^6}{10^2} = 10^{6-2} = 10^4$

Example 4. Divide each of the following powers of 10:

(a) $\dfrac{10^{12}}{10^4} = 10^{12-4} = 10^8$

(b) $\dfrac{10^{-5}}{10^5} = 10^{(-5)-5} = 10^{-10}$

(c) $\dfrac{10^6}{10^{-9}} = 10^{6-(-9)} = 10^{15}$

(d) $10^{-8} \div 10^{-5} = 10^{-8-(-5)} = 10^{-3}$

(e) $10^5 \div 10^9 = 10^{5-9} = 10^{-4}$

Raising a Power of 10 to a Power

To raise a power of 10 to a power, multiply the exponents as follows:

$$(10^a)^b = 10^{ab}$$

Example 5. Find the power $(10^2)^3$

Method 1: $(10^2)^3 = 10^2 \cdot 10^2 \cdot 10^2$

$= 10^{2+2+2}$ (Using product of powers rule.)

$= 10^6$

Method 2: $(10^2)^3 = 10^{(2)(3)} = 10^6$

Example 6. Find each power of 10:

(a) $(10^4)^3 = 10^{(4)(3)} = 10^{12}$

(b) $(10^{-5})^2 = 10^{(-5)(2)} = 10^{-10}$

(c) $(10^{-6})^{-3} = 10^{(-6)(-3)} = 10^{18}$

(d) $(10^4)^{-4} = 10^{(4)(-4)} = 10^{-16}$

(e) $(10^{10})^8 = 10^{(10)(8)} = 10^{80}$

In Sec. 3.1 we claimed that $10^0 = 1$. Let's see why. To show this, we use the substitution principle, which states that if $a = b$ and $a = c$, then $b = c$.

$a = b$

$$\frac{10^n}{10^n} = 10^{n-n} \quad \text{(To divide powers, subtract the exponents.)}$$
$$= 10^0$$

$a = c$

$$\frac{10^n}{10^n} = 1 \quad \text{(Any number other than zero divided by itself equals 1.)}$$

Therefore, $b = c$; that is,

> **Zero Power of 10**
>
> $$10^0 = 1$$

We also have used the fact that $10^{-a} = \frac{1}{10^a}$. To show this we start with $\frac{1}{10^a}$:

$$\frac{1}{10^a} = \frac{10^0}{10^a} \quad (1 = 10^0)$$
$$= 10^{0-a} \quad \text{(To divide powers, subtract exponents.)}$$
$$= 10^{-a}$$

Therefore,

> **Negative Power of 10**
>
> $$10^{-a} = \frac{1}{10^a}$$

In a similar manner we can also show that

> $$\frac{1}{10^{-a}} = 10^a$$

Combinations of multiplications and divisions of powers of 10 can also be done easily using the rules of exponents.

Example 7. Perform the indicated operations. Express the results using positive exponents.

(a)
$$\frac{10^2 \cdot 10^{-3}}{10^4 \cdot 10^{-7}} = \frac{10^{2+(-3)}}{10^{4+(-7)}}$$

$$= \frac{10^{-1}}{10^{-3}}$$

$$= 10^{(-1)-(-3)}$$

$$= 10^2$$

(b)
$$\frac{10^{-5} \cdot 10^8 \cdot 10^{-6}}{10^3 \cdot 10^4 \cdot 10^{-1}} = \frac{10^{(-5)+8+(-6)}}{10^{3+4+(-1)}}$$

$$= \frac{10^{-3}}{10^6}$$

$$= 10^{-3-6}$$

$$= 10^{-9}$$

$$= \frac{1}{10^9}$$

Multiplying Numbers in Scientific Notation

To multiply numbers in scientific notation, multiply the decimals between 1 and 10. Then add the exponents of the powers of 10.

Example 8. Multiply: $(4.5 \times 10^8)(5.2 \times 10^{-14})$. Write the result in scientific notation.

$$(4.5 \times 10^8)(5.2 \times 10^{-14}) = (4.5)(5.2) \times (10^8)(10^{-14})$$

$$= 23.4 \times 10^{-6}$$

$$= (2.34 \times 10^1) \times 10^{-6}$$

$$= 2.34 \times 10^{-5}$$

Notice that 23.4×10^{-6} is not in scientific notation because 23.4 is not between 1 and 10.

To find this product using a calculator that accepts numbers in scientific notation, use the following procedure.

Flow Chart	Buttons Pushed	Display
$\boxed{\text{Enter } 4.5 \times 10^8}$	$\boxed{4} \rightarrow \boxed{.} \rightarrow \boxed{5} \rightarrow \boxed{\text{EXP}}^{*} \rightarrow \boxed{8}$	4.5 08
$\boxed{\text{Push times}}$	$\boxed{\times}$	4.5 08
$\boxed{\text{Enter } 5.2 \times 10^{-14}}$	$\boxed{5} \rightarrow \boxed{.} \rightarrow \boxed{2} \rightarrow \boxed{\text{EXP}} \rightarrow \boxed{1} \rightarrow \boxed{4} \rightarrow \boxed{+/-}^{**}$	5.2 −14
$\boxed{\text{Push equals}}$	$\boxed{=}$	2.34 −05

The product is 2.34×10^{-5}.

*Some calculators have this button marked $\boxed{\text{EE}}$.
**This button is used to change the sign of the *last number* that has been entered.

Dividing Numbers in Scientific Notation

To divide numbers in scientific notation, divide the decimals between 1 and 10. Then subtract the exponents of the powers of 10.

Example 9. Divide $\dfrac{4.8 \times 10^{-7}}{1.6 \times 10^{-11}}$. Write the result in scientific notation.

$$\frac{4.8 \times 10^{-7}}{1.6 \times 10^{-11}} = \frac{4.8}{1.6} \times \frac{10^{-7}}{10^{-11}}$$
$$= 3 \times 10^4$$

Using a calculator, we have

Flow Chart	Buttons Pushed	Display
$\boxed{\text{Enter } 4.8 \times 10^{-7}}$	$\boxed{4} \rightarrow \boxed{.} \rightarrow \boxed{8} \rightarrow \boxed{\text{EXP}} \rightarrow \boxed{7} \rightarrow \boxed{+/-}$	4.8 −07
$\boxed{\text{Push divide}}$	$\boxed{\div}$	4.8 −07
$\boxed{\text{Enter } 1.6 \times 10^{-11}}$	$\boxed{1} \rightarrow \boxed{.} \rightarrow \boxed{6} \rightarrow \boxed{\text{EXP}} \rightarrow \boxed{1} \rightarrow \boxed{1} \rightarrow \boxed{+/-}$	1.6 −11
$\boxed{\text{Push equals}}$	$\boxed{=}$	3. 04 or 30000

The quotient is 3×10^4.

Example 10. Evaluate $\dfrac{(6 \times 10^{-6})(3 \times 10^9)}{(2 \times 10^{-10})(4 \times 10^{-5})}$. Write the result in scientific notation.

$$\frac{(6 \times 10^{-6})(3 \times 10^9)}{(2 \times 10^{-10})(4 \times 10^{-5})} = \frac{(6)(3)}{(2)(4)} \times \frac{(10^{-6})(10^9)}{(10^{-10})(10^{-5})}$$

$$= 2.25 \times 10^{18}$$

Again, use a calculator.

Flow Chart	Buttons Pushed	Display
Enter 6×10^{-6}	$\boxed{6} \to \boxed{EXP} \to \boxed{6} \to \boxed{+/-}$	6. −06
Push times	$\boxed{\times}$	6. −06
Enter 3×10^9	$\boxed{3} \to \boxed{EXP} \to \boxed{9}$	3. 09
Push divide	$\boxed{\div}$	1.8 04 or 18000
Enter 2×10^{-10}	$\boxed{2} \to \boxed{EXP} \to \boxed{1} \to \boxed{0} \to \boxed{+/-}$	2. −10
Push divide	$\boxed{\div}$	9. 13
Enter 4×10^{-5}	$\boxed{4} \to \boxed{EXP} \to \boxed{5} \to \boxed{+/-}$	4. −05
Push equals	$\boxed{=}$	2.25 18

The result is 2.25×10^{18}.

Powers of Numbers in Scientific Notation

To raise a number in scientific notation to a power, multiply the decimal between 1 and 10 times itself the number of times indicated by the power. Then also multiply the exponent of the power of 10 by this same power.

Example 11. Find the power $(4.5 \times 10^6)^2$. Write the result in scientific notation.

$$(4.5 \times 10^6)^2 = (4.5)^2 \times (10^6)^2$$

$$= 20.25 \times 10^{12}$$

$$= 2.025 \times 10^1 \times 10^{12}$$

$$= 2.025 \times 10^{13}$$

Flow Chart	Buttons Pushed	Display
Enter 4.5×10^6	$\boxed{4} \to \boxed{.} \to \boxed{5} \to \boxed{\text{EXP}} \to \boxed{6}$	4.5 06
Push x^2	$\boxed{x^2}$	2.025 13

The result is 2.025×10^{13}.

Example 12. Find the power $(3 \times 10^{-8})^5$. Write the result in scientific notation.

$$(3 \times 10^{-8})^5 = 3^5 \times (10^{-8})^5$$
$$= 243 \times 10^{-40}$$
$$= (2.43 \times 10^2) \times 10^{-40}$$
$$= 2.43 \times 10^{-38}$$

The y^x button is used to raise a number to any power.

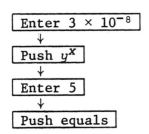

Flow Chart	Buttons Pushed	Display
Enter 3×10^{-8}	$\boxed{3} \to \boxed{\text{EXP}} \to \boxed{8} \to \boxed{+/-}$	3. −08
Push y^x	$\boxed{y^x}$	3. −08
Enter 5	$\boxed{5}$	5.
Push equals	$\boxed{=}$	2.43 −38

The result is 2.43×10^{-38}.

EXERCISES 3.6   *Perform the indicated operations using the laws of exponents. Express the results using positive exponents:*

1. $10^4 \cdot 10^9$

2. $10^4 \div 10^{-6}$

3. $(10^4)^3$

4. $\dfrac{10^4}{10^8}$

5. $10^{-6} \cdot 10^{-4}$

6. $\dfrac{1}{10^{-5}}$

7. $10^{-2} \div 10^{-5}$

8. $(10^3)^{-2}$

9. $(10^{-3})^4$

10. $\dfrac{(10^0)^3}{10^{-2}}$

11. $10^{-15} \cdot 10^{10}$

12. $\dfrac{10^0 \cdot 10^{-3}}{10^{-6} \cdot 10^3}$

13. $\dfrac{10^3 \cdot 10^2 \cdot 10^{-7}}{10^5 \cdot 10^{-3}}$

14. $\dfrac{10^{-2} \cdot 10^{-3} \cdot 10^{-7}}{10^3 \cdot 10^4 \cdot 10^{-5}}$

15. $\dfrac{10^8 \cdot 10^{-6} \cdot 10^{10} \cdot 10^0}{10^4 \cdot 10^{-17} \cdot 10^8}$

16. $\dfrac{(10^{-4})^6}{10^4 \cdot 10^{-3}}$

17. $\dfrac{(10^{-9})^{-2}}{10^{16} \cdot 10^{-4}}$

18. $\left(\dfrac{10^4}{10^{-7}}\right)^3$

19. $\left(\dfrac{10^5 \cdot 10^{-2}}{10^{-4}}\right)^2$

20. $\left(\dfrac{10^{-7} \cdot 10^{-2}}{10^9}\right)^{-3}$

Perform the indicated operations. Write each result in scientific notation:

21. $(4 \times 10^{-6})(6 \times 10^{-10})$ 22. $(3 \times 10^{7})(3 \times 10^{-12})$

23. $\dfrac{4.5 \times 10^{16}}{1.5 \times 10^{-8}}$ 24. $\dfrac{1.6 \times 10^{6}}{6.4 \times 10^{10}}$

25. $\dfrac{(4 \times 10^{-5})(6 \times 10^{-3})}{(3 \times 10^{-10})(8 \times 10^{8})}$

26. $\dfrac{(5 \times 10^{4})(3 \times 10^{-5})(4 \times 10^{6})}{(1.5 \times 10^{6})(2 \times 10^{-11})}$

27. $(1.2 \times 10^{6})^{3}$ 28. $(2 \times 10^{-9})^{4}$

29. $(6.2 \times 10^{-5})(5.2 \times 10^{-6})(3.5 \times 10^{8})$

30. $\dfrac{(5 \times 10^{-6})^{2}}{4 \times 10^{6}}$ 31. $\left(\dfrac{2.5 \times 10^{-4}}{7.5 \times 10^{8}}\right)^{2}$

32. $\left(\dfrac{2.5 \times 10^{-9}}{5 \times 10^{-7}}\right)^{4}$ 33. $(18{,}000)(0.00005)$

34. $\dfrac{2{,}400{,}000}{36{,}000}$ 35. $(4500)(69{,}000)(150{,}000)$

36. $\dfrac{(3500)(0.00164)}{2700}$ 37. $\dfrac{84{,}000 \times 0.0004 \times 142{,}000}{0.002 \times 3200}$

38. $\dfrac{(0.0025)^{2}}{3500}$ 39. $\left(\dfrac{48{,}000 \times 0.0144}{0.0064}\right)^{2}$

40. $\left(\dfrac{0.0027 \times 0.16}{12{,}000}\right)^{3}$

41. $\left(\dfrac{1.3 \times 10^{4}}{(2.6 \times 10^{-3})(5.1 \times 10^{8})}\right)^{5}$

42. $\left(\dfrac{9.6 \times 10^{-3}}{(2.45 \times 10^{-4})(1.1 \times 10^{5})}\right)^{6}$

43. $\left(\dfrac{18.4 \times 2100}{0.036 \times 950}\right)^{8}$

44. $\left(\dfrac{0.259 \times 6300}{866 \times 0.013}\right)^{10}$

3.7 Percent

Percent is the comparison of any number of parts to 100 parts. "Percent" means "per hundred." The symbol for percent is %.

You wish to put milk in a pitcher so that it is 25% "full" (Fig. 3.4). First, imagine a line drawn down the side of the pitcher. Then, imagine the line divided into 100 equal parts. Each mark shows 1%: that is, each mark shows one part of 100 parts. Finally, count 25 marks from the bottom. The amount of milk below the line is 25% of what the pitcher will hold.

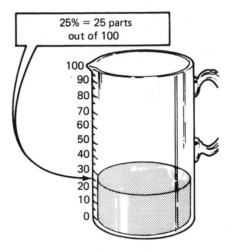

25% = 25 parts out of 100

Figure 3.4

The pitcher in Fig. 3.5 is 83% full.

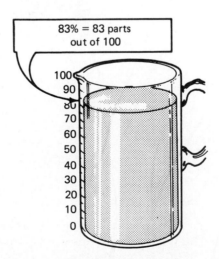

83% = 83 parts out of 100

Figure 3.5

The pitcher in Fig. 3.6 is 100% full.

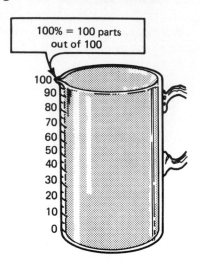

Figure 3.6

Note that 100% is a full, or one whole, pitcher of milk.

One dollar equals 100 cents or 100 pennies. Then, 36% of one dollar equals 36 of 100 parts, or 36 cents or 36 pennies. (See Fig. 3.7.)

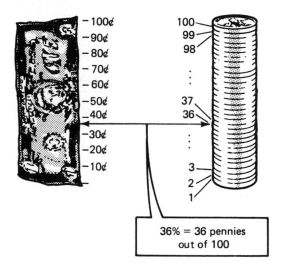

Figure 3.7

To save 10% of your salary, you would have to save $10 out of each $100 earned.

When the United States government spends 40% of its budget on defense, it spends on defense $40 out of every $100 it collects.

When a salesman earns a commission of 8%, he receives $8 out of each $100 of goods he sells.

A car's radiator holds a mixture which is 25% antifreeze. That is, in each hundred parts of mixture, there are 25 parts of pure antifreeze.

A state charges a 5% sales tax. That is, for each $100 of goods that you buy, a tax of $5 is added to your bill. The $5, a 5% tax, is then paid to the state.

Just remember "percent" means "per hundred."

CHANGING A PERCENT TO A DECIMAL

Percent means the number of parts per 100 parts. Any percent can be written as a fraction with 100 as the denominator.

Example 1. Change each percent to a fraction and then to a decimal.

(a) $75\% = \dfrac{75}{100} = 0.75$ (75 hundredths)

(b) $45\% = \dfrac{45}{100} = 0.45$ (45 hundredths)

(c) $16\% = \dfrac{16}{100} = 0.16$ (16 hundredths)

(d) $7\% = \dfrac{7}{100} = 0.07$ (7 hundredths)

As you can see from Example 1,

Changing a Percent to a Decimal

To change a percent to a decimal, move the decimal point two places to the <u>left</u> (divide by 100). Then, remove the percent sign (%).

Example 2. Change each percent to a decimal.

(a) 44% = 0.44
(b) 24% = 0.24
(c) 115% = 1.15
(d) 5.7% = 0.057
(e) 0.25% = 0.0025
(f) 100% = 1

44% = .44.%

= 0.44

Move the decimal point two places to the <u>left</u> and remove the percent sign (%).

If the percent has a fraction in it, write the fraction as a decimal. Then, do as before.

Example 3. Change each percent to a decimal.

(a) $12\frac{1}{2}\% = 12.5\% = 0.125$

(b) $6\frac{3}{4}\% = 6.75\% = 0.0675$

(c) $165\frac{1}{4}\% = 165.25\% = 1.6525$

(d) $\frac{3}{5}\% = 0.6\% = 0.006$

When we work problems using percent, we must use the decimal form of the percent, or its equivalent fraction form.

CHANGING A DECIMAL TO A PERCENT

Changing a decimal to a percent is the reverse of what we did in Example 1.

Example 4. Write 0.75 as a percent.

$0.75 = \dfrac{75}{100}$ (75 hundredths)

$ = 75\%$ ("hundredths" means "percent")

Changing a Decimal to a Percent

To change a decimal to a percent, move the decimal point two places to the <u>right</u> (multiply by 100). Write the percent sign (%) after the number.

Example 5. Change each decimal to a percent.

(a) $0.38 = 38\%$ $0.38 = 0.\underline{38}.\%$
(b) $0.42 = 42\%$
(c) $0.08 = 8\%$ $ = 38\%$
(d) $0.195 = 19.5\%$ Move the decimal point two places
(e) $1.25 = 125\%$ to the <u>right</u>. Write the percent
(f) $2 = 200\%$ sign (%) after the number.

CHANGING A FRACTION TO A PERCENT

In some problems we need to change a fraction to a percent.

To change a fraction to a percent,

(a) first change the fraction to a decimal;

(b) then change this decimal to a percent.

Example 6. Change $\frac{3}{5}$ to a percent.

First, change $\frac{3}{5}$ to a decimal by dividing the numerator by the denominator.

$$
\begin{array}{r}
0.6 \\
5\overline{)3.0} \\
\underline{3.0}
\end{array}
$$

Then change 0.6 to a percent by moving the decimal point two places to the right. Write the percent sign (%) after the number.

$$0.6 = 60\%$$

So, $\frac{3}{5}$ = 0.6 = 60%.

Example 7. Change $\frac{3}{8}$ to a decimal.

$$
\begin{array}{r}
0.375 \\
8\overline{)3.000} \\
\underline{2\ 4} \\
60 \\
\underline{56} \\
40 \\
\underline{40}
\end{array}
$$

Then, change 0.375 to a percent.

$$0.375 = 37.5\%$$

So, $\frac{3}{8}$ = 0.375 = 37.5%.

<u>Example 8</u>. Change $\frac{5}{6}$ to a percent.

First, change $\frac{5}{6}$ to a decimal.

$$
\begin{array}{r}
0.83 \\
6\overline{)5.000} \\
4\ 8 \\
\hline
20 \\
18 \\
\hline
2
\end{array}
\quad \text{r 2 or } 0.83\frac{2}{6} = 0.83\frac{1}{3}
$$

When the division is carried out to the hundredths place and the remainder is not zero, write the remainder in fraction form, with the remainder over the divisor.

Then, change $0.83\frac{1}{3}$ to a percent.

$$0.83\frac{1}{3} = 83\frac{1}{3}\%$$

So, $\frac{5}{6} = 0.83\frac{1}{3} = 83\frac{1}{3}\%$.

<u>Example 9</u>. Change $1\frac{2}{3}$ to a percent.

First, change $1\frac{2}{3}$ to a decimal.

$$
\begin{array}{r}
0.66 \\
3\overline{)2.00} \\
1\ 8 \\
\hline
20 \\
18 \\
\hline
2
\end{array}
\quad \text{r 2}
\qquad \text{That is, } 1\frac{2}{3} = 1.66\frac{2}{3}.
$$

Then, change $1.66\frac{2}{3}$ to a percent.

$$1.66\frac{2}{3} = 166\frac{2}{3}\%$$

So, $1\frac{2}{3} = 1.66\frac{2}{3} = 166\frac{2}{3}\%$.

CHANGING A PERCENT TO A FRACTION

To change a percent to a fraction,

(a) change the percent to a decimal;

(b) then change the decimal to a fraction in lowest terms.

Example 10. Change 25% to a fraction in lowest terms.

First, change 25% to a decimal by moving the decimal point two places to the left. Remove the percent sign (%).

$$25\% = 0.25$$

Then, change 0.25 to a fraction. Reduce it to lowest terms.

$$0.25 = \frac{25}{100} = \frac{1}{4}$$

So, 25% = 0.25 = $\frac{1}{4}$.

Example 11. Change 215% to a mixed number.

First, change 215% to a decimal.

$$215\% = 2.15$$

Then, change 2.15 to a mixed number in lowest terms.

$$2.15 = 2\frac{15}{100} = 2\frac{3}{20}$$

So, 215% = 2.15 = $2\frac{3}{20}$.

To change a percent that contains a mixed number to a fraction,

(a) change the mixed number to an improper fraction,

(b) then multiply this result by $\frac{1}{100}$ * and remove the percent sign (%).

*Multiplying by $\frac{1}{100}$ is the same as dividing by 100. This is what we do to change a percent to a decimal.

<u>Example 12</u>. Change $33\frac{1}{3}\%$ to a fraction.

First, change the mixed number to an improper fraction.

$$33\frac{1}{3}\% = \frac{100}{3}\%$$

Then, multiply this result by $\frac{1}{100}$ and remove the percent sign (%).

$$\frac{100}{3}\% \times \frac{1}{100} = \frac{\cancel{100}^{1}}{3} \times \frac{1}{\cancel{100}_{1}} = \frac{1}{3}$$

So, $33\frac{1}{3}\% = \frac{1}{3}$.

<u>Example 13</u>. Change $83\frac{1}{3}\%$ to a fraction.

First, $83\frac{1}{3}\% = \frac{250}{3}\%$

Then, $\frac{250}{3}\% \times \frac{1}{100} = \frac{\cancel{250}^{5}}{3} \times \frac{1}{\cancel{100}_{2}} = \frac{5}{6}$

So, $83\frac{1}{3}\% = \frac{5}{6}$.

EXERCISES 3.7   *Change each percent to a decimal:*

1. 27%   2. 15%   3. 6%   4. 5%
5. 56%   6. 78%   7. 156%   8. 232%
9. 29.2%   10. 36.2%   11. 8.7%   12. 99.4%
13. 128.7% 14. 32.78% 15. 947.8% 16. 68.29%
17. 0.28%   18. 0.78%   19. 0.068% 20. 0.0093%

21. ·0.0086%   22. 0.00079%   23. $4\frac{1}{4}\%$ 24. $9\frac{1}{2}\%$

25. $\frac{3}{8}\%$ 26. $27\frac{2}{5}\%$ 27. $50\frac{1}{3}\%$ 28. $2\frac{1}{8}\%$

29. $10\frac{1}{6}\%$ 30. $461\frac{2}{9}\%$

Change each decimal to a percent:

31. 0.54   32. 0.25   33. 0.08   34. 0.02
35. 0.62   36. 0.79   37. 0.27   38. 0.04
39. 2.17   40. 0.345   41. 4.35   42. 0.225
43. 0.185   44. 6.25   45. 0.297   46. 7.11
47. 5.19   48. 0.815   49. 0.0187 50. 0.0342
51. 0.0916 52. 0.0029 53. 0.00825 54. 0.00062

Change each fraction to a percent:

55. $\frac{4}{5}$ 56. $\frac{3}{4}$ 57. $\frac{1}{8}$ 58. $\frac{2}{5}$

59. $\frac{1}{6}$ 60. $\frac{1}{3}$ 61. $\frac{4}{9}$ 62. $\frac{3}{7}$

63. $\frac{3}{5}$ 64. $\frac{5}{6}$ 65. $\frac{13}{40}$ 66. $\frac{3}{40}$

67. $\frac{9}{25}$ 68. $\frac{17}{50}$ 69. $\frac{7}{16}$ 70. $\frac{15}{16}$

71. $\frac{20}{30}$ 72. $\frac{42}{50}$ 73. $\frac{96}{40}$ 74. $\frac{100}{16}$

75. $\frac{212}{20}$ 76. $\frac{815}{100}$ 77. $1\frac{3}{4}$ 78. $2\frac{1}{3}$

79. $2\frac{5}{12}$ 80. $1\frac{11}{12}$ 81. $4\frac{2}{3}$ 82. $5\frac{3}{8}$

83. $6\frac{1}{2}$ 84. $9\frac{3}{10}$

Change each percent to a fraction or a mixed number in lowest terms:

85. 75%	86. 45%	87. 16%	88. 80%
89. 60%	90. 15%	91. 93%	92. 32%
93. 275%	94. 325%	95. 125%	96. 150%
97. 216%	98. 610%	99. 120%	100. $12\frac{1}{2}$%
101. $10\frac{3}{4}$%	102. $13\frac{2}{5}$%	103. $10\frac{7}{10}$%	104. $40\frac{7}{20}$%
105. $17\frac{1}{4}$%	106. $6\frac{1}{3}$%	107. $16\frac{1}{6}$%	108. $72\frac{1}{8}$%
109. $7\frac{4}{9}$%	110. $12\frac{3}{16}$%		

3.8 Percentage, Base, and Rate

Any percent problem calls for finding one of three things:

1. the percentage,
2. the rate of percent, or
3. the base.

These problems are solved using one of three percent formulas. In these formulas, we let

$$P = \text{the percentage,}$$
$$R = \text{the rate of percent, and}$$
$$B = \text{the base.}$$

The following may help you identify which letter stands for each given number and the unknown number or quantity in a problem:

1. The rate, R, usually has either a percent sign (%) or the word "percent" with it.
2. The base, B, is usually the whole amount. The base is often the number that follows the word "of."
3. The percentage, P, is usually some fractional part of the base, B. P is the number that is not R or B.

<u>Example 1</u>. Given: "25% of $80 is $20." Identify R, B, and P.

R is 25%. (25 is the number with a percent sign. Remember to change 25% to the decimal 0.25 for use in a formula.)

B is $80. ($80 is the whole amount. It also follows
         the word "of.")
P is $20. ($20 is the part. It is also the number
         that is not R or B.)

Example 2. Given: "72% of the 75 students who took this
course last year are now working." Find how many are
now working. Identify R, B, and P.

R is 72%. (72 is the number with a percent sign.)
B is 75 students. (75 is the whole amount. It also
                   follows the word "of.")
P is the unknown. (The unknown is the number that is
                   some fractional part of the base.
                   It is also the number that is not
                   R or B.)

EXERCISES 3.8   *Identify the rate, R, the base, B, and the percentage, P,
in each:*

1. 60 is 25% of 240.   2. $33\frac{1}{3}$% of 300 is 100.

3. 250 is 10% of 2500.   4. 72 is 15% of 480.

5. 40% of 270 is 108.   6. 107 is 17% of 629.41.

7. 252 is 9% of 2800.   8. 16.2% of 149 is 24.14.

9. 129 is 20% of 645.   10. 26.9% of 7130 is 1917.97.

 11. At plant A, 4% of the tires made are defective. Plant A
    made 28,000 tires. How many tires were defective?

12. On the last test 25 of the 28 students got passing
    grades. What percent of students passed?

13. A girls' volleyball team won 60% of their games. The
    team won 21 games. How many games did they play?

 14. A rancher loses 10% of the herd every winter due to
    weather. He has a herd of 15,000. How many will he
    lose this winter?

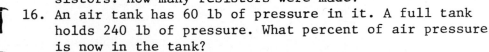 15. An electronic firm finds that 6% of the resistors it
    makes are defective. There were 2050 defective re-
    sistors. How many resistors were made?

16. An air tank has 60 lb of pressure in it. A full tank
    holds 240 lb of pressure. What percent of air pressure
    is now in the tank?

17. A machinist is to allow a 1% tolerance on a shaft of
    12 feet. What is the amount of tolerance?

18. The interest on a $500 loan is $90. What is the percent
    of interest?

PERCENT PROBLEMS: FINDING THE PERCENTAGE

After you have determined which two numbers are known, you find the third or unknown number by using one of three formulas. The first formula is given below.

> To find the <u>percentage</u>, use the formula
>
> $$P = BR$$

<u>Example 3</u>. Find 75% of 180.

$R = 75\% = 0.75$
$B = 180$
$P =$ the unknown
$P = BR$
$\quad = (180)(0.75)$
$\quad = 135$

<u>Example 4</u>. Find $8\frac{3}{4}\%$ of $1200.

$R = 8\frac{3}{4}\% = 8.75\% = 0.0875$
$B = \$1200$
$P =$ the unknown
$P = BR$
$\quad = (\$1200)(0.0875)$
$\quad = \$105$

Example 5. A 20 quart radiator needs to be filled so that it has 25% antifreeze. How many quarts of pure antifreeze must be used?

$R = 25\% = 0.25$
$B = 20$ quarts
$P =$ the unknown
$P = BR$
$\quad = (20 \text{ quarts})(0.25)$
$\quad = 5$ quarts

Example 6. Harry owns an 850 acre farm. Seventy-two percent of his land is tillable (land which can be plowed). How many tillable acres does Harry have?

$R = 72\% = 0.72$
$B = 850$ acres
$P =$ the unknown
$P = BR$
$\quad = (850 \text{ acres})(0.72)$
$\quad = 612$ acres

Example 7. A fuse is a safety device with a core. When too much current flows, the core melts. The circuit is then broken. The size of a fuse is the number of amperes of current the fuse is made to normally carry. A given 50 amp (A) fuse blows at 20% overload. What is the maximum current the fuse will carry?

First, find the amount of current overload:

$$R = 20\% = 0.20$$
$$B = 50 \text{ A}$$
$$P = \text{the overload}$$
$$P = BR$$
$$= (50 \text{ A})(0.20)$$
$$= 10 \text{ A}$$

The maximum current the fuse will carry is the normal current plus the overload:

$$50 \text{ A} + 10 \text{ A} = 60 \text{ A}$$

EXERCISES 3.9

1. Find 20% of 65.

2. Find 30% of 78.

3. Find 70% of 60.

4. Find 50% of 128.

5. Find 17% of 30.

6. Find 28% of 46.

7. Find 45% of 51.

8. Find 35% of 70.

9. Find 29.1% of 600.

10. Find 31.2% of 40.

11. Find 47.8% of 130.

12. Find 19.7% of 80.

13. Find 0.2% of 80.

14. Find 0.9% of 700.

15. Find 0.02% of 128.

16. Find 0.0319% of 750.

17. Find $9\frac{1}{4}\%$ of 64.

18. Find $8\frac{2}{3}\%$ of 120.

19. Find 238% of 48.

20. Find 126% of 720.

 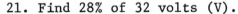

21. Find 28% of 32 volts (V).

22. A voltmeter reads a voltage of 128 V. The voltmeter has a maximum error of 12%. What is its maximum error in volts?

23. The efficiency of an electric motor is 60%. The rated output of the motor is 78 hp. What is the true output?

24. A production line makes 258 calculators a day. If 18% are found defective, how many are defective each day?

25. The materials for a construction job have been estimated at $28,000. Inflation will cause costs to rise 9% a year. What will the cost be one year from now?

26. A manufacturer must increase production by 12%. The current production is 2050 units a week. Find the number of units after the increase.

27. An automobile dealer's overhead costs are 12% of his automobile sales. His sales for one month are $322,210. What are his overhead costs?

 28. The coolant capacity of a radiator is 14 quarts. A mixture of 30% water and 70% coolant is needed for proper protection. What amount of water in the radiator is needed for proper protection?

 29. An automobile salesman receives a commission of 2% on each new car he sells. What is his commission on an $8032 car?

 30. A beef half has a hanging weight of 280 lb. This beef loses 30% in processing. What is the weight of the processed beef?

 31. A farmer has 880 acres he is going to plant in corn. The government has a set-aside program in which he does not plant 20% of his acreage and he receives a subsidy. How many acres could he put in the program?

 32. A cattle feed is 8% protein solids. How many pounds of protein solids are in 2 tons of feed?

 33. You take 50% of a 70 mg tablet. How many mg would you take?

 34. An antiseptic solution should be 15% antiseptic. How much antiseptic is needed for 28 fl oz of solution?

 35. Mary needs to give 75% of a 4.5 grain tablet. How many grains must she give?

PERCENT PROBLEMS: FINDING THE BASE

> To find the **base**, use the formula
>
> $$B = \frac{P}{R}$$

Example 8. The number 10 is 20% of what number?

$R = 20\% = 0.20$
$B = $ the unknown (You must find what number follows the word "of.")
$P = 10$
$B = \dfrac{P}{R}$
$\quad = \dfrac{10}{0.20} = 50$

Example 9. $45 is $9\frac{3}{4}\%$ of what amount?

$$R = 9\frac{3}{4}\% = 9.75\% = 0.0975$$
$$B = \text{the unknown}$$
$$P = \$45$$
$$B = \frac{P}{R}$$
$$= \frac{\$45}{0.0975}$$
$$= \$461.54$$

Example 10. Aluminum is 12% of the mass of a given car. This car has 186 kg of aluminum in it. What is the total mass of the car?

$$R = 12\% = 0.12$$
$$B = \text{the unknown}$$
$$P = 186 \text{ kg}$$
$$B = \frac{P}{R}$$
$$= \frac{186 \text{ kg}}{0.12}$$
$$= 1550 \text{ kg}$$

Example 11. Martha paid $475 as a 20% down payment on a used car. How much did the car cost?

$$R = 20\% = 0.20$$
$$B = \text{the unknown}$$
$$P = \$475$$
$$B = \frac{P}{R}$$
$$= \frac{\$475}{0.20}$$
$$= \$2375$$

EXERCISES 3.10

1. The number 360 is 72% of what number?
2. The number 20 is 25% of what number?
3. The number 16.2 is 18% of what number?
4. The number 125 is 50% of what number?
5. The number 40 is 30% of what number?
6. The number 540 is 20% of what number?
7. The number 620 is 31% of what number?
8. The number 140 is 35% of what number?
9. The number 2040 is 7.5% of what number?
10. The number 3850 is 11.2% of what number?

11. $72 is 4.5% of what amount?
12. $72 is 5.25% of what amount?
13. $2028 is 8.2% of what amount?
14. $4060 is 11.7% of what amount?
15. $20,080 is 21% of what amount?
16. $18,730 is 28% of what amount?
17. $37,500 is 30% of what amount?
18. $27,750 is $33\frac{1}{3}$% of what amount?
19. $128,650 is 40% of what amount?
20. $241,250 is 60% of what amount?

21. A steer needs 1.8 lb of protein per day. The feed is 6% protein. How many pounds of feed does the steer need each day?

22. The corn yield on a farm was 20,048 bushels after drying. The drying loss was 6%. What was the yield before drying?

23. A milk tanker carries milk which is 3.8% butterfat. There are 2052 lb of butterfat in the tanker. How many pounds of milk are in the tanker?

24. Martha received a discount of 18% on her new car. This amounted to $1780. What was the original cost of the car?

25. The profit on a mechanic's labor is 20% of the total labor bill. His profit on a valve job was $22. What was the total labor bill?

26. A tire has worn 72% of its useful tread. The tire has been used 22,000 miles. What would be the normal useful wear of the tire?

27. The output of a generator is 42,000 watts (W). The generator has an efficiency rating of 70%. What is the input?

$$\text{Efficiency} = \frac{\text{output}}{\text{input}} \times 100\%.$$

28. A motor draws 52% of the total amperage of a circuit. The motor draws 18 amps (A). What is the total amperage of the circuit?

29. The line loss of a circuit is 18% of the original voltage. The loss is 21.6 volts (V). What is the original voltage?

30. A machinist found 4% of his stock faulty. He found 18 feet faulty. How much stock did he have?

31. A certain alloy is 12% tin. There are 207 lb of tin. What is the number of pounds of alloy?

32. A construction project is bid with a 12% profit margin. The profit should be $17,250. What was the bid on the project?

33. 27 mL is 18% of what amount?
34. A solution is 4% salt. The solution has 0.8 L of salt in it. What is the amount of the total solution?

35. 0.35 grain is 4% of what amount?

PERCENT PROBLEMS: FINDING THE RATE

To find the <u>rate</u> of percent, use the formula

$$R = \frac{P}{B}$$

<u>Example 12.</u> The number 36 is what percent of 60?

R = the unknown
B = 60
P = 36
$R = \dfrac{P}{B}$
$\quad = \dfrac{36}{60}$
$\quad = 0.60 = 60\%$

<u>Example 13.</u> What percent of 20 metres is 5 metres?

R = the unknown
B = 20 m
P = 5 m
$R = \dfrac{P}{B}$
$\quad = \dfrac{5 \text{ m}}{20 \text{ m}}$
$\quad = 0.25 = 25\%$

<u>Example 14.</u> Harry had 3600 acres of timber. A lumber company cut 900 acres. What percent of his timber land was cut?

R = the unknown
B = 3600 acres
P = 900 acres
$R = \dfrac{P}{B}$
$\quad = \dfrac{900 \text{ acres}}{3600 \text{ acres}}$
$\quad = 0.25 = 25\%$

Example 15. Georgia's salary was $360 per week. Then she was given a raise of $30 per week. What percent raise did she get?

$$R = \text{the unknown}$$
$$B = \$360$$
$$P = \$30$$
$$R = \frac{P}{B}$$
$$= \frac{\$30}{\$360}$$
$$= 0.08\frac{1}{3} = 8\frac{1}{3}\%$$

Example 16. Pedro bought a used car for $2100. He paid $300 as a down payment and borrowed the rest for one year. He paid $225 in interest on the balance. What percent interest did he pay?

The amount of money Pedro borrowed was

$$\$2100 - \$300 = \$1800$$

So,
$$R = \text{the unknown}$$
$$B = \$1800$$
$$P = \$225$$
$$R = \frac{P}{B}$$
$$= \frac{\$225}{\$1800}$$
$$= 0.125 = 12.5\%$$

EXERCISES 3.11

Find each percent rounded to the nearest tenth of a percent, when necessary:

1. The number 24 is what percent of 48?
2. The number 18 is what percent of 72?
3. What percent of 30 is 18?
4. What percent of 52 is 69?
5. What percent of 50 is 14?
6. What percent of 12,000 is 375?
7. What percent of 7.15 is 3.5?
8. What percent of $\frac{1}{8}$ is $\frac{1}{15}$?
9. What percent of $7\frac{1}{2}$ is 3.75?
10. What percent of 20 is 4?
11. What percent of 207 m is 69 m?
12. What percent of 510 cm is 170 cm?
13. What percent of $520 is $130?
14. What percent of $880 is $110?
15. What percent of $1150 is $230?
16. What percent of $78 is $156?

17. What percent of $2080 is $208?
18. What percent of 5280 ft is 880 yd?
19. What percent of 1050 cm is 2.15 m?
20. What percent of 432 in. is 2.9 ft?

21. What percent of 50 amps (A) is 12 A?

22. A generator has an input of 45,000 watts (W) and an output of 37,500 W. What is the efficiency of the generator?

23. The voltage on one end of the line is 120 volts (V) and on the other end is 115.6 V. What is the percent of the lost voltage?

24. A machinist found that 208 stove bolts out of 25,100 were defective. What is the percent of defective bolts?

25. A shipbuilder increased the total number of workers in his yards by 250. The number of workers was 28,100. What was the percent of increase?

26. An alloy of 5050 lb has 230 lb of lead in it. What percent of the alloy is lead?

27. An engine has an input of 90 hp and output of 68 hp. What percent of the input is the output?

28. A mechanic finds that labor on a repair job is $62. Parts were $18. What percent of the total bill is the labor cost?

29. A live steer that weighs 1100 lb dresses out at 620 lb. What percent of the live weight is the dressed weight?
30. A feed is mixed so that 22 lb of corn is used for every 80 lb of feed. What percent of the feed is corn?
31. A new tire is guaranteed for 40,000 miles. At 18,000 miles the tire goes bad. What percent of the guaranteed 40,000 miles was not used?
32. What percent of 0.650 grain is 0.250 grain?
33. What percent of 0.320 grain is 0.075 grain?
34. What percent of 540 mL is 25 mL?

When you work percent problems that are not separated into the three types (P, B, or R), you must decide which of the three types a given problem is. Then, you must use the correct formula.

The triangle in Fig. 3.8 can be used to help you remember the three formulas for percent.

B and R are next to each other on a horizontal line, an arrangement which indicates multiplication. To find P, you multiply the numbers which replace B and R.

To find B, you see P is over R, which is the fraction $\frac{P}{R}$ and indicates division.

To find R, you see the fraction $\frac{P}{B}$, which also indicates division.

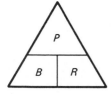

Figure 3.8

Example 17. A cow needs 2.4 lb of protein per day. The feed is 8% protein. How many pounds of feed does she need each day?

$$R = 8\% = 0.08$$
$$P = 2.4 \text{ lb}$$
$$B = \text{the unknown (the amount of feed)}$$

Since B is to be found, use the formula $B = \dfrac{P}{R}$.

$$B = \frac{2.4 \text{ lb}}{0.08}$$
$$= 30 \text{ lb of feed per day}$$

Example 18. Castings are listed at $9.50 each. A 12% discount is given if 50 or more are bought at one time. We buy 60 castings.

(a) What is the discount on one casting?
(b) What is the cost of one casting?
(c) What is the total cost?

(a) Discount equals 12% of $9.50.

$$R = 12\% = 0.12$$
$$B = \$9.50$$
$$P = \text{the unknown (the discount)}$$
$$P = BR$$
$$= (\$9.50)(0.12)$$
$$= \$1.14 \text{ (the discount on one casting)}$$

(b) Cost (of one casting) = list - discount
$$= \$9.50 - \$1.14$$
$$= \$8.36$$

(c) Total cost = cost of one casting times the number of castings
$$= (\$8.36)(60)$$
$$= \$501.60$$

You may also need to find the percent increase or percent decrease in a given quantity.

Example 19. Mary's hourly wages changed from $10.40 to $11.05. Find the percent increase in her wages.

First, let's find the change in her wages:

$$\$11.05 - \$10.40 = \$0.65$$

Then, this change is what percent of her original wage?

$$R = \frac{P}{B} = \frac{\$0.65}{\$10.40} = 0.0625 = 6.25\%$$

This process of finding the percent increase or percent decrease may be summarized by the following formula:

$$\text{percent increase} \atop \text{(or percent decrease)} = \frac{\text{the change}}{\text{the original value}} \times 100\%$$

Example 20. Normal ac line voltage is 115 volts (V). Find the percent decrease if the line voltage drops to 109 V.

$$\text{percent decrease} = \frac{\text{the change}}{\text{the original value}} \times 100\%$$

$$= \frac{115 \text{ V} - 109 \text{ V}}{115 \text{ V}} \times 100\%$$

$$= 5.22\%$$

EXERCISES 3.12

Find each percent rounded to the nearest tenth of a percent, when necessary.

1. A generator delivers 120 volts (V) to a line. But, the voltage available at a motor some distance away is only 116 V. What percent of delivered voltage has been lost?

2. What percent of 38 amps (A) is 8 A?

3. A motor draws 45% of its total amperage of 23 A. How many amperes does the motor draw?

4. The efficiency of a generator is 81%. The input is 65,000 watts (W). What is the output?

   $\text{Efficiency} = \dfrac{\text{output}}{\text{input}} \times 100\%.$

5. The reading on a vernier scale is 0.004 in. The total measurement is 0.604 in. What percent (to the nearest hundredth) of the total measurement is the vernier reading?

6. In making a number of square-head bolts, 144 were found defective. This was 3% of the total number made. How many bolts were made in all?

7. A machinist got a raise in pay of $34 per week. This was 8% of his old pay. How much did he make before he got the raise?

8. A tire maker made 32 truck tires on a certain day. This was only 64% of his normal production. How many tires does he normally make in one day?

9. On a certain welding job, $75 was spent on materials and $125 on labor. What percent of the total cost was spent on materials? What percent was spent on labor?

10. An airplane manufacturer employs 25,000 people. He must increase the number of employees by $11\frac{1}{2}\%$. How many more people must he hire?

11. A certain alloy is made of 12% antimony, 5% tin, and 83% lead. How many pounds of each metal are needed to make 450 pounds of this alloy?

12. Suppose 2000 lb of ore yield 160 lb of iron. What percent of the ore is iron?

13. Of 1800 workers in a plant, 97% have a perfect safety record. How many workers have had accidents? How many workers have perfect safety records?

14. The allowance for shrinkage when casting iron pipe is $\frac{1}{8}$ inch per foot. What percent of shrinkage is this?

15. Four square inches is what percent of the area of a square four inches on a side?

16. The potential of a motor is 90% of the power needed to run it. The potential of a motor is stated at 9 hp (horsepower). What horsepower is needed to run it?

17. A motor with an input of 9.6 hp has an output of 6.4 hp. What percent of the output is the input?

18. A contractor borrows $18,000 at 10.2% interest per year. If he pays the debt at the end of two years, how much interest does he pay? If he pays the debt at the end of 20 months, how much interest does he pay?

19. A material loses 4% by drying. It weighs 8.4 lb when dry. What was its weight before drying?

20. An engine has an input of 90 hp. It uses 12% of its input power to overcome friction and other energy losses. Find the actual power that the engine is capable of delivering; that is, find its output.

21. The brake lining on a new car is $\frac{7}{32}$ in. thick. After 15,000 miles, the brake lining is $\frac{9}{64}$ in. thick. What percent of the brake lining has worn? Assuming the same rate of wear, how many more miles can the car be driven before the brake lining is completely worn through?

22. A truck tire sells for $215 new with an original tread depth of $\frac{19}{32}$ in. The tread breaks with $\frac{23}{64}$ in. remaining. What is the cost of adjustment to the truck owner for a new tire?

23. In one year, cattle in the U.S. numbered 131,800,000 head. During the next year, producers reduced their numbers of cattle by 2.9%. How many cattle were being raised the next year?

24. A live 240 lb hog weighs 165 lb dressed. How much weight is waste? Dressing percent is the ratio of the dressed weight to the live weight written as a percent. Find the dressing percent.

25. The application rate of Aatrex 80W is $2\frac{3}{4}$ lb/acre.

How many pounds are needed for 160 acres of corn? The "80W" means that it is 80% active ingredients by weight. How many pounds of active ingredients will be applied? How many pounds of inert ingredients will be applied?

 26. U.S. soybeans average 39% protein. A bushel of soybeans weighs 60 lb. How many pounds of protein are in a bushel? A 120 acre field yields 45 bu/acre. How many pounds of protein are yielded from that field?

 27. A dairy cow produced 7525 lb of milk in a year. A gallon of milk weighs 8.6 lb. How many gallons of milk did the cow produce? The milk tested at 4.2% butterfat. How many gallons of butterfat did the cow produce?

 28. You need 15% of a 60-mg tablet. How many mg would you take?

 29. Mary needs to give 40% of a 0.75 grain tablet. How many grains does she give?

 30. You need 0.15% of 2000 mL. How many mL do you need?

 31. What percent of 0.600 grain is 0.150 grain?

 32. What percent of 0.250 grain is 0.750 grain?

33. 42 mL is 35% of what quantity?

 34. During a line voltage surge, the normal ac voltage increased from 115 V to 128 V. Find the percent increase.

 35. During manufacturing the pressure in a hydraulic line increases from 75 lb/in^2 to 115 lb/in^2. What is the percent increase in pressure?

36. The value of Caroline's house decreased from $72,500 to $58,500 when the area's major employer closed and moved to another state. Find the percent decrease in the value of her house.

37. Due to wage concessions, Bill's hourly wages dropped from $12.65 to $10.85. Find the percent decrease in his wages.

38. A building has 28,000 ft^2 of floor space. When an addition of 6500 ft^2 is built, what is the percent increase in floor space?

39. A business computer package that usually sells for $675 is on sale for $595. Find the percent decrease in its selling price.

An *invoice* is an itemized list of goods and services specifying the price and terms of sale.

40. Complete the following invoice for parts and labor for the addition to a home for the week indicated.

BILL'S PLUMBING
120 East Main Street
Poughkeepsie, New York 12600

Date: 6/25—6/29

Name: Gary Jones

Address: 2630 E. Elm St.

City: Poughkeepsie, NY 12600

Quantity	Item	Cost/Unit	Total Cost
22 ea	$\frac{3}{4}$" fittings	$0.59	
14 ea	$\frac{3}{4}$" nozzles	2.09	
12 ea	$\frac{3}{4}$" 90° ells	0.77	
6 ea	$\frac{3}{4}$" faucets	4.69	
6 ea	$\frac{3}{4}$" valves	6.45	
6 ea	$\frac{3}{4}$" unions	1.96	
5 ea	$\frac{3}{4}$" T joints	1.09	
4 ea	$\frac{3}{4}$" 45° ells	1.29	
120 ft	$\frac{3}{4}$" type M copper pipe	0.59/ft	
32 hr	Labor	18.00/hr	
		TOTAL	
		Less 5% Cash Discount Net 30 Days	
		NET TOTAL	

41. Complete the following feed invoice.

SOUTH COUNTY FEED COMPANY
Highway 51 South
Janesville, Wisconsin 53545

CUSTOMER: Rolling Acres, Inc.   DATE: 8/5

QUANTITY	ITEM	PRICE	DISCOUNT	AMOUNT
600 lb	Pig starter	$24.00/100 lb	$0.25/100 lb	
500 lb	Beef grower	10.60/100 lb	0.11/100 lb	
14,000 lb	Bulk feed	14.75/100 lb	0.15/100 lb	
800 lb	Molasses	6.20/100 lb	—	
6	Salt blocks	0.85/50# block	—	
		Subtotal		
		2% Discount (if paid in 30 days)		
		TOTAL		

42. Complete the following invoice for grain sold at a local elevator.

FARMERS TERMINAL GRAIN COMPANY   Beardstown, Illinois 62618

CUSTOMER NAME: ___Shaw Farms, Inc.___   ACCOUNT NO. ___3786___

Date	Gross wt— pounds	Weight of empty truck	Net wt— pounds	Type of grain	No. of bushels*	Price/bu	Amount
7/2	21560	9160	12400	Wheat	207	$3.17	$656.19
7/3	26720	9240		Wheat		3.23	
7/5	20240	7480		Wheat		3.20	
7/6	28340	9200		Wheat		3.22	
7/8	26760	9160		Wheat		3.25	
7/8	17880	7485		Wheat		3.25	
10/1	25620	9080		Soybeans		6.20	
10/1	21560	7640		Soybeans		6.20	
10/2	26510	9060		Soybeans		6.24	
10/2	22630	7635		Soybeans		6.24	
10/4	22920	9220		Soybeans		6.35	
10/5	20200	7660		Soybeans		6.28	
10/6	25880	9160		Soybeans		6.40	
10/7	21300	7675		Soybeans		6.41	
10/8	18200	7665		Soybeans		6.45	
10/12	26200	9150		Corn		2.31	
10/12	22600	7650		Corn		2.31	
10/13	27100	9080		Corn		2.37	
10/15	22550	7635		Corn		2.25	
10/15	23600	7680		Corn		2.25	
10/17	26780	9160		Corn		2.36	
10/18	28310	9200		Corn		2.43	
10/21	21560	7665		Corn		2.33	
10/22	25750	9160		Corn		2.31	
					TOTAL		

*Round to nearest bushel. Note: Corn weighs 56 lb/bu; soybeans weigh 60 lb/bu; wheat weighs 60 lb/bu.

43. Complete the following electronics parts invoice.

APPLIANCE DISTRIBUTORS INCORPORATED			1400 WEST ELM STREET ST. LOUIS, MISSOURI 63100	
Sold to: Ed's TV Kampsville, IL 62053			Date: 9/26	
Quantity	Description	Unit price	Discount	Net amount
3	67A76-1	$ 8.58	40%	
5	A8934-1	35.10	25%	
5	A8935-1	33.95	25%	
8	A8922-2	33.90	25%	
2	A8919-2X	24.60	20%	
5	700A256	18.80	15%	
		SUBTOTAL		
		Less 5% if paid in 30 days		
		TOTAL		

44. Many lumber yards write invoices for their lumber by the piece. Complete the following invoice for the rough framing of the shell of a home.

| KURT'S LUMBER | 400 WEST OAK | | AKRON, OHIO 44300 |

SOLD TO Robert Bennett   32 Park Pl   E. Akron 44305   DATE 5/16

QUANTITY	DESCRIPTION	UNIT PRICE	TOTAL
66	2" × 4" × 16', fir, plate mat.	$ 4.89	
30	2" × 4" × 10', fir, studs	2.79	
14	2" × 4" × 8', fir, knee wall studs	1.99	
17	2" × 6" × 12', fir, kit. ceiling joists	5.29	
4	2" × 12" × 12', fir, kitchen girders	11.10	
9	2" × 6" × 10', fir, kitchen rafters	4.29	
7	2" × 4" × 12', fir, collar beams	3.59	
10	2" × 8" × 12', fir, 2nd floor joists	6.69	
6	2" × 8" × 16', fir, 2nd floor joists	8.58	
11	2" × 8" × 20', fir, 2nd floor joists	11.88	
15	4' × 8' × 3/4" T & G plywood	19.85	
27	2" × 8" × 18', fir, kit. & liv. rm. raft.	11.35	
7	2" × 8" × 16', fir, kit. & liv. rm. raft.	8.95	
1	2" × 10" × 22', fir, kit. & liv. rm. ridge	18.90	
10	1" × 8" × 14', #2 w. pine, sub facia	5.85	
27	2" × 8" × 22', fir, bedroom rafters	13.55	
7	2" × 8" × 16', fir, dormer rafters	8.95	
1	2" × 10" × 20', fir, bedroom ridge	15.85	
7	2" × 6" × 20', fir, bedroom ceil. joists	10.19	
8	2" × 6" × 8', fir, bedroom ceil. joists	3.29	
3	2" × 12" × 14", fir, stair stringers	12.89	
80	4' × 8' × 1/2", COX roof deck plywood	6.99	
7	rolls #15 felt building paper	9.95	
1	keg 50 lb #16 cement coated nails	34.50	
1	keg 50 lb simplex roofing nails	64.25	
1	keg 50 lb gold roofing nails	39.75	
250	precut fir studs	1.15	

Subtotal _____

Less 2% cash discount _____

Subtotal _____

$4\frac{3}{4}$% sales tax _____

NET TOTAL _____

142

Chapter 3 Review

Write each common fraction as a decimal:

1. $\dfrac{9}{16}$ 2. $\dfrac{5}{12}$

Change each decimal to a common fraction or a mixed number:

3. 0.45   4. 19.625

Perform the indicated operations:

5. 8.6 + 140 + 0.048 + 19.63
6. 25 + 16.3 − 40 + 0.05 − 26.1
7. 86.7 − 18.035   8. 34 − 0.28
9. 0.605 × 5300   10. 18.05 × 0.109
11. 74.73 ÷ 23.5   12. 9.27 ÷ 0.45
13. Round 248.1563 to: (a) the nearest hundred, (b) the nearest tenth, and (c) the nearest ten.
14. Round 5.64908 to: (a) the nearest tenth, (b) the nearest hundredth, and (c) the nearest ten-thousandth.

Write each number in scientific notation:

15. 476,000   16. 0.0014

Write each number in decimal form:

17. 5.35×10^{-5} 18. 6.1×10^{7}

Find the larger number:

19. 0.00063; 0.00105   20. 0.056; 0.06

Find the smaller number:

21. 0.000075; 0.0006   22. 0.04; 0.00183

Perform the indicated operations using the laws of exponents and express the results using positive exponents:

23. $10^{9} \cdot 10^{-14}$ 24. $10^{6} \div 10^{-3}$

25. $(10^{-4})^{3}$ 26. $\dfrac{(10^{-3} \cdot 10^{5})^{3}}{10^{6}}$

Perform the indicated operations and write each result in scientific notation:

27. $(9.5 \times 10^{10})(4.6 \times 10^{-13})$ 28. $\dfrac{8.4 \times 10^8}{3 \times 10^{-6}}$

29. $\dfrac{(50,000)(640,000,000)}{(0.0004)^2}$

30. $(4.5 \times 10^{-8})^2$ 31. $(2 \times 10^9)^4$

Change each percent to a decimal:

32. 15%   33. $8\frac{1}{4}\%$

Change each decimal to a percent:

34. 0.065   35. 1.2

36. What is $8\frac{3}{4}\%$ of $12,000?

37. In a small electronics business the overhead is $16,000 and the gross income is $42,000. What percent is the overhead?

38. 180 represents 8% of what number?

39. A new tire has a tread depth of $\frac{13}{32}$ in. At 16,000 miles the tread depth is $\frac{11}{64}$ in. What percent of the tread is left?

40. A farmer bales 60 tons of hay which are 20% moisture. How many tons of dry matter does he harvest?

THE METRIC SYSTEM

 Introduction to the Metric System

In early recorded time, parts of the human body were used for standards of measurement. However, these standards were neither uniform nor acceptable to all. The next step was to define the various standards, such as the inch, the foot, the rod, and so on. But each country introduced or defined its own standards, which were often not related to those in other countries. Then, in 1670, as the need for a single worldwide measurement system became recognized, Gabriel Mouton, a Frenchman, proposed a uniform decimal measurement system. Then, by the 1800s, metric standards were first adopted worldwide. In 1960 the International System of Units, with SI as the international abbreviation, was adopted for the modern metric system.

The U.S. Congress legalized the use of the metric system throughout the United States on July 28, 1866, which is over 100 years ago. The United States has signed various metric treaties and has participated in many international conferences and conventions dealing with the metric system. In the early 1970s, the United States found that it was the only industrial nation not committed to converting to the metric system. On December 23, 1975, President Gerald R. Ford signed the Metric Conversion Act of 1975, the purpose of which is "to declare a national policy of coordinating the increasing use of the metric system in the United States."

At present, only the United States and one other country are left to make the move toward world uniformity in measurement. This huge imbalance makes it imperative that the United States convert as soon as possible so that it can compete more favorably for international trade. Most countries have restrictions on allowing non-metric products to enter. Metric countries just naturally want metric products.

The SI metric system has seven base units:

Basic unit	SI abbreviation	For measuring
metre*	m	length
kilogram	kg	mass
second	s	time
ampere	A	electric current
kelvin	K	temperature
candela	cd	light intensity
mole	mol	molecular substance

*At present, there is some difference of opinion on the spelling of metre and litre. We have chosen the "re" spelling because it is the internationally accepted spelling and because it distinguishes the unit from other meters, such as parking and electric meters.

Other commonly used metric units are:

Unit	SI abbreviation	For measuring
litre	L	volume
cubic metre	m^3	volume
square metre	m^2	area
newton	N	force
metre per second	m/s	speed
joule	J	energy
watt	W	power
radian	rad or (...r)	plane angle

Since the metric system is a decimal or base 10 system, it is very similar to our decimal number system. It is an easy system to use because calculations are based on the number 10 and its multiples. The SI system has special prefixes which name multiples and submultiples, and which may be used with most all SI units. The following table shows the prefixes and the corresponding symbols.

Prefixes for SI Units

Multiple or submultiple* decimal form	Power of 10	Prefix	Prefix symbol	Pronun- ciation	Meaning
1,000,000,000,000	10^{12}	tera	T	tĕr´ă	one trillion times
1,000,000,000	10^9	giga	G	jĭg´ă	one billion times
1,000,000	10^6	mega	M	mĕg´ă	one million times
1,000	10^3	kilo**	k	kĭl´ō	one thousand times
100	10^2	hecto	h	hĕk´tō	one hundred times
10	10^1	deka	da	dĕk´ă	ten times
0.1	10^{-1}	deci	d	dĕs´ĭ	one tenth of
0.01	10^{-2}	centi**	c	sĕnt´ĭ	one hundredth of
0.001	10^{-3}	milli**	m	mĭl´ĭ	one thousandth of
0.000001	10^{-6}	micro	μ	mī´krō	one millionth of
0.000000001	10^{-9}	nano	n	năn´ō	one billionth of
0.000000000001	10^{-12}	pico	p	pē´kō	one trillionth of

*Factor by which the unit is multiplied.
**Most commonly used prefixes.

Since the same prefixes are used with most all SI metric units, memorizing long lists or many tables is not necessary.

Example 1. Write the SI abbreviation for 45 kilometres.

The symbol for the prefix *kilo* is k.
The symbol for the unit *metre* is m.
The SI abbreviation for 45 kilometres is 45 km.

Example 2. Write the SI unit for the abbreviation 50 mg.

The prefix for m is *milli*.
The unit for g is *gram*.
The SI unit for 50 mg is 50 milligrams.

In summary, the English system is an ancient one based on standards initially determined by parts of the human body, which is why there is no consistent relationship between units. In the metric system, however, standard units are subdivided in multiples of 10, similar to our number system, and the names associated with each subdivision have prefixes that indicate a multiple of 10.

EXERCISES 4.1   *Give the metric prefix for each value:*

1. 1000   2. 100   3. 0.01   4. 0.1
5. 0.001   6. 10   7. 1,000,000   8. 0.000001

Give the SI symbol, or abbreviation, for each prefix:

9. hecto   10. kilo   11. deci   12. milli
13. centi   14. deka   15. micro   16. mega

Write the abbreviation for each quantity:

17. 65 milligrams   18. 125 kilolitres   19. 82 centimetres
20. 205 millilitres   21. 36 microamps
22. 75 kilograms   23. 19 hectolitres   24. 5 megawatts

Write the SI unit for each abbreviation:

25. 18 m   26. 15 L   27. 36 kg   28. 85 mm   29. 24 ps
30. 9 dam   31. 135 mL   32. 45 dL   33. 45 mA   34. 75 MW
35. The basic SI unit of length is _____.
36. The basic SI unit of mass is _____.
37. The basic SI unit of electric current is _____.
38. The basic SI unit of time is _____.
39. The common SI units of volume are _____ and _____.
40. The common SI unit of power is _____.

4.2 Length

The basic SI unit of length is the metre (m). The height of a door knob is about 1 m. (See Fig. 4.1.) One metre is a little more than 1 yd. (See Fig. 4.2.)

Long distances are measured in kilometres (km). (1 km = 1000 m.) The length of five city blocks is about 1 km. (See Fig. 4.3.)

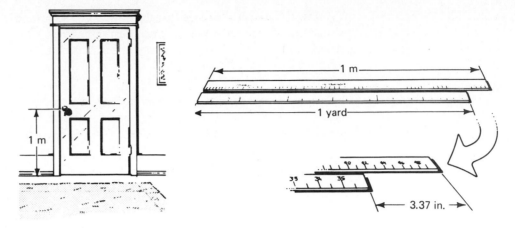

Figure 4.1   Figure 4.2

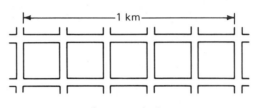

Figure 4.3

The centimetre (cm) is used to measure short distances, such as the length of this book (about 28 cm), or the width of a board. The width of your small fingernail is about 1 cm. (See Fig. 4.4.)

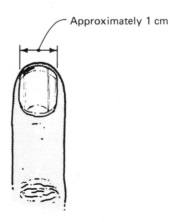

Figure 4.4

The millimetre (mm) is used to measure very small lengths, such as the thickness of a sheet of metal or the depth of a tire tread. The thickness of a dime is about 1 mm. (See Fig. 4.5.)

Figure 4.5

Each of the large numbered divisions on the metric ruler in Fig. 4.6 marks off one centimetre (cm). The smaller lines indicate halves of centimetres, and the smallest divisions show tenths of centimetres, or millimetres.

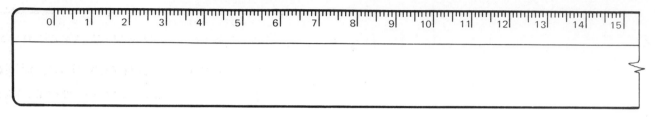

Figure 4.6

Example 1. Read *A*, *B*, *C*, and *D* on the metric ruler in Fig. 4.7. Give each result in millimetres, centimetres, and metres.

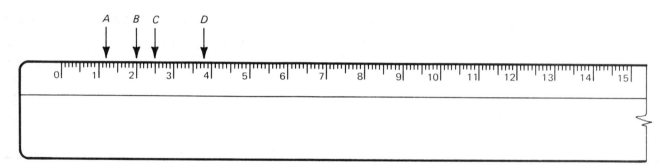

Figure 4.7

Answers. A = 12 mm, 1.2 cm, 0.012 m

           B = 20 mm, 2.0 cm, 0.020 m

           C = 25 mm, 2.5 cm, 0.025 m

           D = 38 mm, 3.8 cm, 0.038 m

To convert from one unit to another in the metric system we could use the same conversion factor procedure that we used in the English system in Sec. 2.6. That is,

---

Conversion Factor

The correct conversion factor is equal to one (1) in fraction form. The numerator is expressed in the new units and the denominator is expressed in the old units.

Example 2. Change 3.6 km to metres.

Since kilo means 10^3 or 1000, 1 km = 1000 m. The two possible conversion factors are $\dfrac{1 \text{ km}}{1000 \text{ m}}$ and $\dfrac{1000 \text{ m}}{1 \text{ km}}$. We want to choose the one whose numerator is expressed in the new units (m) and whose denominator is expressed in the old units (km). This is $\dfrac{1000 \text{ m}}{1 \text{ km}}$.

Thus, $3.6 \text{ km} \times \dfrac{1000 \text{ m}}{1 \text{ km}} = 3600 \text{ m}$

Example 3. Change 4 m to centimetres.

First, centi means 10^{-2} or 0.01 and 1 cm = 10^{-2} m. Choose the conversion factor with metres in the denominator and centimetres in the numerator.

That is, $4 \text{ m} \;\; \dfrac{1 \text{ cm}}{10^{-2} \text{ m}} = 400 \text{ cm}$

Note: Conversions within the metric system only involve moving the decimal point.

EXERCISES 4.2   *Which is longer?*

1. 1 metre or 1 millimetre   2. 1 metre or 1 centimetre
3. 1 metre or 1 kilometre   4. 1 millimetre or 1 kilometre
5. 1 centimetre or 1 millimetre
6. 1 kilometre or 1 centimetre

Fill in each blank:

7. 1 m = _____ mm   8. 1 km = _____ m

9. 1 cm = _____ m   10. 1 m = _____ cm

11. 1 m = _____ km   12. 1 hm = _____ m

13. 1 cm = _____ mm   14. 1 dam = _____ dm

Which metric unit (km, m, cm, or mm) should you use to measure each item?

15. diameter of an automobile tire
16. thickness of sheet metal
17. metric wrench sizes
18. length of an auto race
19. length of a discus throw in track and field
20. length and width of a table top
21. distance between Chicago and St. Louis
22. thickness of plywood
23. thread size of a pipe
24. length and width of a house lot

Fill in each blank with the most reasonable unit (km, m, cm, or mm):

25. A standard film size for cameras is 35 ___ .

26. The diameter of a wheel on a ten-speed bicycle is 56 ___ .

27. A jet plane generally flies about 8-9 ___ high.

28. The width of a door in our house is 910 ___ .

29. The length of the ridge on our roof is 24 ___ .

30. John's waist size is 95 ___ .

31. The steering wheel on Brenda's car is 36 ___ in diameter.

32. Jan drives 12 ___ to school.

33. The standard metric size for plywood is 1200 ___ wide and 2400 ___ long.

34. The distance from home plate to centerfield in a base-ball park is 125 ___ .

35. Read the measurements indicated by the letters on the metric ruler below and give each result in millimetres, centimetres, and metres.

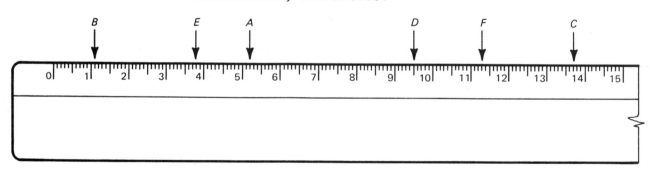

Use a metric ruler to measure each line segment. Give each result in millimetres, centimetres, and metres:

36. ├─────────────┤

37. ├──────────────────┤

38. ├─────────────────────────────┤

39. ├──────────────────────────┤

40. ├──────────────┤

41. ├───────────────────────┤

42. ├──────────┤

43. ├──────────────────────────────────┤

44. ├─────┤

45. Change 675 m to km.   46. Change 450 cm to m.
47. Change 1540 mm to m.   48. Change 3.2 km to m.
49. Change 65 cm to m.   50. Change 1.4 m to mm.
51. Change 7.3 m to cm.   52. Change 48 cm to mm.
53. Change 125 mm to cm.   54. Change 0.75 m to μm.
55. What is your height in metres?
56. What is your height in centimetres?
57. What is your height in millimetres?

4.3 Mass and Weight

The *mass* of an object is the quantity of material making up
the object. One unit of mass in the SI system is the gram
(g). The gram is defined as the mass contained in one cubic
centimetre (cm^3) of water, at its maximum density.

A common paper clip has a   Three aspirin have a mass of
mass of about 1 g. (See   about 1 g. (See Fig. 4.9.)
Fig. 4.8.)

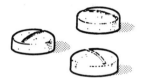

Figure 4.8   Figure 4.9

Because the gram is so small, the kilogram is the basic
unit of mass in the SI system. One kilogram is defined as
the mass contained in one cubic decimetre (dm^3) of water,
at its maximum density.

For very large quantities, such as a trainload of coal
or grain or a shipload of ore, the metric ton (1000 kg) is
used.

The milligram (mg) is used to measure very, very small
masses such as medicine dosages. One grain of salt has a
mass of about 1 mg.

The *weight* of an object is a measure of the earth's
gravitational force—or pull—acting on the object. The SI
unit of weight is the newton (N). As you are no doubt
aware, the terms "mass" and "weight" are commonly used
interchangeably by the general public. We have presented
them as technical terms, as they are used in the technical,
engineering, and scientific professions. To further illustrate the difference, the mass of an astronaut remains
relatively constant while his/her weight varies (the weight
decreases as the distance from the earth increases). If the
spaceship is in orbit or farther out in space, we say the
crew is "weightless" because they seem to float freely in

space. Their mass has not changed while their weight is near zero. (See Fig. 4.10.)

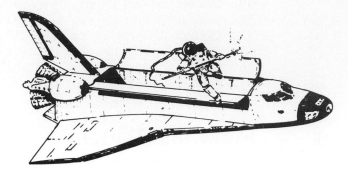

Figure 4.10

<u>Example 1</u>. Change 12 kg to grams.

First, kilo means 10^3 or 1000 and 1 kg = 1000 g. Choose the conversion factor which has kilograms in the denominator and grams in the numerator.

Thus, $12 \ \cancel{kg} \times \dfrac{1000 \ g}{1 \ \cancel{kg}} = 12,000 \ g$

<u>Example 2</u>. Change 250 mg to grams.

First, milli means 10^{-3} or 0.001 and 1 mg = 10^{-3} g. Choose the conversion factor which has milligram in the denominator and gram in the numerator.

That is, $250 \ \cancel{mg} \times \dfrac{10^{-3} \ g}{1 \ \cancel{mg}} = 0.25 \ g$

EXERCISES 4.3   *Which is larger?*

1. 1 gram or 1 milligram   2. 1 gram or 1 kilogram
3. 1 milligram or 1 kilogram
4. 1 metric ton or 1 kilogram
5. 1 milligram or 1 microgram
6. 1 kilogram or 1 microgram

Fill in each blank:

7. 1 g = ___ mg   8. 1 kg = ___ g

9. 1 cg = ___ g   10. 1 mg = ___ g

11. 1 metric ton = ___ kg   12. 1 g = ___ cg

13. 1 mg = ___ µg   14. 1 µg = ___ mg

Which metric unit (kg, g, mg, or metric ton) should you use to measure each item?

15. a bar of handsoap
16. a vitamin capsule
17. a bag of flour
18. a 4-wheel drive tractor
19. a pencil
20. your mass
21. a trainload of soy-beans
22. a bag of potatoes
23. a contact lens
24. an apple

Fill in each blank with the most reasonable unit (kg, g, mg, or metric ton):

25. A slice of bread has a mass of about 25 ___.

26. Elevators in the college have a load limit of 2200 ___.

27. I take 1000 ___ of vitamin C every day.

28. My uncle's new truck can haul a load of 4 ___.

29. Postage rates for letters are based on the ___.

30. I take 1 ___ of vitamin C every day.

31. My best friend has a mass of 65 ___.

32. A jar of peanut butter contains 1200 ___.

33. The local grain elevator shipped 20,000 ___ of wheat last year.

34. One common aspirin size is 325 ___.

35. Change 875 g to kg.
36. Change 127 mg to g.
37. Change 85 g to mg.
38. Change 1.5 kg to g.
39. Change 3.6 kg to g.
40. Change 430 g to mg.
41. Change 270 mg to g.
42. Change 18 mg to ug.
43. Change 2.5 metric tons to kg.
44. Change 18,000 kg to metric tons.
45. What is your mass in kilograms?

4.4 Volume and Area

A common unit of volume in the metric system is the litre (L). One litre of milk is a little more than 1 quart of milk. The litre is commonly used for liquid volume. (See Fig. 4.11.)

The cubic metre (m^3) is used to measure large volumes. One cubic metre is the volume contained in a cube 1 m on an edge. The cubic centimetre (cm^3) is used to measure small volumes. It is the volume contained in a cube 1 cm on an edge.

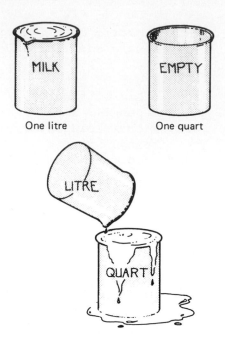

Figure 4.11

The relationship between the litre and the cubic centimetre needs to be made. The litre is defined as the volume in 1 cubic decimetre (dm^3). That is, 1 L of liquid fills a cube 1 dm (10 cm) on an edge. (See Fig. 4.12.)

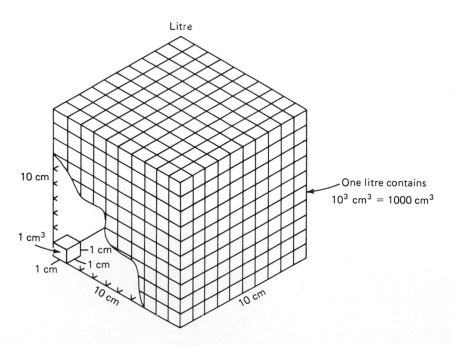

Figure 4.12

The volume of this cube can also be found by the formula

$$V = \ell wh$$
$$V = (10 \text{ cm})(10 \text{ cm})(10 \text{ cm})$$
$$= 1000 \text{ cm}^3$$

Thus, $1 \text{ L} = 1000 \text{ cm}^3$

Dividing each by 1000, we have

$$\frac{1}{1000} \text{ L} = 1 \text{ cm}^3$$

or $\boxed{1 \text{ mL} = 1 \text{ cm}^3}$ $\left(1 \text{ mL} = \frac{1}{1000} \text{ L} \right)$.

Milk, soft drinks, and gasoline are sold by the litre. Liquid medicine and eye drops are sold by the millilitre. Large quantities of liquid are sold by the kilolitre (1000 L).

In Sec. 4.3, the kilogram was defined as the mass of 1 dm^3 of water. Since $1 \text{ dm}^3 = 1$ L, one liter of water has a mass of 1 kg.

Example 1. Change 0.5 L to millilitres.

$$0.5 \text{ L} \times \frac{1000 \text{ mL}}{1 \text{ L}} = 500 \text{ mL}$$

Example 2. Change 4.5 cm^3 to mm^3.

$$4.5 \text{ cm}^3 \times \left(\frac{10 \text{ mm}}{1 \text{ cm}} \right)^3 = 4500 \text{ mm}^3$$

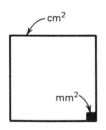

Figure 4.13

The basic unit of area in the metric system is the square metre (m^2), the area contained in a square whose sides are each 1 m long. The square centimetre (cm^2) and the square millimetre (mm^2) are smaller units of area. (See Fig. 4.13.) The larger-area units are the square kilometre (km^2) and the hectare (ha).

Example 3. Change 2400 cm^2 to m^2.

$$2400 \text{ cm}^2 \times \left(\frac{1 \text{ m}}{100 \text{ cm}} \right)^2 = 0.24 \text{ m}^2$$

Example 4. Change 1.2 km^2 to m^2.

$$1.2 \text{ km}^2 \times \left(\frac{1000 \text{ m}}{1 \text{ km}} \right)^2 = 1,200,000 \text{ m}^2$$

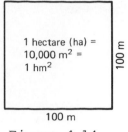

1 hectare (ha) =
10,000 m² =
1 hm²

100 m

100 m

Figure 4.14

In the metric system the hectare (ha) is the basic land area unit. The area of one hectare is equal to the area of a square 100 m on a side, which is equal to 10,000 m² or one square hectometre (hm²). (See Fig. 4.14.)

The hectare is used because it is more convenient to say and use than "square hectometre." The metric prefixes are *not* to be used with the hectare unit. That is, instead of saying "kilohectare," we say "1000 hectares."

<u>Example 5</u>. How many hectares are contained in a rectangular field 360 m by 850 m?

The area in m² is

$$(360 \text{ m})(850 \text{ m}) = 306{,}000 \text{ m}^2$$

$$306{,}000 \text{ m}^2 \times \frac{1 \text{ ha}}{10{,}000 \text{ m}^2} = 30.6 \text{ ha}$$

EXERCISES 4.4   *Which is larger?*

1. 1 litre or 1 millilitre
2. 1 millilitre or 1 kilolitre
3. 1 cubic millimetre or 1 cubic centimetre
4. 1 cubic metre or 1 litre
5. 1 square kilometre or 1 hectare
6. 1 square centimetre or 1 square millimetre

Fill in each blank:

7. 1 L = ____ mL   8. 1 mL = ____ L

9. 1 m³ = ____ cm³   10. 1 mm³ = ____ cm³

11. 1 cm² = ____ mm²   12. 1 km² = ____ ha

13. 1 m³ = ____ L   14. 1 cm³ = ____ mL

Which metric unit (m^3, L, mL, m^2, cm^2, or ha) should you use to measure the following?

15. oil in your car's crankcase
16. cough syrup
17. floor space in a warehouse
18. size of a farm
19. cross section of a piston
20. piston displacement in an engine
21. cargo space in a truck
22. paint needed to paint a house
23. eye drops
24. page size of this book
25. size of an industrial park
26. gasoline in your car's gas tank

Fill in each blank with the most reasonable unit (m^3, L, mL, m^2, cm^2, or ha):

27. Gus ordered 12 ___ of concrete for his new driveway.

28. I drink 250 ___ of orange juice each morning.

29. Bill, a farmer, owns a 2500 ___ storage tank for diesel fuel.

30. Dwight planted 75 ___ of wheat this year.

31. Our house has 195 ___ of floor space.

32. We must heat 520 ___ of living space in our house.

33. When I was a kid, I mowed 6 ___ of lawns each week.

34. Our community's water tower holds 650 ___ of water.

35. The cross section of a log is 2500 ___.

36. Don bought a 25-___ tarpaulin for his truck.

37. I need copper tubing with a cross section of 4 ___.

38. We should each drink 2 ___ of water each day.

39. Change 1500 mL to L.   40. Change 0.60 L to mL.
41. Change 1.5 m^3 to cm^3.   42. Change 450 mm^3 to cm^3.
43. Change 85. cm^3 to mL.   44. Change 650 L to m^3.
45. Change 85,000 m^2 to km^2.
46. Change 18 m^2 to cm^2.
47. Change 85,000 m^2 to ha.
48. Change 250 ha to km^2.
49. What is the mass of 500 mL of water?
50. What is the mass of 1 m^3 of water?
51. How many hectares are contained in a rectangular field which measures 75 m by 90 m?
52. How many hectares are contained in a rectangular field which measures $\frac{1}{4}$ km by $\frac{1}{2}$ km?

4.5 Time, Current, and Other Units

The basic SI unit of time is the second (s), which is the same in all units of measurement. Time is also measured in minutes (min), hours (h), days, and years.

$$1 \text{ min} = 60 \text{ s}$$
$$1 \text{ h} = 60 \text{ min}$$
$$1 \text{ day} = 24 \text{ h}$$
$$1 \text{ year} = 365\frac{1}{4} \text{ day (approx.)}$$

<u>Example 1</u>. Change 4 h 15 min to seconds.

First, $4 \, h \times \dfrac{60 \text{ min}}{1 \, h} = 240$ min

Then,   4 h 15 min = 240 min + 15 min
                                   = 255 min

Finally, $255 \, min \times \dfrac{60 \text{ s}}{1 \, min} = 15{,}300$ s

Very short periods of time are commonly used in electronics. These are measured in parts of a second, given with the appropriate metric prefix.

<u>Example 2</u>. What is the meaning of each unit?

(a) 1 ms = 1 millisecond = 10^{-3} s and means one-thousandth of a second.
(b) 1 μs = 1 microsecond = 10^{-6} s and means one-millionth of a second.
(c) 1 ns = 1 nanosecond = 10^{-9} s and means one-billionth of a second.
(d) 1 ps = 1 picosecond = 10^{-12} s and means one-trillionth of a second.

<u>Example 3</u>. Change 25 ms to seconds.

First, milli means 10^{-3} and 1 ms = 10^{-3} s. So,

$$25 \, ms \times \frac{10^{-3} \text{ s}}{1 \, ms} = 0.025 \text{ s}$$

<u>Example 4</u>. Change 0.00000025 s to nanoseconds.

First, nano means 10^{-9} and 1 ns = 10^{-9} s. So,

$$0.00000025 \, s \times \frac{1 \text{ ns}}{10^{-9} \, s} = 250 \text{ ns}$$

The basic SI unit of electric current is the ampere (A), or sometimes called amp. This same unit is used in the English system. The ampere is a fairly large amount of current, so smaller currents are measured in parts of an ampere and are given the appropriate SI prefix.

<u>Example 5</u>. What is the meaning of each unit?

(a) 1 mA = 1 milliampere = 10^{-3} A and means one-thousandth of an ampere.
(b) 1 μA = 1 microampere = 10^{-6} A and means one-millionth of an ampere.

Example 6. Change 275 μA to amperes.

First, micro means 10^{-6} and 1 μA = 10^{-6} A. So,

$$275 \; \cancel{μA} \times \frac{10^{-6} \; A}{1 \; \cancel{μA}} = 0.000275 \; A$$

Example 7. Change 0.045 A to milliamps.

First, milli means 10^{-3} and 1 mA = 10^{-3} A. So,

$$0.045 \; \cancel{A} \times \frac{1 \; mA}{10^{-3} \; \cancel{A}} = 45 \; mA$$

The common metric unit for both electrical and mechanical power is the watt (W).

Example 8. What is the meaning of each unit?

(a) 1 mW = 1 milliwatt = 10^{-3} W and means one-thousandth of a watt.
(b) 1 kW = 1 kilowatt = 10^3 W and means one thousand watts.
(c) 1 MW = 1 megawatt = 10^6 W and means one million watts.

Example 9. Change 0.025 W to milliwatts.

First, milli means 10^{-3} and 1 mW = 10^{-3} W. So,

$$0.025 \; \cancel{W} \times \frac{1 \; mW}{10^{-3} \; \cancel{W}} = 25 \; mW$$

Example 10. Change 2.3 MW to watts.

First, mega means 10^6 and 1 MW = 10^6 W. So,

$$2.3 \; \cancel{MW} \times \frac{10^6 \; W}{1 \; \cancel{MW}} = 2.3 \times 10^6 \; W \; or \; 2,300,000 \; W$$

A few other units commonly used in electronics deserve mention here.

Unit	Symbol	Used to measure
volt	V	voltage
ohm	Ω	resistance
hertz	Hz	frequency
farad	F	capacitance
henry	H	inductance
coulomb	C	charge

The metric prefixes are used with each of these units in the same way we used them with the previous metric units.

EXERCISES 4.5   *Which is larger?*

1. 1 amp or 1 milliamp
2. 1 microsecond or 1 picosecond
3. 1 second or 1 nanosecond
4. 1 megawatt or 1 milliwatt
5. 1 kilovolt or 1 megavolt
6. 1 volt or 1 millivolt

Write the abbreviation for each unit:

7. 43 kilowatts   8. 7 millivolts
9. 17 picoseconds   10. 1.2 amperes
11. 3.2 megawatts   12. 55 microfarads
13. 450 ohms   14. 70 nanoseconds

Fill in each blank:

15. 1 kW = ___ W   16. 1 mA = ___ A

17. 1 ns = ___ s   18. 1 day = ___ s

19. 1 A = ___ μA   20. 1 F = ___ μF

21. 1 V = ___ MV   22. 1 Hz = ___ kHz

23. Change 0.35 A to mA.   24. Change 18 kW to W.
25. Change 350 ms to s.   26. Change 1 h 25 min 16 s to s.
27. Change 13,950 s to h, min, and s.
28. Change 15 MV to kV.
29. Change 175 μF to mF.   30. Change 145 ps to ns.
31. Change 1500 kHz to MHz. 32. Change 5×10^{12} W to MW.

 Temperature

The basic SI unit for temperature is kelvin (K), which is used mostly in scientific and engineering work. Everyday temperatures are measured in degrees Celsius (°C). The United States also measures temperatures in degrees Fahrenheit (°F).

On the Celsius scale, water freezes at 0° and boils at 100°. Each degree Celsius is 1/100 of the difference between the boiling temperature and the freezing temperature of water. Fig. 4.15 shows some approximate temperature readings in degrees Celsius and Fahrenheit and compares them with some related activity.

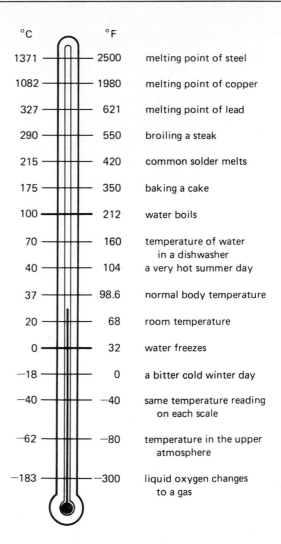

°C	°F	
1371	2500	melting point of steel
1082	1980	melting point of copper
327	621	melting point of lead
290	550	broiling a steak
215	420	common solder melts
175	350	baking a cake
100	212	water boils
70	160	temperature of water in a dishwasher
40	104	a very hot summer day
37	98.6	normal body temperature
20	68	room temperature
0	32	water freezes
−18	0	a bitter cold winter day
−40	−40	same temperature reading on each scale
−62	−80	temperature in the upper atmosphere
−183	−300	liquid oxygen changes to a gas

Figure 4.15

The formulas for changing from degrees Celsius and degrees Fahrenheit are:

$$C = \frac{5}{9}(F - 32) \text{ or } C = 0.56(F - 32)$$

$$F = \frac{9}{5}C + 32 \quad \text{or } F = 1.8C + 32$$

Example 1. Change 68° F to degrees Celsius.

$$C = \frac{5}{9}(F - 32)$$

$$C = \frac{5}{9}(68° - 32°)$$

$$= \frac{5}{9}(36°)$$

$$= 20°$$

Thus, 68°F = 20°C.

Example 2. Change 35°C to degrees Fahrenheit.

$$F = \frac{9}{5}C + 32$$

$$F = \frac{9}{5}(35°) + 32°$$

$$= 63° + 32°$$

$$= 95°$$

That is, 35°C = 95°F.

Example 3. Change 10°F to degrees Celsius.

$$C = \frac{5}{9}(F - 32)$$

$$C = \frac{5}{9}(10° - 32°)$$

$$= \frac{5}{9}(-22°)$$

$$= -12.2°$$

So, 10°F = -12.2°C.

Example 4. Change -60°C to degrees Fahrenheit.

$$F = \frac{9}{5}C + 32$$

$$F = \frac{9}{5}(-60°) + 32°$$

$$= -108° + 32°$$

$$= -76°$$

SO, -60°C = -76°F.

EXERCISES 4.6   *Use Fig. 4.15 to choose the most reasonable answer for each statement:*

1. The boiling temperature of water is
   (a) 212°C   (b) 100°C   (c) 0°C   (d) 50°C

2. The freezing temperature of water is
   (a) 32°C   (b) 100°C   (c) 0°C   (d) -32°C

3. Normal body temperature is
   (a) 100°C   (b) 50°C   (c) 37°C   (d) 98.6°C

4. The body temperature of a person who has a fever is
   (a) 102°C   (b) 52°C   (c) 39°C   (d) 37°C

5. The temperature on a hot summer day in the California desert is
   (a) 108°C   (b) 43°C   (c) 60°C   (d) 120°C

6. The temperature on a cold winter day in Chicago is

    (a) 20°C   (b) 10°C   (c) 30°C   (d) -10°C

7. The thermostat in your home should be set at

    (a) 70°C   (b) 50°C   (c) 19°C   (d) 30°C

8. Solder melts at

    (a) 215°C   (b) 420°C   (c) 175°C   (d) 350°C

9. Freezing rain is most likely to occur at

    (a) 32°C   (b) 25°C   (c) -18°C   (d) 0°C

10. The weather forecast calls for a low temperature of 3°C. What should you plan to do?
    (a) Sleep with the windows open.
    (b) Protect your plants from frost.
    (c) Sleep with the air conditioner on.
    (d) Sleep with an extra blanket.

Fill in each blank:

11. 77°F = ___ °C   12. 45°C = ___ °F

13. 325°C = ___ °F   14. 140°F = ___ °C

15. -16°C = ___ °F   16. 5°F = ___ °C

17. -16°F = ___ °C   18. -40°C = ___ °F

19. -78°C = ___ °F   20. -10°F = ___ °C

4.7 Metric and English Conversion

Sometimes in technical work you must change from one system of measurement to another. The approximate conversions between metric units and English units are found in the Metric and English Conversion Table, Table 3. Most numbers are rounded to three or four significant digits. Due to this rounding and your choice of conversion factors, there may be a small difference in the last digit(s) of the answers involving conversion factors. This small difference is acceptable. In this section, round each result to three significant digits, when necessary. You may review significant digits in Section 5.1.

Fig. 4.16 shows the relative sizes of each of four sets of common metric and English units of area.

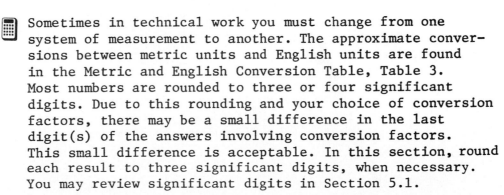
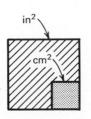
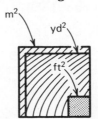
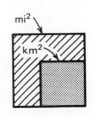
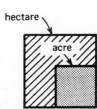

Figure 4.16

<u>Example 1</u>. Change 17 in. to centimetres.

$$17 \; \cancel{in.} \times \frac{2.54 \; cm}{1 \; \cancel{in.}} = 43.2 \; cm$$

<u>Example 2</u>. Change 1950 g to pounds.

$$1950 \; \cancel{g} \times \frac{1 \; lb}{454 \; \cancel{g}} = 4.30 \; lb$$

<u>Example 3</u>. Change 0.85 qt to millilitres.

$$0.85 \; \cancel{qt} \times \frac{0.946 \; \cancel{L}}{1 \; \cancel{qt}} \times \frac{10^3 \; mL}{1 \; \cancel{L}} = 804 \; mL$$

<u>Example 4</u>. Change 5 yd^2 to ft^2.

$$5 \; \cancel{yd^2} \times \frac{9 \; ft^2}{1 \; \cancel{yd^2}} = 45 \; ft^2$$

<u>Example 5</u>. How many square inches are in a metal plate 14 cm^2 in area?

$$14 \; \cancel{cm^2} \times \frac{0.155 \; in^2}{1 \; \cancel{cm^2}} = 2.17 \; in^2$$

<u>Example 6</u>. Change 147 ft^3 to cubic yards.

$$147 \; \cancel{ft^3} \times \frac{1 \; yd^3}{27 \; \cancel{ft^3}} = 5.44 \; yd^3$$

<u>Example 7</u>. How many cubic yards are in 12 m^3?

$$12 \; \cancel{m^3} \times \frac{1.31 \; yd^3}{1 \; \cancel{m^3}} = 15.7 \; yd^3$$

In the English system the acre is the basic unit of land area. Historically, the acre was the amount of ground that a yoke of oxen could plow in one day.

$$1 \; acre = 43{,}560 \; ft^2$$
$$1 \; mi^2 = 640 \; acres = 1 \; section$$

<u>Example 8</u>. How many acres are in a rectangular field which measures 1350 ft by 2750 ft?

The area in ft^2 is

$$(1350 \; ft)(2750 \; ft) = 3{,}712{,}500 \; ft^2$$

$$3{,}712{,}500 \; \cancel{ft^2} \times \frac{1 \; acre}{43{,}560 \; \cancel{ft^2}} = 85.2 \; acres$$

Professional journals and publications in nearly all scientific areas, including agronomy and animal science, have been metric for several years now so that scientists around the world could better understand and benefit from U.S. research. Land areas in the United States are measured mostly in the English system.

When converting between metric and English land area units, we use the following relationship:

$$1 \text{ hectare} = 2.47 \text{ acres}$$

However, a good approximation is

$$1 \text{ hectare} = 2.5 \text{ acres}$$

Example 9. How many acres are in 30.6 ha?

$$30.6 \text{ ha} \times \frac{2.47 \text{ acres}}{1 \text{ ha}} = 75.6 \text{ acres}$$

Example 10. How many hectares are in the rectangular field in Example 8?

$$85.2 \text{ acres} \times \frac{1 \text{ ha}}{2.47 \text{ acres}} = 34.5 \text{ ha}$$

Considerable patience and education will be necessary as well as many man-hours spent before the hectare becomes the common unit of land area in the United States. The mammoth task of changing all property documents is only one of many obstacles.

EXERCISES 4.7   *Fill in each blank: (Round each result to three significant digits, when necessary.)*

1. 8 lb = ___ kg
2. 16 ft = ___ m
3. 38 cm = ___ in.
4. 81 m = ___ ft
5. 4 yd = ___ cm
6. 17 qt = ___ L
7. 30 kg = ___ lb
8. 15 mi = ___ km
9. 3.2 in. = ___ mm
10. 2 lb 4 oz = ___ g
11. A road sign reads "75 km to Chicago." What is this distance in miles?
12. A camera uses 35-mm film. How many inches wide is it?
13. The diameter of a bolt is 0.425 in. Express this diameter in mm.
14. Change $3\frac{13}{32}$ in. to cm.
15. A tank contains 8 gal of fuel. How many litres of fuel are in the tank?

16. A satellite weighs 150 kg. How many pounds does it weigh?
17. A football field 100 yd long is (a) how many feet long? (b) How many metres long?
18. A micro wheel weighs 0.045 oz. What is its weight in mg?
19. A hole is to be drilled in a metal plate 5 in. in diameter. What is the diameter (a) in cm and (b) in mm?
20. A can contains 15 oz of tomato sauce. How many grams does the can contain?
21. Change 3 yd^2 to m^2.
22. Change 12 cm^2 to in^2.
23. How many m^2 are in 140 yd^2?
24. How many m^2 are in 15 yd^2?
25. Change 18 in^2 to cm^2.
26. How many ft^2 are in a rectangle 12.6 yd long and 8.6 yd wide? ($A = \ell w$)
27. How many ft^2 are in a rectangle 12.6 m long and 8.6 m wide?
28. Find the area of the figure below in in^2.

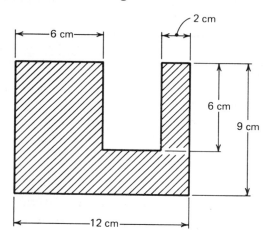

29. Change 15 yd^3 to m^3.
30. Change 5473 in^3 to cm^3.
31. How many mm^3 are in 17 in^3?
32. How many in^3 are in 25 cm^3?
33. Change 84 ft^3 to cm^3.
34. How many cm^3 are in 98 in^3?
35. A commercial lot 80 ft wide and 180 ft deep sold for $32,400. What was the price per square foot? What was the price per frontage foot?

36. A concrete sidewalk is to be built as shown below around the outside of a corner lot that measures 140 ft by 180 ft. The sidewalk is to be 5 ft wide. What is the surface area of the sidewalk? The sidewalk is to be 4 in. thick. How many yards (actually cubic yards) of concrete are needed? Concrete costs $58/yd^3 delivered. How much will the sidewalk cost?

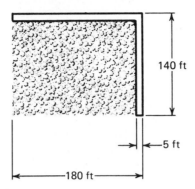

37. How many acres are in a rectangular field which measures 2400 ft by 625 ft?

38. How many acres are in a rectangular field which measures 820 yd by 440 yd?

39. How many hectares are in the field in Exercise 37?

40. How many hectares are in the field in Exercise 38?

41. A house lot measures 145 ft by 186 ft. What part of an acre is the lot?

42. How many acres are in $\frac{1}{4}$ mi^2?

43. How many acres are in $\frac{1}{8}$ section?

44. How many acres are in 520 square rods?

45. A corn yield of 10,550 kg/ha is equivalent to how many lb/acre? To how many bu/acre? (1 bu of corn weighs 56 lb.)

46. A soybean yield of 45 bu/acre is equivalent to how many kg/ha? To how many metric tons/hectare? (1 bu of soybeans weighs 60 lb.)

47. How many acres are in eight 30-in. rows 440 yards long?

48. How many rows 30 in. apart can be planted in a rectangular field 3300 ft long and 2600 ft wide? Suppose seed corn is planted at 20 lb/acre. (a) How many bushels of seed corn will be needed to plant the field? (b) Suppose 1 lb of seed corn contains 1200 kernels. How many bags, each containing 80,000 kernels, will be needed?

Chapter 4 Review

Give the metric prefix for each value:

1. 0.001
2. 1000

Give the SI abbreviation for each prefix:

3. mega
4. micro

Write the SI abbreviation for each quantity:

5. 42 millilitres
6. 8.3 nanoseconds

Write the SI unit for each abbreviation:

7. 18 km
8. 350 mA
9. 50 µs

Which is larger?

10. 1 L or 1 mL
11. 1 kW or 1 MW
12. 1 km^2 or 1 ha
13. 1 m^3 or 1 L

Fill in each blank:

14. 650 m = ___ km
15. 750 mL = ___ L
16. 6.1 kg = ___ g
17. 4.2 A = ___ µA
18. 18 MW = ___ W
19. 25 µs = ___ ns
20. 250 cm^2 = ___ mm^2
21. 25,000 m^2 = ___ ha
22. 0.6 m^3 = ___ cm^3
23. 250 cm^3 = ___ mL
24. 72°F = ___ °C
25. −25°C = ___ °F
26. Water freezes at ___ °C.
27. Water boils at ___ °C.
28. 180 lb = ___ kg
29. 126 ft = ___ m
30. 360 cm = ___ in.
31. 275 in^2 = ___ cm^2
32. 18 yd^2 = ___ ft^2
33. 5 m^3 = ___ ft^3

Choose the most reasonable quantity:

34. Jan and Chuck drive (a) 1600 cm (b) 470 m (c) 12 km (d) 2400 mm to college each day.
35. Chuck's mass is (a) 80 kg (b) 175 kg (c) 14 µg (d) 160 Mg.
36. Their car's gas tank holds (a) 18 L (b) 15 kL (c) 240 mL (d) 60 L of gasoline.
37. Jan, being of average height, is (a) 5.5 m (b) 325 mm (c) 55 cm (d) 165 cm tall.
38. Their car averages (a) 320 km/L (b) 15 km/L (c) 35 km/L (d) 0.75 km/L.
39. On Illinois winter mornings, the temperature sometimes dips to (a) −50°C (b) −30°C (c) 30°C (d) −80°C.
40. Jan drives (a) 85 km/h (b) 50 km/h (c) 150 km/h (d) 25 km/h on the interstate.

MEASUREMENT

RPM
Hundreds

Figure 5.1

RPM
Thousands

Figure 5.2

5.1 Approximate Numbers and Accuracy

APPROXIMATE NUMBERS (MEASUREMENTS) VERSUS EXACT NUMBERS

A *tachometer* is used to measure the number of revolutions an object makes with respect to some unit of time. Since the unit of time is usually minutes, a tachometer usually measures revolutions per minute (rpm). This measurement is usually given in integral units, such as 10 rpm or 255 rpm. Tachometers are used in industry to test motors to see whether or not they turn at a specified rate. In the shop, both wood and metal lathes have specified rpm rates. Tachometers are also commonly used in sports cars to help drivers shift gears at the appropriate engine rpm.

Consider the diagram of the tachometer in Fig. 5.1. Each of the printed integral values on the dial indicates hundreds of rpm. That is, if the dial indicator is at 10, then the reading is 10 hundred rpm, or 1000 rpm. If the dial indicator is at 70, then the reading is 70 hundred rpm, or 7000 rpm. Each of the subdivisions between 0 and 10, 10 and 20, 20 and 30, and so forth, represents an additional one hundred rpm.

Example 1. Read the tachometer in Fig. 5.2.

The indicator is at the sixth division past 30; so the reading is 36 thousand, or 36,000 rpm.

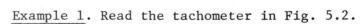

When reading tachometers, the readings are only approximate. Tachometers are calibrated in tens, hundreds, or thousands of revolutions per minute, and it is impossible to read the exact number of rpm.

Next consider the measurement of the length of a metal block like the one in Fig. 5.3.

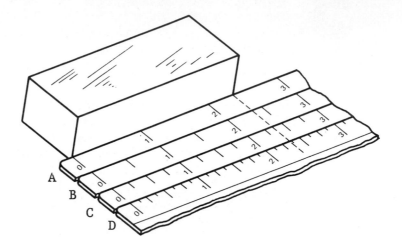

Figure 5.3

1. First, measure the block with ruler A graduated only in inches. This means measurements will be to the nearest inch. The measured length is 2 in.
2. Measure the same block with ruler B graduated in half-inches. Measurements now are to the nearest half-inch. The measured length is $2\frac{1}{2}$ in. to the nearest half-inch.
3. Measure the block again with ruler C graduated in fourths of an inch. To the nearest $\frac{1}{4}$ in. the measurement is $2\frac{1}{4}$ in.
4. Measure the block again with ruler D graduated in eighths of an inch. To the nearest $\frac{1}{8}$ in. the measurement is $2\frac{3}{8}$ in.

If you continue this process, by using finer and finer graduations on the ruler, will you ever find the "exact" length? All measurements are only approximations, so the "exact" length cannot be found. A measurement is only as good as the measuring instrument you use. It would be rather difficult for you to measure the diameter of a pinhead with a ruler.

Up to this time in your study of mathematics, probably all numbers and all measurements have been treated as exact numbers. An *exact number* is a number which has been determined as a result of counting, such as 24 students enrolled in this class. Or by some definition, such as 1 hour (h) = 60 minutes (min), or 1 in. = 2.54 cm. (These are conversion definitions agreed to by government bureaus of standards the world over.) Addition, subtraction, multiplication, and division of exact numbers are usually the main content of elementary school mathematics.

However, nearly all data of a technical nature involve *approximate numbers*; that is, numbers which have been determined by some measurement process. This process may possibly be direct, as with a ruler, and possibly indirect, as with a surveying transit. Before studying how to perform calculations with approximate numbers (measurements), we must determine the "correctness" of an approximate number. First, we must realize that no measurement can be found exactly. The length of the cover of this book can be found using many instruments. The better the measuring device used, the better the measurement.

In summary,

1. only counting numbers are exact;
2. all measurements are approximations.

ACCURACY AND SIGNIFICANT DIGITS

The *accuracy* of a measurement refers to the number of digits, called *significant digits*, which it contains. These indicate the number of units we are reasonably sure of having counted when making a measurement. The greater the number of significant digits given in a measurement, the better the accuracy, and vice versa.

Example 2. The average distance between the moon and the earth is 239,000 mi. This measurement indicates measuring 239 thousands of miles. Its accuracy is indicated by 3 significant digits.

Example 3. A measurement of 10,900 m indicates measuring 109 hundreds of metres. Its accuracy is 3 significant digits.

Example 4. A measurement of 0.042 cm indicates measuring 42 thousandths of a centimetre. Its accuracy is 2 significant digits.

Notice that sometimes a zero is significant and sometimes it is not. In order to determine whether a digit is significant or not, we give the following rules:

Significant Digits

1. All nonzero digits are significant.

   For example, the measurement 1765 kg has 4 significant digits. (This measurement indicates measuring 1765 units of kilograms.)

2. All zeros between significant digits are significant.

   For example, the measurement 30,060 m has 4 significant digits. (This measurement indicates measuring 3006 tens of metres.)

3. A zero in a whole-number measurement that is specially tagged, such as by a bar above it, is significant.

   For example, the measurement $3\overline{0},000$ ft has 2 significant digits. (This measurement indicates measuring $3\overline{0}$ thousands of feet.)

4. All zeros to the right of a significant digit <u>and</u> a decimal point are significant.

   For example, the measurement 6.100 L has 4 significant digits. (This measurement indicates measuring $610\overline{0}$ thousandths of litres.)

5. Zeros to the right in a whole-number measurement that is not tagged are <u>not</u> significant.

   For example, the measurement 4600 V has 2 significant digits. (This measurement indicates measuring 46 hundreds of volts.)

6. Zeros to the left in a decimal measurement that is less than one are <u>not</u> significant.

   For example, the measurement 0.00360 s has 3 significant digits. (This measurement indicates measuring $36\overline{0}$ hundred-thousandths of a second.)

Example 5. Determine the accuracy of each measurement.

Measurement	Accuracy (significant digits)
(a) 109.006 m	6
(b) 0.000589 kg	3
(c) 75 V	2
(d) 239,000 mi	3
(e) 239,00$\overline{0}$ mi	6
(f) 239,$\overline{0}$00 mi	5
(g) 0.03200 mg	4
(h) 1.20 cm	3
(i) 9.020 µA	4
(j) 100.050 km	6

EXERCISES 5.1   *Give the number of significant digits for each measurement:*

1. 115 V	2. 47,000 lb	3. 7009 ft
4. 420 m	5. 6972 m	6. 320,070 ft
7. 440$\overline{0}$ ft	8. 4400 Ω	9. 44$\overline{0}$0 m
10. 0.0040 g	11. 173.4 m	12. 2070 ft
13. 41,$\overline{0}$00 mi	14. 0.025 A	15. 0.0350 in.
16. 6700 g	17. 173 m	18. 8060 ft
19. 240,$\overline{0}$00 V	20. 2500 g	21. 72,00$\overline{0}$ mi
22. 137 V	23. 0.047000 A	24. 7.009 g
25. 0.20 mi	26. 69.72 m	27. 32.0070 g
28. 61$\overline{0}$ L	29. 15,0$\overline{0}$0 mi	30. 0.07050 mL
31. 100.020 in.	32. 250.0100 m	33. 900,200 ft
34. 15$\overline{0}$ cm	35. 16,000 W	36. 0.001005 m

5.2 Precision and Greatest Possible Error

The *precision of a measurement* refers to the smallest unit with which a measurement is made; that is, the position of the last significant digit.

Example 1. The precision of the measurement 239,000 mi is 1000 mi. (The position of the last significant digit is in the thousands place.)

Example 2. The precision of the measurement 10,900 m is 100 m. (The position of the last significant digit is in the hundreds place.)

Example 3. The precision of the measurement 6.90 L is 0.01 L. (The position of the last significant digit is in the hundredths place.)

Example 4. The precision of the measurement 0.0016 A is 0.0001 A. (The position of the last significant digit is in the ten-thousandths place.)

Example 5. Determine the precision of each measurement given in Example 5 in the previous section, on page 175.

	Measurement	Precision	Accuracy (significant digits)
(a)	109.006 m	0.001 m	6
(b)	0.000589 kg	0.000001 kg	3
(c)	75 V	1 V	2
(d)	239,000 mi	1000 mi	3
(e)	239,00͞0 mi	1 mi	6
(f)	239,0͞0͞0 mi	10 mi	5
(g)	0.03200 mg	0.00001 mg	4
(h)	1.20 cm	0.01 cm	3
(i)	9.020 μA	0.001 μA	4
(j)	100.050 km	0.001 km	6

The precision of a measuring instrument is determined by the smallest unit or calibration on the instrument. The precision of the tachometer on the left in Fig. 5.4 is 100 rpm. The precision of the tachometer on the right is 1000 rpm.

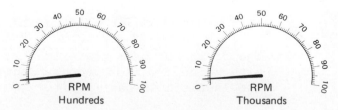

Figure 5.4

The precision of a ruler graduated in eighths of an inch is $\frac{1}{8}$ in. The precision of one graduated in fourths of an inch is $\frac{1}{4}$ in.

However, if a measurement is given as $4\frac{5}{8}$ in., you have no way of knowing what ruler was used. Therefore you cannot tell whether the precision is $\frac{1}{8}$ in., $\frac{1}{16}$ in., $\frac{1}{32}$ in., or what. The measurement could have been $4\frac{5}{8}$ in., $4\frac{10}{16}$ in., $4\frac{20}{32}$ in., or a similar measurement of some other precision. Unless it is stated otherwise, you should assume that the smallest unit used is the one that is recorded. In this case the precision is $\frac{1}{8}$ in.

Now study closely the enlarged portions of the tachometer (tach) readings in Fig. 5.5, given in hundreds of rpm. Note that in each case the measurement is 4100 rpm, although the locations of the pointer are slightly different. Any actual rpm between 4050 and 4150 is read 4100 rpm on the scale of the tachometer.

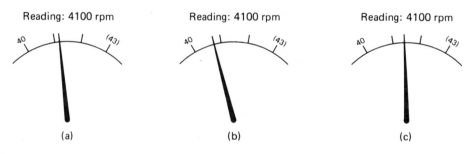

Figure 5.5

The *greatest possible error* is one-half of the smallest unit on the scale on which the measurement is read. We see this in the tach readings above where any reading within 50 rpm of 4100 rpm is read 4100 rpm. Therefore the greatest possible error is 50 rpm.

If you have a tach reading of 5300 rpm, the greatest possible error is $\frac{1}{2}$ of 100 rpm, or 50 rpm. This means the actual rpm is between 5300 − 50 and 5300 + 50, or between 5250 rpm and 5350 rpm.

Next consider the measurements of the three metal rods shown in Fig. 5.6. Note that in each case the measurement of the length is $4\frac{5}{8}$ in., although it is obvious that the rods are of different lengths. Any rod that measures $4\frac{9}{16}$ in. to $4\frac{11}{16}$ in. will measure $4\frac{5}{8}$ in. on this scale. The greatest possible error is one-half of the smallest unit on the scale with which the measurement is made.

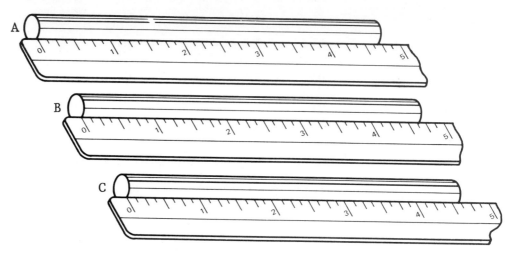

Figure 5.6

If the length of a metal rod is given as $3\frac{3}{16}$ in., the greatest possible error is $\frac{1}{2}$ of $\frac{1}{16}$ in., or $\frac{1}{32}$ in. This means that the actual length of the rod is between $3\frac{3}{16}$ in. $- \frac{1}{32}$ in. and $3\frac{3}{16}$ in. $+ \frac{1}{32}$ in., or between $3\frac{5}{32}$ in. and $3\frac{7}{32}$ in.

Greatest Possible Error

The greatest possible error of a measurement is equal to one half its precision.

Example 6. Find the precision and greatest possible error of the measurement 8.00 kg.

The position of the last significant digit is in the hundredths place; therefore, the precision is 0.01 kg. The greatest possible error is one half the precision.

$$(0.5)(0.01 \text{ kg}) = 0.005 \text{ kg}$$

Example 7. Find the precision and the greatest possible error of the measurement 26,000 gal.

The position of the last significant digit is in the thousands place; therefore, the precision is 1000 gal.

The greatest possible error is one-half the precision, or

$$\frac{1}{2} \times 1000 \text{ gal} = 500 \text{ gal}.$$

Example 8. Find the precision and greatest possible error of the measurement 0.0460 mg.

Precision: 0.0001 mg

Greatest possible error: (0.5)(0.0001 mg) = 0.00005 mg

Example 9. Find the precision and the greatest possible error of the measurement $5\frac{5}{8}$ in.

The smallest unit is $\frac{1}{8}$ in., which is the precision.

The greatest possible error is $\frac{1}{2} \times \frac{1}{8}$ in. $= \frac{1}{16}$ in., which means the actual length is within $\frac{1}{16}$ in. of $5\frac{5}{8}$ in.

Many calculations with measurements are performed by people who do not make the actual measurements. Therefore, it is often necessary to agree on a method of recording measurements to indicate the precision of the instrument used.

EXERCISES 5.2   *Find (a) the precision and (b) the greatest possible error of each measurement.*

1. 2.70 A	2. 13.0 ft	3. 14.00 cm
4. 1.000 in.	5. 15 km	6. 1.010 cm
7. 17.50 mi	8. 6.100 m	9. 0.040 A
10. 0.0001 in.	11. 0.0805 W	12. 10,$\bar{0}$00 W
13. 14$\bar{0}$0 Ω	14. 301,000 Hz	15. 3$\bar{0}$,000 L
16. 7,0$\bar{0}$0,000 g	17. 428.0 cm	18. 60.0 cm
19. 120 V	20. 30$\bar{0}$ km	21. 67.500 m
22. $1\frac{7}{8}$ in.	23. $3\frac{2}{3}$ yd	24. $3\frac{3}{4}$ yd
25. $9\frac{7}{32}$ in.	26. $4\frac{5}{8}$ mi	27. $9\frac{5}{16}$ mi
28. $5\frac{13}{64}$ in.	29. $9\frac{4}{9}$ in^2	30. $18\frac{4}{5}$ in^3

5.3 The Vernier Caliper

In your use of English and metric rulers for making measurements, you have seen that it is difficult to get very precise results. When it is necessary to make more precise measurements, you must have a more precise instrument. One such instrument is the *vernier caliper*, which is used by technicians in machine shops, plant assembly lines, and many other related industries. This instrument is a slide-type caliper used to take inside, outside, and depth measurements. It has two metric scales and two English scales and is shown in detail in Fig. 5.7.

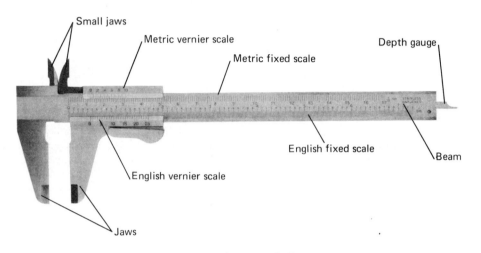

Figure 5.7

To make an outside measurement, the jaws are closed snugly around the outside of an object, as in Fig. 5.8.

For an inside measurement the smaller jaws are placed inside the object to be measured, as in Fig. 5.9.

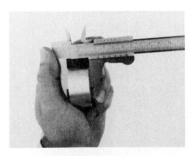

Figure 5.8

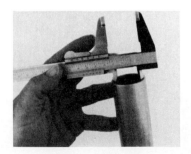

Figure 5.9

For a depth measurement insert the depth gauge into the opening to be measured, as in Fig. 5.10.

Let's first consider the metric scales (Fig. 5.11). One of them is fixed and located on the upper part of the beam. This fixed scale is divided into centimetres and subdivided into millimetres, so record all readings in millimetres (mm). The other metric scale, called the vernier scale, is the upper scale on the slide. The vernier scale is divided into tenths of millimetres (0.1 mm) and subdivided into halves of tenths of millimetres ($\frac{1}{20}$ mm or 0.05 mm). The precision of the vernier scale is therefore 0.05 mm.

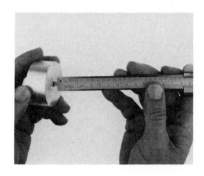

Figure 5.10

Figure 5.11

The figures in the examples and exercises have been computer generated for easier reading.

To read this vernier caliper in millimetres:

Step 1: Determine the number of whole millimetres in a measurement by counting—on the fixed scale—the number of millimetre graduations that are to the left of the zero graduation on the vernier scale. (Remember that each numbered graduation on the fixed scale represents 10 mm.) The zero graduation on the vernier scale may be directly in line with a graduation on the fixed scale. If so, read the total measurement directly from the fixed scale. Write it in millimetres, followed by a decimal point and two zeros.

Step 2: The zero graduation on the vernier scale may not be directly in line with a graduation on the fixed scale. In that case, find the graduation on the vernier scale that is most nearly in line with any graduation on the fixed scale.
(a) If the vernier graduation is a long graduation, it represents the number of tenths of millimetres between the last graduation on the fixed scale and the zero graduation on the vernier scale. Then insert a zero in the hundredths place.
(b) If the vernier graduation is a short graduation, add 0.05 mm to the vernier graduation which is on the immediate left of the short graduation.

Step 3: Add the numbers from Steps 1 and 2 to determine the total measurement.

Example 1. Read the measurement on the vernier caliper in Fig. 5.12 in millimetres.

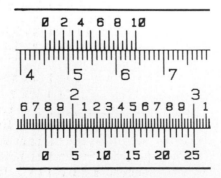

Figure 5.12

Step 1: The first mark to the left of zero
mark is   45.00 mm

Step 2: The mark on vernier scale that most
nearly lines up with a mark on fixed scale   0.20 mm

Step 3: Total measurement   45.20 mm

Example 2. Read the measurement on the vernier caliper
in Fig. 5.13 in millimetres.

The total measurement is 21.00 mm, because the zero
graduation on the vernier scale most nearly lines up
with a mark on the fixed scale.

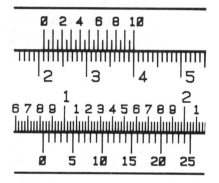

Figure 5.13   Figure 5.14

Example 3. Read the measurement on the vernier caliper
in Fig. 5.14 in millimetres.

Step 1: The first mark to the left of zero
mark is   104.00 mm

Step 2: The mark on vernier scale that most
nearly lines up with a mark on fixed scale   0.80 mm

Step 3: Total measurement   104.80 mm

EXERCISES 5.3   *Read the measurement shown on each vernier caliper in
millimetres:*

1. 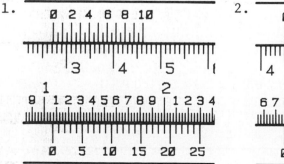   2.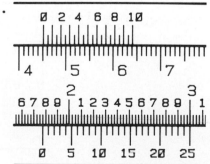

3.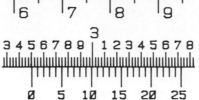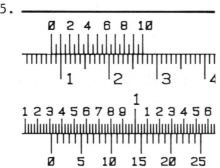

4.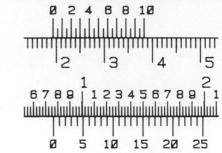

5.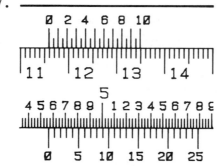

6.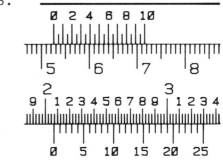

7.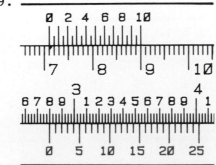

8.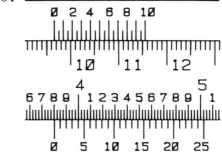

9.

10.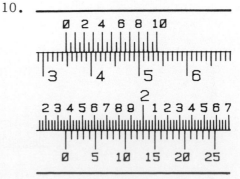

11—20. Read in millimetres the measurement shown on each
        vernier caliper on pages 187—188.

Now, consider the two English scales on the vernier
caliper in Fig. 5.15. One is fixed, the other movable. The
fixed scale is located on the lower part of the beam where
each inch is divided into tenths (0.100 in., 0.200 in.,
0.300 in., and so on). Each tenth is subdivided into four
parts, each of which represents 0.025 in. The vernier scale
is divided into 25 parts, each of which represents thousandths (0.001 in.). This means that the precision of this
scale is 0.001 in.

To read this vernier caliper in thousandths of
an inch:

Step 1: Determine the number of inches and tenths
of inches by reading the first numbered division
that is to the left of the zero graduation on the
vernier scale.

Step 2: Add 0.025 in. to the number from Step 1 for
each graduation between the last numbered division
on the fixed scale and the zero graduation on the
vernier scale. (If this zero graduation is directly
in line with a graduation on the fixed scale, read
the total measurement directly from the fixed
scale.)

Step 3: If the zero graduation on the vernier scale
is not directly in line with a graduation on the
fixed scale:
(a) Find the graduation on the vernier scale that
    is most nearly in line with any graduation on
    the fixed scale. This graduation determines
    the number of thousandths of inches in the
    measurement.
(b) Add the numbers from Steps 1, 2, and 3a to
    determine the total measurement.

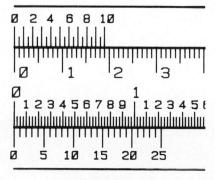

Figure 5.15

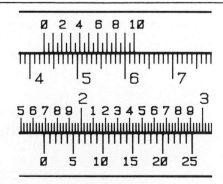

Figure 5.16

Example 1. Read the measurement shown on the vernier caliper in Fig. 5.16 in inches.

Step 1: The first numbered mark to left of zero mark   1.600 in.

Step 2: The number of 0.025-in. graduations (3 × 0.025 in.)   0.075 in.

Step 3a: The mark on vernier scale that most nearly lines up with a mark on fixed scale   0.015 in.

Step 3b: Total measurement   1.690 in.

Example 2. Read the measurement shown on the vernier caliper in Fig. 5.17 in inches.

Step 1: The first numbered mark to left of zero mark   0.800 in.

Step 2: The number of 0.025-in. graduations (1 × 0.025 in.)   0.025 in.

Step 3b: Total measurement   0.825 in.

Note: The zero graduation is directly in line with a graduation on the fixed scale.

Example 3. Read the measurement shown on the vernier caliper in Fig. 5.18 in inches.

Step 1: The first numbered mark to left of zero mark   1.200 in.

Step 2: The number of 0.025-in. graduations (2 × 0.025 in.)   0.050 in.

Step 3a: The mark on vernier scale that most nearly lines up with a mark on fixed scale   0.008 in.

Step 3b: Total measurement   1.258 in.

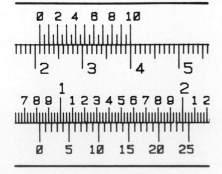

Figure 5.17

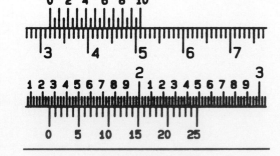

Figure 5.18

EXERCISES 5.3A   *Read the measurement shown on each vernier caliper in inches:*

1.

2.

3.

4.

5.

6.

7.

8.

9.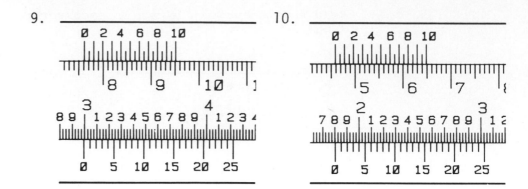

10.

11—20. Read in inches the measurement shown on each vernier caliper on pages 183—184.

5.4 The Micrometer Caliper

The *micrometer caliper* (micrometer or "mike") is a more precise instrument than the vernier caliper and is used in technical fields where fine precision is required. Micrometers are available in metric units and English units. The metric "mike" is graduated and read in hundredths of a millimetre (0.01 mm); the English "mike" is graduated and read in thousandths of an inch (0.001 in.)

The parts of a micrometer are labeled in Fig. 5.19. Place the object to be measured between the anvil and spindle, and turn the thimble until the object fits snugly. *Do not force the turning of the thimble, since this may damage the very delicate threads on the spindle located inside the thimble.* Some calipers have a ratchet to protect the instrument; the ratchet prevents the thimble from being turned with too much force. Fig. 5.20 shows a metric micrometer in which the basic parts are labeled.

The barrels of most metric micrometers are graduated in millimetres. The micrometer in Fig. 5.20 also has graduations of halves of millimetres, which are indicated by the lower set of graduations on the barrel. The threads on the spindle are made so it takes two complete turns of the thimble for the spindle to move precisely one millimetre. The head is divided into 50 equal divisions—each division indicating 0.01 mm, which is the precision.

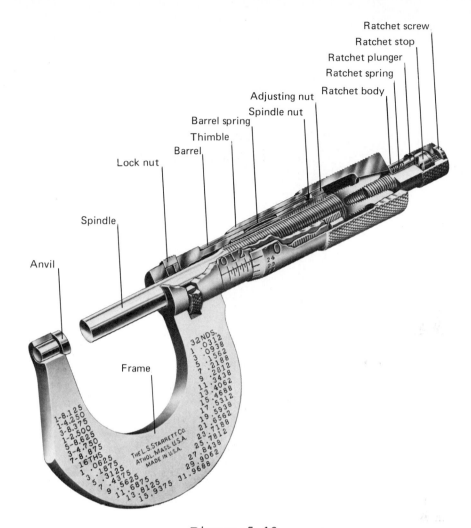

Figure 5.19

Reading a Metric Micrometer in Millimetres

Step 1: Find the whole number of mm in the measurement by counting the number of mm graduations on
the barrel to the left of the head.

Step 2: Find the decimal part of the measurement by
reading the graduation on the head (see Fig. 5.20)
that is most nearly in line with the center line on
the barrel. Then multiply this reading by 0.01. If
the head is at, or immediately to the right of, the
half-mm graduation, add 0.50 mm to the reading on
the head.

Step 3: Add the numbers found in Step 1 and Step 2.

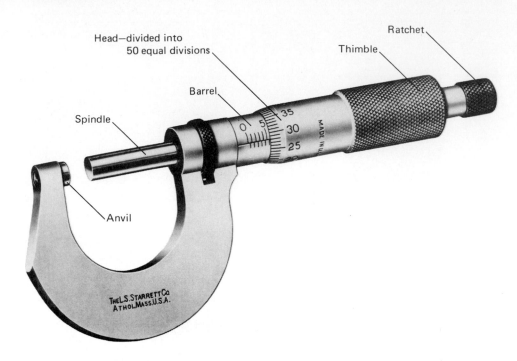

Figure 5.20

<u>Example 1.</u> Read the measurement shown on the metric micrometer in Fig. 5.21.

<u>*Step 1*</u>: The barrel reading is   6.00 mm

<u>*Step 2*</u>: The head reading is   <u>0.24 mm</u>

<u>*Step 3*</u>: The total measurement is   6.24 mm

<u>Example 2.</u> Read the measurement shown on the metric micrometer in Fig. 5.22.

<u>*Step 1*</u>: The barrel reading is   14.00 mm

<u>*Step 2*</u>: The head reading is   <u>0.08 mm</u>

<u>*Step 3*</u>: The total measurement is   14.08 mm

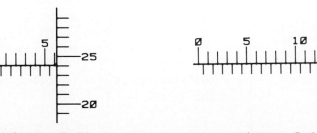

Figure 5.21   *Figure 5.22*

Example 3. Read the measurement shown on the metric micrometer in Fig. 5.23.

Step 1: The barrel reading is   8.00 mm

Step 2: The head reading is   0.65 mm
(Note that the head is past
half-mm mark.)

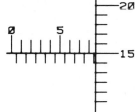

Step 3: The total measure-
ment is   8.65 mm   Figure 5.23

EXERCISES 5.4   *Read the measurement shown on each metric micrometer:*

1.   2.   3.

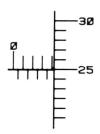   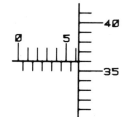   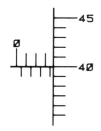

4.   5.

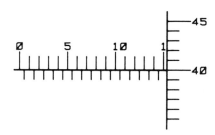   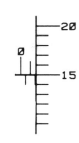

6.   7.   8.

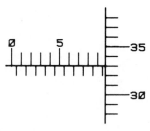   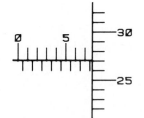   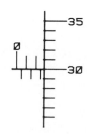

9.

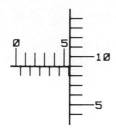

10.

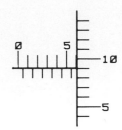

11.

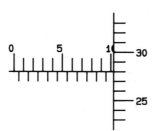

12.

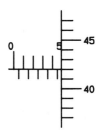

13.

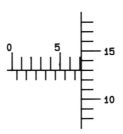

14.

15.

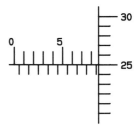

16.

17.

18.

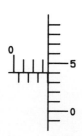

19.

20.

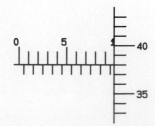

The barrel of the English micrometer shown in Fig. 5.24 is divided into tenths of an inch and subdivided into 0.025 in. The threads on the spindle allow the spindle to move 0.025 in. in one complete turn of the thimble and 4 × 0.025 in., or 0.100 in., in four complete turns. The head is divided into 25 equal divisions—each division indicating 0.001 in., which is the precision.

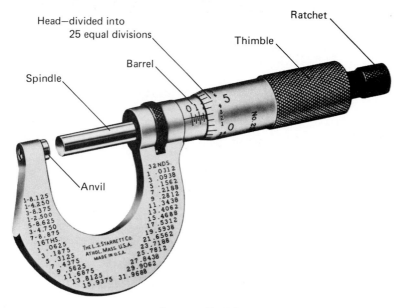

Figure 5.24

Reading an English Micrometer in 0.001 in.

Step 1: Read the last numbered graduation showing on the barrel. Multiply this number by 0.100 in.

Step 2: Find the number of smaller graduations between the last numbered graduation and the head. Multiply this number by 0.025 in.

Step 3: Find the graduation on the head which is most nearly in line with the center line on the barrel. Multiply the number represented by this graduation by 0.001 in.

Step 4: Add the numbers found in Steps 1, 2, and 3.

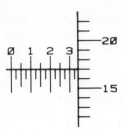

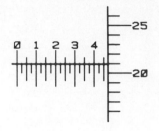

Figure 5.25   *Figure 5.26*

Example 4. Read the measurement shown on the English micrometer in Fig. 5.25.

Step 1: 3 numbered divisions on barrel
(3 × 0.100 in.)   0.300 in.

Step 2: 1 small division on barrel
(1 × 0.025 in.)   0.025 in.

Step 3: The head reading is 17
(17 × 0.001 in.)   0.017 in.

Step 4: The total measurement is   0.342 in.

Example 5. Read the measurement shown on the English micrometer in Fig. 5.26.

Step 1: 4 numbered divisions on barrel
(4 × 0.100 in.)   0.400 in.

Step 2: 2 small divisions on barrel
(2 × 0.025 in.)   0.050 in.

Step 3: The head reading is 21
(21 × 0.001 in.)   0.021 in.

Step 4: The total measurement is   0.471 in.

EXERCISES 5.4A   *Read the measurement shown on each English micrometer:*

1.   2.

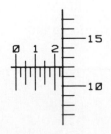

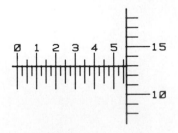

3.

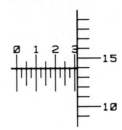

4.

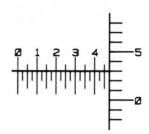

5.

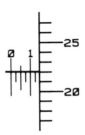

6.

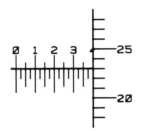

7.

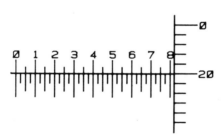

8.

9.

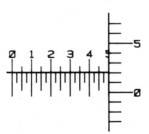

10.

11.

12.

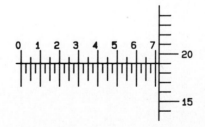

13.

14.

15.

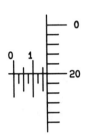

16.

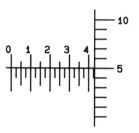

17.

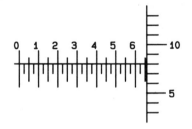

18.

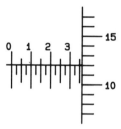

19.

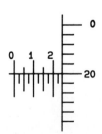

20.

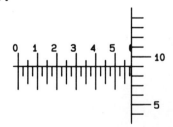

OTHER MICROMETERS

By adding a vernier scale on the barrel (as shown in
Fig. 5.27) of each of the micrometers we just studied,
we can increase the precision by one more decimal place.
That is, the metric micrometer with vernier scale has
precision 0.001 mm. The English micrometer with vernier
scale has precision 0.0001 in. Of course, these micrometers
cost more because they require more precise threading than
the ones previously discussed. Nevertheless, many jobs
require this precision.

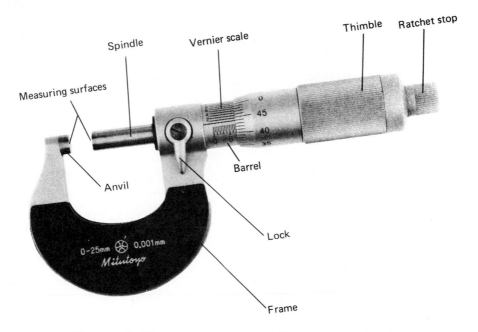

Figure 5.27 Micrometer with Vernier Scale

Micrometers are basic, useful, and important tools of the technician. Fig. 5.28 shows just a few of their uses.

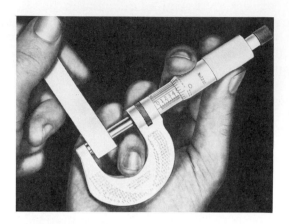

(a) Measuring a piece of die steel.

(b) Measuring the diameter of a crankshaft bearing.

(c) Measuring tubing wall thickness with a round anvil micrometer.

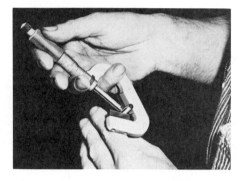

(d) Checking out-of-roundness on centerless grinding work.

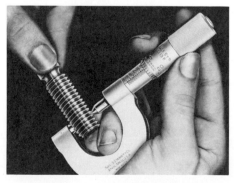

(e) Measuring the pitch diameter of a screw thread.

(f) Measuring the depth of a shoulder with a micrometer depth gauge.

Figure 5.28

5.5 Precision versus Accuracy

Recall that the *precision* of a measurement refers to the smallest unit with which a measurement is made; that is, the *position* of the last significant digit. And, the *accuracy* of a measurement refers to the *number* of digits, called significant digits, which indicate the number of units we are reasonably sure of having counted when making a measurement. Unfortunately, some people tend to use the terms *precision* and *accuracy* interchangeably. Each term expresses a different aspect of a given measurement.

Example 1. Compare the precision and the accuracy of the measurement 0.0004 mm.

Since the precision is 0.0001 mm, its precision is relatively good. However, since the accuracy is only one significant digit, its accuracy is relatively poor.

Example 2. Given the measurements 13.00 m, 0.140 m, 3400 m, and 0.006 m, find the measurement which is (a) the least precise, (b) the most precise, (c) the least accurate, and (d) the most accurate.

First, let's find the precision and the accuracy of each measurement.

Measurement	Precision	Accuracy (significant digits)
13.00 m	0.01 m	4
0.140 m	0.001 m	3
3400 m	100 m	2
0.006 m	0.001 m	1

From the table we find:

(a) the least precise measurement: 3400 m
(b) the most precise measurement: 0.140 m and 0.006 m
(c) the least accurate measurement: 0.006 m
(d) the most accurate measurement: 13.00 m

EXERCISES 5.5   *In each set of measurements, find the measurement which is (a) the most accurate, and (b) the most precise:*

1. 14.7 in.; 0.017 in.; 0.09 in.

2. 459 ft; 600 ft; 190 ft

3. 0.737 mm; 0.94 mm; 16.01 mm

4. 4.5 cm; 9.3 cm; 7.1 cm

5. 0.0350 A; 0.025 A; 0.00050 A; 0.041 A

6. 134.00 g; 5.07 g; 9.000 g; 0.04 g

7. 145 cm; 73.2 cm; 2560 cm; 0.391 cm

8. 15.2 km; 631.3 km; 20.0 km; 37.7 km

9. 205,000 Ω; 45,000 Ω; 5$\overline{0}$0,000 Ω; 90,000 Ω

10. 1,500,000 V; 65,000 V; 30,$\overline{0}$00 V; 20,000 V

In each set of measurements, find the measurement which is (a) the least accurate, and (b) the least precise:

11. 15.5 in.; 0.053 in.; 0.04 in.

12. 635 ft; 400 ft; 240 ft

13. 43.4 cm; 0.48 cm; 14.05 cm

14. 4.9 kg; 670 kg; 0.043 kg

15. 0.0730 A; 0.043 A; 0.00008 A; 0.91 A

16. 197.0 m; 5.43 m; 4.000 m; 0.07 m

17. 2.1 m; 31.3 m; 461.5 m; 0.6 m

18. 295 m; 91.3 m; 1920 m; 0.360 m

19. 405,000 Ω; 35,000 Ω; 8$\overline{0}$0,000 Ω; 500,000 Ω

20. 1,600,000 V; 36,000 V; 40,$\overline{0}$00 V; 60,000 V

5.6 Addition and Subtraction of Measurements

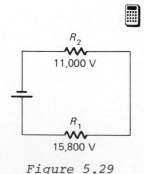

Figure 5.29

In a series circuit the electromagnetic force (emf) of the source equals the sum of the separate voltage drops across each resistor in the circuit. Suppose that someone measures the voltage across the first resistor R_1 in Fig. 5.29. He uses a voltmeter calibrated in hundreds of volts, and measures 15,800 V. Across the second resistor R_2, he uses a voltmeter in thousands of volts, and measures 11,000 V. Does the total emf equal 26,800 V? Note that the first voltmeter and its reading indicate a precision of 100 V and a greatest possible error of 50 V. This means the actual reading lies between 15,750 V and 15,850 V.

The second voltmeter and its reading indicate a precision of 1000 V and a greatest possible error of 500 V. The actual reading, therefore, lies between 10,500 V and 11,500 V. This means that we are not very certain of the digit in the hundreds place in the sum 26,800 V.

To be consistent when adding or subtracting measurements of different precision, the sum or difference can be no more *precise* than the least precise measurement. That is,

---

To add or subtract measurements of different precision:

1. Make certain that all measurements are expressed in the same units. If they are not, change them all to any common unit.
2. Next, round each measurement to the same precision as the least precise measurement.
3. Then, add or subtract.

In the above discussion (of Fig. 5.29) the sum is, then:

Measurement   *Rounded*

R_1: 15,800 V → 16,000 V
R_2: 11,000 V → 11,000 V
                        ──────────
                         27,000 V

Example 1. Use the rules for addition of measurements to add 13,800 ft, 14,020 ft, 19,864 ft, and 14,700 ft.

The least precise measurement—either 13,800 ft or 14,700 ft—has precision 100 ft. Round each measurement to the nearest hundred feet and then add.

Measurement *Rounded*

13,800 ft → 13,800 ft
14,020 ft → 14,000 ft
19,864 ft → 19,900 ft
14,700 ft → 14,700 ft
                  ──────────
                   62,400 ft

Example 2. Use the rules for addition of measurements to add 735,000 V, 490,000 V, 86,000 V, 1,300,000 V, and 200,000 V.

The least precise measurement is 1,300,000 V, which has precision 100,000 V. Round each measurement to the nearest hundred thousand volts and then add.

Measurement   Rounded

$$
\begin{array}{rcl}
735,000 \text{ V} & \to & 700,000 \text{ V} \\
490,000 \text{ V} & \to & 500,000 \text{ V} \\
86,000 \text{ V} & \to & 100,000 \text{ V} \\
1,300,000 \text{ V} & \to & 1,300,000 \text{ V} \\
200,000 \text{ V} & \to & 200,000 \text{ V} \\
\hline
& & 2,800,000 \text{ V}
\end{array}
$$

Example 3. Use the rules for addition of measurements to add 13.8 m, 140.2 cm, 1.853 m, and 29.95 cm.

First, change each of the measurements to a common unit, say m. Next, round each measurement to the same precision as the least precise measurement, which in this case is 13.8 m. That is, round to tenths of metres. Then, add.

$$
\begin{array}{rcrcl}
13.8 \text{ m} & \to & 13.8 \text{ m} & \to & 13.8 \text{ m} \\
140.2 \text{ cm} & \to & 1.402 \text{ m} & \to & 1.4 \text{ m} \\
1.853 \text{ m} & \to & 1.853 \text{ m} & \to & 1.9 \text{ m} \\
29.95 \text{ cm} & \to & 0.2995 \text{ m} & \to & 0.3 \text{ m} \\
& & & & \hline
& & & & 17.4 \text{ m}
\end{array}
$$

Example 4. Use the rules for subtraction of measurements to subtract 24.35 cm from 41.8 cm.

Each measurement is expressed in the same unit, cm. Round to the precision of the least precise measurement, 41.8 cm. Then subtract.

$$
\begin{array}{rcl}
41.8 \text{ cm} & \to & 41.8 \text{ cm} \\
24.35 \text{ cm} & \to & 24.4 \text{ cm} \\
& & \hline
& & 17.4 \text{ cm}
\end{array}
$$

EXERCISES 5.6 *Use the rules for addition of measurements to find the sum of each set of measurements:*

1. 14.7 m; 3.4 m
2. 168 in.; 34.7 in.; 61 in.
3. 42.6 cm; 16.41 cm; 1.417 cm; 34.4 cm
4. 407 g; 1648.5 g; 32.74 g; 98.1 g
5. 26,000 W; 19,600 W; 8450 W; 42,500 W
6. 5420 km; 1926 km; 850 km; 2000 km
7. 140,000 V; 76,200 V; 4700 V; 254,000 V; 370,000 V
8. 19,200 m; 8930 m; 50,040 m; 137 m
9. 14 V; 1.005 V; 0.017 V; 3.6 V
10. 120.5 cm; 16.4 cm; 1.417 m
11. 10.555 cm; 9.55 mm; 13.75 cm
12. 1350 cm; 1476 mm; 2.876 m; 4.82 m

Use the rules for subtraction of measurements to subtract the second measurement from the first:

13. 140.2 cm
       13.8 cm

14. 14.02 mm
       13.8 mm

15. 9200 mi
       627 mi

16. 1,900,000 V
        645,000 V

17. 167   mm
       13.2 cm

18. 16.41 oz
     11.372 oz

19. 98.1 g
     32.74 g

20. 4.000 in.
     2.006 in.

21. 0.54361 in.
     0.214   in.

 22. Four pieces of metal are to be bolted together. They have thicknesses of 0.136 in., 0.408 in., 1.023 in., and 0.88 in. What is the total thickness of the four pieces?

 23. Five pieces of metal are clamped together. Their thicknesses are 2.38 mm, 10.5 mm, 3.50 mm, 1.455 mm, and 8.200 mm. What is the total thickness of the five pieces?

24. What is the current going through R_5 in the circuit below? (Hint: In a parallel circuit, the current is divided among its branches. That is, $I_T = I_1 + I_2 + I_3$.)

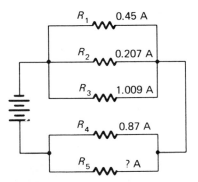

5.7 Multiplication and Division of Measurements

Suppose that you want to find the area of a plot of ground that measures 206 m by 84 m. The product, 17,304 m², shows five significant digits. The original measurements have three and two significant digits, respectively.

To be consistent when multiplying or dividing measurements, the product or quotient can be no more *accurate* than the least accurate measurement. That is,

To multiply or divide measurements:

1. First, multiply and/or divide the measurements as given.
2. Then, round the result to the same number of significant digits as the measurement which has the least number of significant digits.

Using this procedure, the area of the plot of ground 206 m by 84 m is 17,000 m^2.

Example 1. Use the rules for multiplication of measurements to multiply 10.42 g × 3.2 cm.

Step 1: 10.42 g × 3.2 cm = 33.344 g cm

Step 2: Round this product to two significant digits, which is the accuracy of the least accurate measurement, 3.2 cm. That is,

$$10.42 \text{ g} \times 3.2 \text{ cm} = 33 \text{ g cm}$$

Example 2. Use the rules for multiplication of measurements to multiply 125 m × 345 m × 204 m.

Step 1: 125 m × 345 m × 204 m = 8,797,500 m^3

Step 2: Round this product to three significant digits, which is the accuracy of the least accurate measurement (which is the accuracy of each measurement in this example).

That is, 125 m × 345 m × 204 m = 8,8$\overline{0}$0,000 m^3.

Example 3. Use the rules for division of measurements to divide 288,000 ft^3 by 216 ft.

Step 1: $\dfrac{288,000 \text{ ft}^3}{216 \text{ ft}} = 1333.333\ldots \text{ ft}^2$

Step 2: Round this quotient to three significant digits, which is the accuracy of the least accurate measurement (which is the accuracy of each measurement in this example).

That is, $\dfrac{288,000 \text{ ft}^3}{216 \text{ ft}} = 1330 \text{ ft}^2$.

Example 4. Use the rules for multiplication and division of measurements to evaluate $\dfrac{4750 \text{ N} \times 4.82 \text{ m}}{1.6 \text{ s}}$.

Step 1: $\dfrac{4750 \text{ N} \times 4.82 \text{ m}}{1.6 \text{ s}} = 14,309.375 \dfrac{\text{N m}}{\text{s}}$

Step 2: Round this result to two significant digits, which is the accuracy of the least accurate measurement, 1.6 s. That is,

$$\dfrac{4750 \text{ N} \times 4.82 \text{ m}}{1.6 \text{ s}} = 14,000 \dfrac{\text{N m}}{\text{s}}$$

There are even more sophisticated methods for dealing with the calculations of measurements. The method one uses (and indeed whether one should even follow any given procedure), depends on the number of measurements and the sophistication needed for a particular situation.

The procedures for addition, subtraction, multiplication, and division of measurements are based on methods followed and presented by the American Society for Testing and Materials.

Note. To multiply or divide measurements, the units do not need to be the same. (They must be the same in addition and subtraction of measurements.) Also, note that the units are multiplied and/or divided in the same manner as the corresponding numbers.

EXERCISES 5.7   *Use the rules for multiplication and/or division of measurements to evaluate:*

1. 126 m × 35 m

2. 470 mi × 1200 mi

3. 1463 cm × 838 cm

4. 2.4 A × 3600 Ω

5. 18.7 m × 48.2 m

6. 560 cm × 28.0 cm

7. 4.7 Ω × 0.0281 A

8. 5.2 km × 6.71 km

9. 24.2 cm × 16.1 cm × 18.9 cm

10. 0.045 m × 0.0292 m × 0.0365 m

11. 2460 m × 960 m × 1970 m

12. 46$\overline{0}$ in. × 235 in. × 368 in.

13. $(0.480 \text{ A})^2 (150 \ \Omega)$

14. $360 \text{ ft}^2 \div 12 \text{ ft}$

15. $62{,}500 \text{ in}^3 \div 25 \text{ in.}$

16. $9180 \text{ yd}^3 \div 36 \text{ yd}^2$

17. $1520 \text{ m}^2 \div 40 \text{ m}$

18. $18.4 \text{ m}^3 \div 9.2 \text{ m}^2$

19. 4800 V ÷ 14.2 A

20. $\dfrac{4800 \text{ V}}{6.72 \ \Omega}$

21. $\dfrac{5.63 \text{ km}}{2.7 \text{ s}}$

22. $\dfrac{0.497 \text{ N}}{1.4 \text{ m} \times 8.0 \text{ m}}$

23. $\dfrac{(120 \text{ V})^2}{47.6 \ \Omega}$

24. $\dfrac{19 \text{ kg} \times (3.0 \text{ m/s})^2}{2.46 \text{ m}}$

25. $\dfrac{140 \text{ g}}{3.2 \text{ cm} \times 1.7 \text{ cm} \times 6.4 \text{ cm}}$

26. Find the area of a rectangle measured as 6.5 cm by 28.3 cm. ($A = \ell w$)

27. $V = \ell wh$ is the formula for the volume of a rectangular solid, where ℓ = length, w = width, h = height. Find the volume of a rectangular solid when $\ell = 16.4$ ft, $w = 8.6$ ft, and $h = 6.4$ ft.

28. Find the volume of a cube with each edge 8.10 cm long. (Volume of a cube = e^3, where e is the length of each edge.)

29. The formula $s = 4.90t^2$ gives the distance, s, in metres, that a body falls in a given time, t. Find the distance a ball falls in 2.4 seconds.

30. Given K.E. = $\frac{1}{2}mv^2$, $m = 2.87 \times 10^6$ kg, and $v = 13.4$ m/s. Find K.E.

31. A formula for finding the horsepower of an engine is $p = \dfrac{d^2 n}{2.50}$, where d is the diameter of each cylinder in inches and n is the number of cylinders. What is the horsepower of an eight-cylinder engine if each cylinder has a diameter of 3.00 in.? (<u>Note</u>: Eight is an exact number. The number of significant digits in an exact number has no bearing on the number of significant digits in the product or quotient.)

32. Six pieces of metal, each of thickness 2.48 mm, are fitted together. What is the total thickness of the six pieces?

33. Find the volume of a cylinder having a radius of 6.2 m and a height of 8.5 m. The formula for the volume of a cylinder is $V = \pi r^2 h$.

34. In 1970 in the United States 4,200,000,000 bu of corn were harvested from 66,800,000 acres. In 1975 there were 5,800,000,000 bu harvested from 77,900,000 acres. What was the yield in bu/acre for each year? What was the increase in yield?

5.8   Relative Error and Percent of Error

Technicians must determine the importance of measurement error, which may be expressed in terms of relative error. The *relative error* of a measurement is found by comparing the greatest possible error with the measurement itself.

$$\text{Relative error} = \frac{\text{Greatest possible error}}{\text{Measurement}}$$

Example 1. Find the relative error of the measurement 0.08 cm.

The precision is 0.01 cm. The greatest possible error is one half the precision, which is 0.005 cm.

$$\text{Relative error} = \frac{0.005 \ \cancel{cm}}{0.08 \ \cancel{cm}} = 0.0625$$

Note that the units will always cancel, which means that the relative error is expressed as a unitless decimal. When this decimal is expressed as a percent, we have the percent of error. That is,

> The *percent of error* is the relative error expressed as a percent.

Percent of error may be used to compare different measurements because it, being a percent, compares each error in terms of 100. (The percent of error in Example 1 is 6.25%.)

Example 2. Find the relative error and percent of error of the measurement 13.8 m.

The precision is 0.1 m and the greatest possible error is then 0.05 m. Therefore,

$$\text{Relative error} = \frac{0.05 \ \cancel{m}}{13.8 \ \cancel{m}} = 0.00362$$

Percent of error = 0.362%

Example 3. Compare the measurements $3\frac{3}{4}$ in. and 16 mm;

that is, which one is better? (Which one has the smaller percent of error?)

Measurement	$3\frac{3}{4}$ in.	16 mm
Precision	$\frac{1}{4}$ in.	1 mm
Greatest possible error	$\frac{1}{2} \times \frac{1}{4}$ in. $= \frac{1}{8}$ in.	$\frac{1}{2} \times 1$ mm $= 0.5$ mm

Relative error:

$$\frac{\frac{1}{8} \text{ in.}}{3\frac{3}{4} \text{ in.}} = \frac{1}{8} \div 3\frac{3}{4}$$

$$= \frac{1}{8} \div \frac{15}{4}$$

$$= \frac{1}{8} \times \frac{4}{15}$$

$$= \frac{1}{30}$$

$$= 0.0333$$

$$\frac{0.5 \text{ mm}}{16 \text{ mm}} = 0.03125$$

Percent of error	3.33%	3.125%

Therefore, 16 mm is the better measurement because its percent of error is smaller.

EXERCISES 5.8   *For each measurement find the precision, the greatest possible error, the relative error, and the percent of error (to the nearest hundredth percent):*

1. 1400 lb
2. 0.188 cm
3. 240,000 Ω

4. 12,500 V
5. 875 rpm
6. 2 g

7. 2.2 g
8. 2.22 g
9. 18,0$\overline{0}$0 W

10. 1.00 kg
11. 0.041 A
12. $11\frac{7}{8}$ in.

13. $1\frac{3}{4}$ in.
14. 12 ft 8 in.
15. 4 lb 13 oz

Compare each set of measurements by indicating which measurement is better or best:

16. 13.5 cm; $8\frac{3}{4}$ in.
17. 364 m; 36.4 cm

18. 16 mg; 19.7 g; $12\frac{3}{16}$ oz
19. 68,000 V; 3450 Ω; 3.2 A

5.9 Tolerance

In industry, the *tolerance* of a part or component is the acceptable amount that the part or component may vary from a given size. For example, a steel rod may be specified as $14\frac{3}{8}$ in. $\pm \frac{1}{32}$ in. The symbol "$\pm$" is read "plus or minus." This means that the rod may be as long as $14\frac{3}{8}$ in. $+ \frac{1}{32}$ in., or $14\frac{13}{32}$ in., which is called the *upper limit*. Or, it may be as short as $14\frac{3}{8}$ in. $- \frac{1}{32}$ in., or $14\frac{11}{32}$ in., which is called the *lower limit*. That is, any rod between $14\frac{11}{32}$ in. and $14\frac{13}{32}$ in. would be acceptable. We say the tolerance is $\pm\frac{1}{32}$ in. and the tolerance interval is $\frac{2}{32}$ in., or $\frac{1}{16}$ in.

A simple way to check the tolerance of the length of a rod would be to carefully mark off lengths which represent the lower limit and upper limit, as shown in Fig. 5.30. To check the acceptability of a rod, place one end of the rod flush against the metal barrier on the left. If the other end is between the upper and lower limit marks, the part is acceptable. If the rod is longer than the upper limit, it can then be cut to the acceptable limits. If the rod is shorter than the lower limit, it can be melted down for another try.

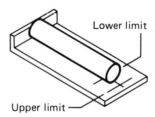

Figure 5.30

Note: The tolerance interval is the difference between the upper limit and the lower limit.

Example 1. The specifications for a stainless steel cylindrical piston are given as:

Diameter: 10.200 cm ±0.001 cm

Height: 14.800 cm ±0.005 cm

Find the upper limit, the lower limit, and the tolerance interval for each dimension.

	Given length	Tolerance	Upper limit	Lower limit	Tolerance interval
Diameter:	10.200 cm	±0.001 cm	10.201 cm	10.199 cm	0.002 cm
Height:	14.800 cm	±0.005 cm	14.805 cm	14.795 cm	0.010 cm

Tolerance may also be expressed as a percent. For example, resistors are color-coded to indicate the tolerance of a given resistor. If the fourth band is silver, this indicates that the acceptable tolerance is ±10% of the given resistance. If the fourth band is gold, this indicates that the acceptable tolerance is ±5% of the given resistance. This is discussed fully in the next section.

Many times bids may be accepted under certain conditions. They may be accepted, for example, when they are less than 10% over the architect's estimate.

Example 2. If the architect's estimate for a given project is $356,200, and bids may be accepted which are less than 10% over the estimate, what is the maximum acceptable bid?

10% of $356,200 = (0.10)($356,200) = $35,620

The upper limit or maximum acceptable bid is $356,200 + $35,620 = $391,820.

EXERCISES 5.9   *Complete the table:*

	Given measurement	Tolerance	Upper limit	Lower limit	Tolerance interval
1.	$3\frac{1}{2}$ in.	$\pm\frac{1}{8}$ in.	$3\frac{5}{8}$ in.	$3\frac{3}{8}$ in.	$\frac{1}{4}$ in.
2.	$5\frac{3}{4}$ in.	$\pm\frac{1}{16}$ in.			
3.	$6\frac{5}{8}$ in.	$\pm\frac{1}{32}$ in.			
4.	$7\frac{7}{16}$ in.	$\pm\frac{1}{32}$ in.			
5.	$3\frac{7}{16}$ in.	$\pm\frac{1}{64}$ in.			
6.	$\frac{9}{64}$ in.	$\pm\frac{1}{128}$ in.			
7.	$3\frac{3}{16}$ in.	$\pm\frac{1}{128}$ in.			
8.	$9\frac{3}{16}$ mi	$\pm\frac{1}{32}$ mi			
9.	1.19 cm	±0.05 cm			
10.	1.78 m	±0.05 m			
11.	0.0180 A	±0.0005 A			
12.	9.437 L	±0.001 L			
13.	24,000 V	±2000 V			
14.	375,000 W	±10,000 W			
15.	10.31 km	±0.05 km			
16.	21.30 kg	±0.01 kg			

	Architect's estimate	Maximum rate above estimate	Maximum acceptable bid
17.	$48,250	10%	
18.	$259,675	7%	
19.	$1,450,945	8%	
20.	$8,275,625	5%	

5.10 Color Code of Electrical Resistors

The resistance of most electrical resistors is usually given in a color code, in a series of four colored bands. Each color on any of the first three bands stands for a digit or number as given in the table below:

Color on any of the first three bands	Digit or number
Black	0
Brown	1
Red	2
Orange	3
Yellow	4
Green	5
Blue	6
Violet	7
Gray	8
White	9
Gold on the third band	Multiply the value by 0.1
Silver on the third band	Multiply the value by 0.01

The fourth band indicates the tolerance of the resistor as given in the following table:

Color of the fourth band	Tolerance
Gold	±5%
Silver	±10%
Black	±20%

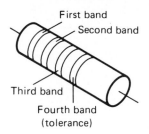

Electrical Resistor

First band

Second band

Third band

Fourth band
(tolerance)

Figure 5.31

The value of each resistor is in ohms, Ω, and given in two significant digits. The color bands are read from left to right when the resistor is in the position below. (See Fig. 5.31.)

The value of each resistor is found by reading the colors as follows:

Step 1. The digit corresponding to the color of the first band is the first digit of the resistance.

Step 2. The digit corresponding to the color of the second band is the second digit of the resistance.

Step 3. (a) The third band indicates the *number* of zeros to be written *after* the first two digits from Steps 1 and 2.

(b) If the third band is gold, multiply the number corresponding to the digits from Steps 1 and 2 by 0.1. That is, place the decimal point *between* the two digits.

(c) If the third band is silver, multiply the number corresponding to the digits from Steps 1 and 2 by 0.01. That is, place the decimal point *before* the two digits.

Step 4. The fourth band indicates the tolerance written as a percent. The tolerance is:
(a) ±5% if the fourth band is gold.
(b) ±10% if the fouth band is silver.
(c) ±20% if the fourth band is black.
If there is no fourth band, the tolerance is ±20%.

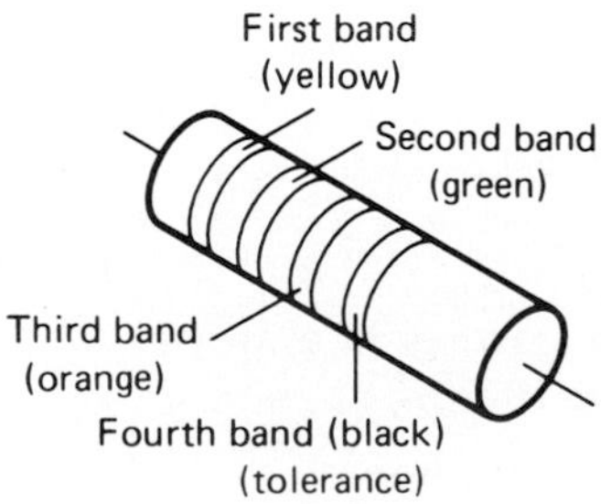

First band
(yellow)

Second band
(green)

Third band
(orange)

Fourth band (black)
(tolerance)

Figure 5.32

Example 1. Find the resistance of the resistor. (See Fig. 5.32.)

Step 1. The first digit is 4—the digit that corresponds to *yellow*.

Step 2. The second digit is 5—the digit that corresponds to *green*.

Step 3a. *Orange* on the third band means that there are three zeros to be written *after* the digits from Steps 1 and 2.

So, the resistance is 45,000 Ω.

Example 2. Find the tolerance, the upper limit, the lower limit, and the tolerance interval, for the resistor in Example 1.

The black fourth band indicates a tolerance of ±20%.
20% of 45,000 Ω = (0.20)(45,000 Ω) = 9000 Ω.
That is, the tolerance is ±9000 Ω.

The upper limit is 45,000 Ω + 9000 Ω = 54,000 Ω.
The lower limit is 45,000 Ω - 9000 Ω = 36,000 Ω.
The tolerance interval is then 18,000 Ω.

Example 3. Find the resistance of the resistor in
Fig. 5.33.

Step 1. The first digit is 3—the digit that corresponds to *orange*.

Step 2. The second digit is 0—the digit that corresponds to *black*.

Step 3a. *Red* on the third band means that there are two
zeros to be written *after* the digits from Steps
1 and 2.

So, the resistance is 3000 Ω.

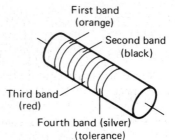

First band
(orange)

Second band
(black)

Third band
(red)

Fourth band (silver)
(tolerance)

Figure 5.33

Example 4. Find the tolerance, the upper limit, the
lower limit, and the tolerance interval, for the
resistor in Example 3.

The silver fourth band indicates a tolerance of ±10%.
10% of 3000 Ω = (0.10)(3000 Ω) = 300 Ω.
That is, the tolerance is ±300 Ω.

The upper limit is 3000 Ω + 300 Ω = 3300 Ω.
The lower limit is 3000 Ω - 300 Ω = 2700 Ω.
The tolerance interval is then 600 Ω.

Example 5. Find the resistance of the resistor
in Fig. 5.34.

Step 1. The first digit is 7—the digit that corresponds
to *violet*.

Step 2. The second digit is 9—the digit that corresponds to *white*.

Step 3b. *Gold* on the third band means to place the decimal point *between* the digits from Steps 1 and 2.

So, the resistance is 7.9 Ω.

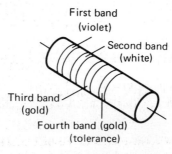

First band
(violet)

Second band
(white)

Third band
(gold)

Fourth band (gold)
(tolerance)

Figure 5.34

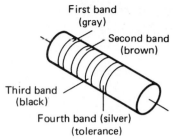

First band
(gray)

Second band
(brown)

Third band
(black)

Fourth band (silver)
(tolerance)

Figure 5.35

Example 6. Find the resistance of the resistor in Fig. 5.35.

Step 1. The first digit is 8—the digit that corresponds to *gray*.

Step 2. The second digit is 1—the digit that corresponds to *brown*.

Step 3a. *Black* on the third band means that there are *no* zeros to be included after the digits from Steps 1 and 2.

So, the resistance is 81 Ω.

Example 7. Find the resistance of the resistor in Fig. 5.36.

Step 1. The first digit is 6—the digit that corresponds to *blue*.

Step 2. The second digit is 5—the digit that corresponds to *green*.

Step 3c. *Silver* on the third band means to place the decimal point *before* the digits from Steps 1 and 2.

So, the resistance is 0.65 Ω.

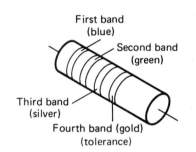

First band
(blue)

Second band
(green)

Third band
(silver)

Fourth band (gold)
(tolerance)

Figure 5.36

Example 8. A television serviceperson needs a 680,000 Ω-resistor. What color code on the first three bands is needed?

Step 1. The color that corresponds to the first digit, 6, is *blue*.

Step 2. The color that corresponds to the second digit, 8, is *gray*.

Step 3a. The color that corresponds to four zeros is *yellow*.

So, the colors that he is looking for are blue, gray, and yellow.

EXERCISES 5.10   *For each resistor below find its resistance and its*
                       tolerance, written as a percent:

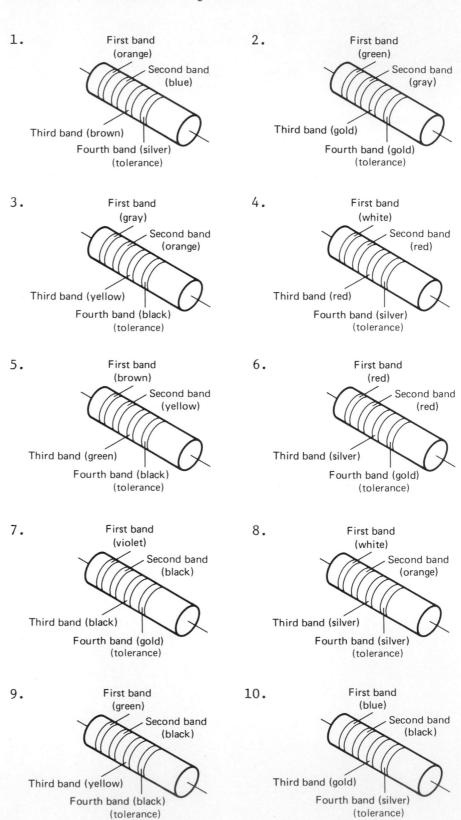

1.   First band
          (orange)
                    Second band
                    (blue)

Third band (brown)
          Fourth band (silver)
          (tolerance)

2.   First band
          (green)
                    Second band
                    (gray)

Third band (gold)
          Fourth band (gold)
          (tolerance)

3.   First band
          (gray)
                    Second band
                    (orange)

Third band (yellow)
          Fourth band (black)
          (tolerance)

4.   First band
          (white)
                    Second band
                    (red)

Third band (red)
          Fourth band (silver)
          (tolerance)

5.   First band
          (brown)
                    Second band
                    (yellow)

Third band (green)
          Fourth band (black)
          (tolerance)

6.   First band
          (red)
                    Second band
                    (red)

Third band (silver)
          Fourth band (gold)
          (tolerance)

7.   First band
          (violet)
                    Second band
                    (black)

Third band (black)
          Fourth band (gold)
          (tolerance)

8.   First band
          (white)
                    Second band
                    (orange)

Third band (silver)
          Fourth band (silver)
          (tolerance)

9.   First band
          (green)
                    Second band
                    (black)

Third band (yellow)
          Fourth band (black)
          (tolerance)

10.   First band
          (blue)
                    Second band
                    (black)

Third band (gold)
          Fourth band (silver)
          (tolerance)

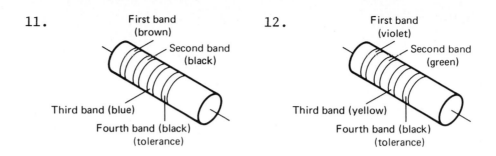

11. First band (brown)
Second band (black)
Third band (blue)
Fourth band (black) (tolerance)

12. First band (violet)
Second band (green)
Third band (yellow)
Fourth band (black) (tolerance)

What color code on the first three bands is needed for each resistance?

13. 4800 Ω 14. 95 Ω 15. 72,000 Ω 16. 3.1 Ω

17. 650,000 Ω 18. 100 Ω 19. 0.25 Ω 20. 9000 Ω

21. 4,500,000 Ω 22. 40 Ω 23. 7.6 Ω 24. 0.34 Ω

Find (a) the tolerance in ohms, Ω, (b) the upper limit, (c) the lower limit, and (d) the tolerance interval for each resistor:

25. Exercise 1   26. Exercise 2   27. Exercise 3

28. Exercise 5   29. Exercise 7   30. Exercise 12

5.11 Reading Circular Scales

In reading dials on meters, you must first determine which scale is being used and what the basic unit of that scale is. Consider, for example, the face of the water meter in Fig. 5.37. Each dial has ten graduations on its scale. Each dial represents a different power of 10. You read the dial starting with the 100,000 scale and continue clockwise to the one-cubic-foot scale. If the indicator is between two digits, read the smaller. The reading is 475,603 ft^3.

Example 1. Read the water meter shown in Fig. 5.38. The reading is 228,688 ft^3.

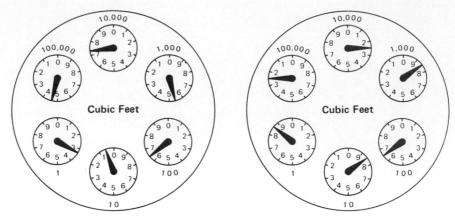

Figure 5.37   Figure 5.38

Consider the face of the electric meter in Fig. 5.39, which measures electricity in kilowatt hours (kWh). The four dials on the meter are read from left to right; the first dial is in thousands, next hundreds, the third in tens, and the fourth in units. Each dial has nine graduations dividing the scale into ten equal parts. If the indicator is between two digits, always read the smaller one. The reading of the dial in Fig. 5.39 is 2473 kWh.

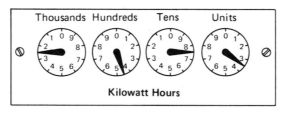

Figure 5.39

Figure 5.40

Example 2. What is the reading on the first dial of the electric meter if the indicator is as in Fig. 5.40?

The reading is 3000 kWh.

Example 3. What is the reading on the electric meter shown in Fig. 5.41?

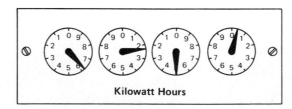

Figure 5.41

The reading is 6250 kWh.

Example 4. The January water meter reading was 312,017 ft^3. The March water meter reading is 314,830 ft^3. The rate for water is:

$$\begin{array}{ll} \text{First 1800 ft}^3, & \text{\$1.41 per 100 ft}^3, \\ \text{Next 4000 ft}^3, & \text{\$0.97 per 100 ft}^3, \\ \text{Next 14,000 ft}^3, & \text{\$0.65 per 100 ft}^3. \end{array}$$

Find the amount of the water bill.

Subtract the old reading from the new reading.

$$\begin{array}{r} 314,830 \text{ ft}^3 \\ \underline{312,017 \text{ ft}^3} \\ 2,813 \text{ ft}^3 \end{array}$$

Round to the lowest hundred and divide by 100. The result is 28.

$$\begin{array}{rll} 18 \times \$1.41 = & \$25.38 & \text{cost of first 1800 ft}^3 \\ 10 \times \$0.97 = & \underline{9.70} & \text{cost of next 1000 ft}^3 \\ \text{Subtotal} & \$35.08 & \\ 3\% \text{ municipal tax} & \underline{1.05} & \\ \text{Net bill} & \$36.13 & \end{array}$$

A sample bill for this account is shown in Fig. 5.42.

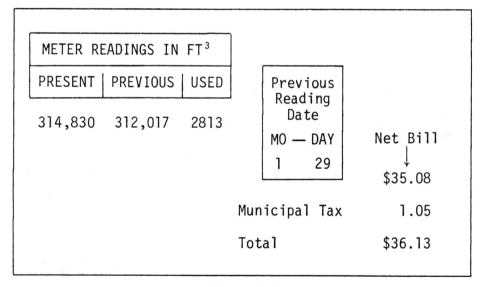

Figure 5.42

Dial gauges are useful for making very precise comparisons between a known measurement and some measurement that must be checked for precision. They are used for inspection operations, in toolrooms, and in machine shops in a wide variety of applications. Some of these uses are shown in Fig. 5.43.

Let's first study the metric dial indicator as shown in Fig. 5.44. Each graduation represents 0.01 mm. If the needle deflects six graduations to the right (+) of zero, the object being measured is 6 × 0.01 mm. This is 0.06 mm larger than the desired measurement. If the needle deflects 32 graduations to the left (−) of zero, the object being

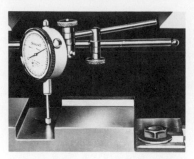

(a) To insure positive vise alignment on milling machine, operator uses indicator against parallel clamped in vise jaws.

(b) Precise alignment of cutting bar is assured by using a dial indicator.

(c) Planer operator uses dial test indicator to check depth of cut on flat casting.

(d) Lathe operator uses dial indicator to check total indicator runout (TIR).

Figure 5.43

measured is 32 × 0.01 mm. This is 0.32 mm smaller than the desired measurement.

Note the smaller dial on the lower left portion of the dial in Fig. 5.44. This small needle records the number of complete revolutions that the large needle makes. Each complete revolution of the large needle corresponds to 1.00 mm.

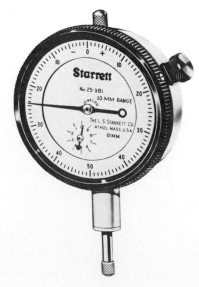

Figure 5.44

Example 5. Read the metric dial in Fig. 5.45.

Figure 5.45

The small needle reads +3 × 1.00 mm = +3.00 mm
The large needle reads +36 × 0.01 mm = +0.36 mm
The total reading is   +3.36 mm

This measurement is 3.36 mm more than the desired
measurement.

Now look closely at the English dial indicator in Fig.
5.46. Each graduation represents 0.001 in. If the needle
deflects 7 graduations to the right (+) of zero, the object
being measured is 7 × 0.001 in. This is 0.007 in. larger
than the desired measurement. If the needle deflects to the
left (−) 14 graduations, the object being measured is
14 × 0.001 in. This is 0.014 in. smaller than the desired
measurement.

Note the smaller dial on the lower left portion of the
dial in Fig. 5.46. This small needle records the number of
complete revolutions that the large needle makes. Each
complete revolution of the large needle corresponds to
0.100 in. Other dials are read in a similar manner.

Figure 5.46

Example 6. Read the English dial in Fig. 5.47.

Figure 5.47

The small needle reads −2 × 0.100 in. = −0.200 in.
The large needle reads −23 × 0.001 in. = −0.023 in.
The total reading is   −0.223 in.

This measurement is 0.223 in. less than the desired measurement.

EXERCISES 5.11   *Read each water meter:*

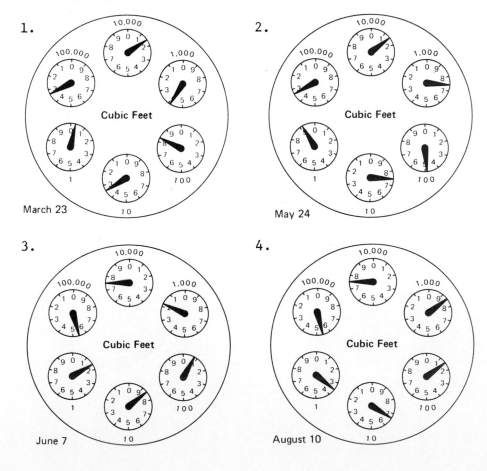

5. **6.**

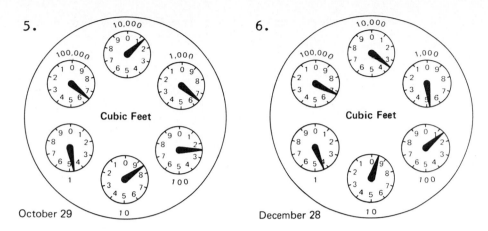

October 29   December 28

In the next three exercises, use the water rates and the municipal tax rate given in Example 4, page 219.

7. Find the net water bill for the water used between the readings of Exercises 1 and 2.
8. Find the net water bill for the water used between the readings of Exercises 3 and 4.
9. Find the net water bill for the water used between the readings of Exercises 5 and 6.

Read each electric meter:

10.   **11.**

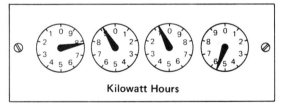 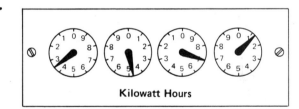

12.   **13.**

14.

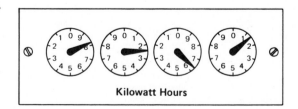

Read each metric dial: (The arrow near the zero indicates the initial deflection of the needle.)

15.

16.

17.

18.

19.

20.

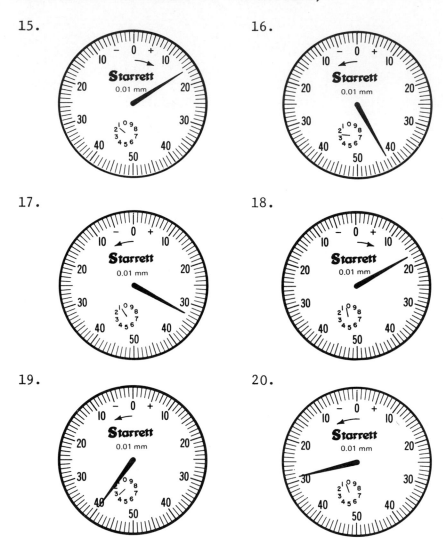

21.   22.

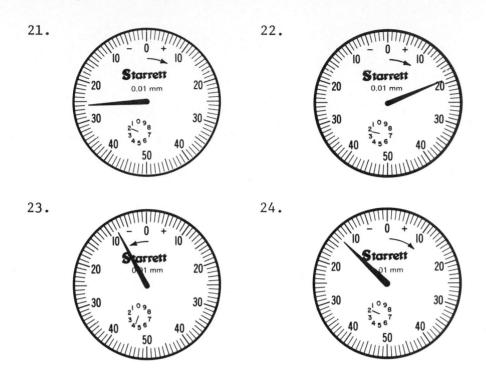

23.   24.

Read each English dial: (*The arrow near the zero indicates
the initial deflection of the needle.*)

25.   26.

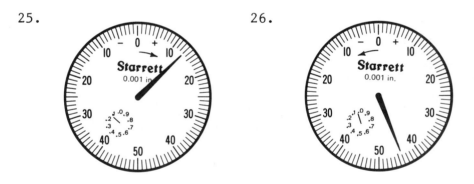

27.

28.

29.

30.

31.

32.

33.

34.

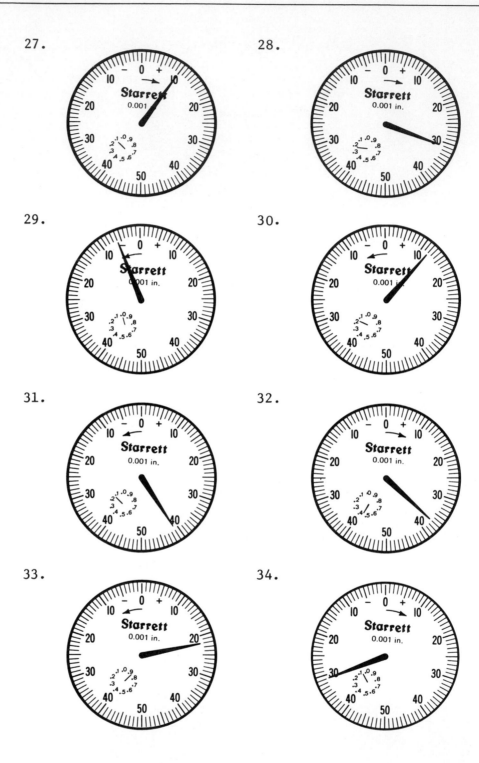

5.12 Reading Uniform Scales

Figure 5.48 shows some of the various scales that may be found on a volt-ohm meter (VOM). This instrument is used to measure voltage (measured in volts, V) and resistance (measured in ohms, Ω) in electrical circuits. Note that the voltage scales are uniform, while the resistance scale is nonuniform. On a given voltage scale, the graduations are equally spaced, and each subdivision represents the number of volts. On the resistance scale, the graduations are not equally spaced, and subdivisions on various intervals represent a different number of ohms. To make things clear in the examples and exercises that follow, we show only one of the VOM scales at a time in a given figure.

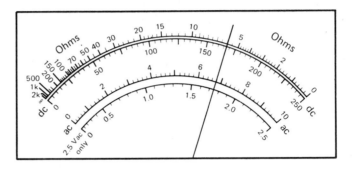

Figure 5.48

The first uniform voltage scale that we study is shown in Fig. 5.49. This scale has a range of 0—10 V. There are 10 large divisions, each representing 1 V. Each large division has 4 graduations dividing it into 5 subdivisions. Each of these subdivisions is $\frac{1}{5}$ V, or 0.2 V.

Example 1. Read the scale shown in Fig. 5.49.

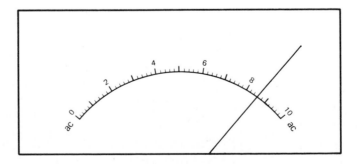

Figure 5.49

The needle is on the third graduation to the right of 8. Each subdivision is 0.2 V. Therefore, the reading is 8.6 V.

Figure 5.50 shows a voltage scale that has a range of 0–2.5 V. There are 5 large divisions, each representing 0.5 V. Each division has 4 graduations dividing it into 5 subdivisions. Each of these subdivisions is $\frac{1}{5} \times 0.5$ V = 0.1 V.

Example 2. Read the scale shown in Fig. 5.50.

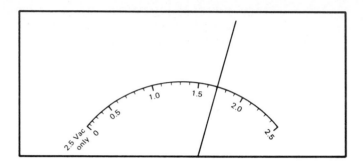

Figure 5.50

The needle is on the second graduation to the right of 1.5. Each subdivision is 0.1 V. Therefore, the reading is 1.7 V.

Figure 5.51 shows a voltage scale that has a range of 0–250 V. There are 10 large divisions, each representing 25 V. Each division has 4 graduations dividing it into 5 subdivisions. Each of these subdivisions is $\frac{1}{5} \times 25$ V = 5 V.

Example 3. Read the scale shown in Fig. 5.51.

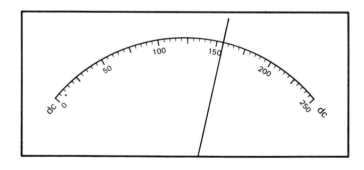

Figure 5.51

The needle is on the first graduation to the right of 150. Each subdivision is 5 V. Therefore, the reading is 155 V.

EXERCISES 5.12   *Read each voltage scale:*

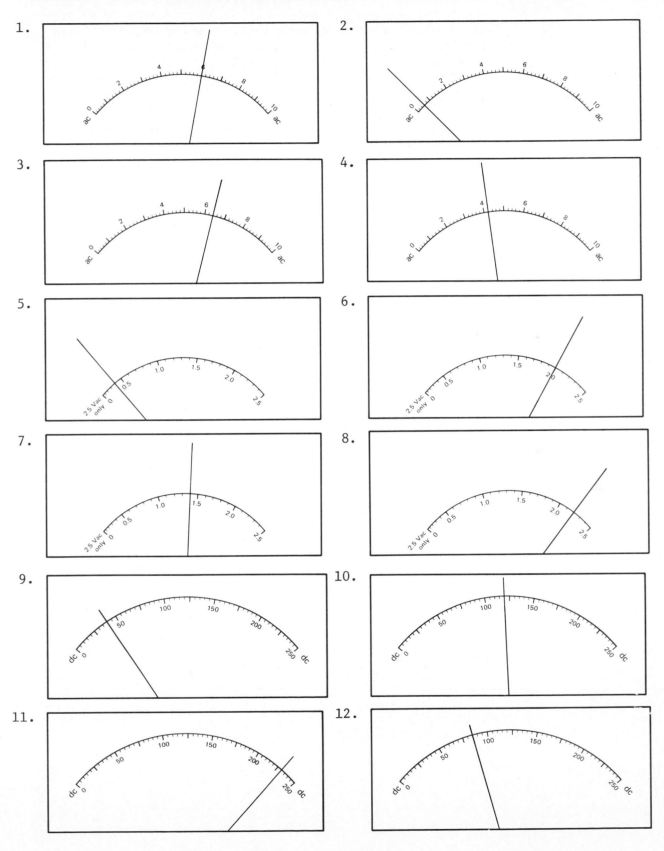

5.13 Reading Nonuniform Scales

Figure 5.52 shows a nonuniform ohm scale usually found on a VOM. First, consider that part of the scale between 0 and 5. Each large division represents 1 ohm (Ω). Each large division is divided into 5 subdivisions. Therefore, each subdivision is $\frac{1}{5} \times 1 \ \Omega$, or 0.2 Ω.

Consider that part of the scale between 5 and 20. Between 5 and 10 each subdivision is divided into 2 sub-subdivisions. Each sub-subdivision is $\frac{1}{2} \times 1 \ \Omega$, or 0.5 Ω. Between 10 and 20 each division represents 1 Ω.

Consider that part of the scale between 20 and 100. Each large division represents 10 Ω. Between 20 and 30 there are 5 subdivisions. Therefore, each subdivision is $\frac{1}{5} \times 10 \ \Omega$, or 2 Ω. Between 30 and 100 each large division has 2 subdivisions. Each subdivision is $\frac{1}{2} \times 10 \ \Omega$, or 5 Ω.

Consider that part of the scale between 100 and 200. Each large division represents 50 Ω. Between 100 and 150 there are 5 subdivisions. Each subdivision is $\frac{1}{5} \times 50 \ \Omega$, or 10 Ω.

Consider that part of the scale between 200 and 500. There are 3 subdivisions. Each subdivision is $\frac{1}{3} \times 300 \ \Omega$, or 100 Ω.

The subdivisions on each part of the ohm scale may be summarized as follows:

Range	Each subdivision represents:
0—5 Ω	0.2 Ω
5—10 Ω	0.5 Ω
10—20 Ω	1 Ω
20—30 Ω	2 Ω
30—100 Ω	5 Ω
100—150 Ω	10 Ω
200—500 Ω	100 Ω

Example 1. Read the scale shown in Fig. 5.52.

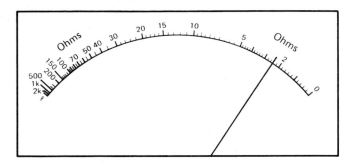

Figure 5.52

The needle is on the second subdivision to the left of 2, where each subdivision represents 0.2 Ω. Therefore, the reading is 2.4 Ω.

Example 2. Read the scale shown in Fig. 5.53.

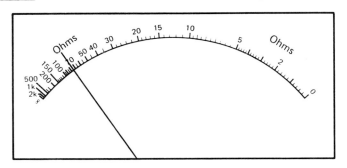

Figure 5.53

The needle is on the subdivision between 70 and 80. A subdivision represents 5 Ω on this part of the scale. Therefore, the reading is 75 Ω.

EXERCISES 5.13   *Read each ohm scale:*

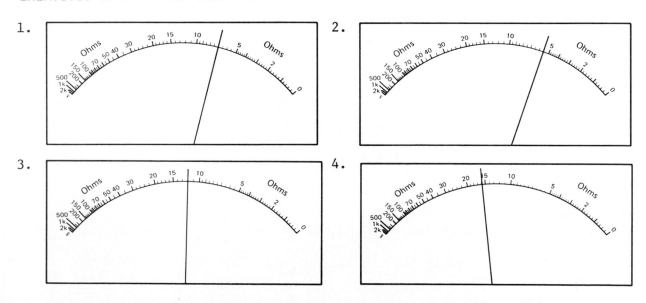

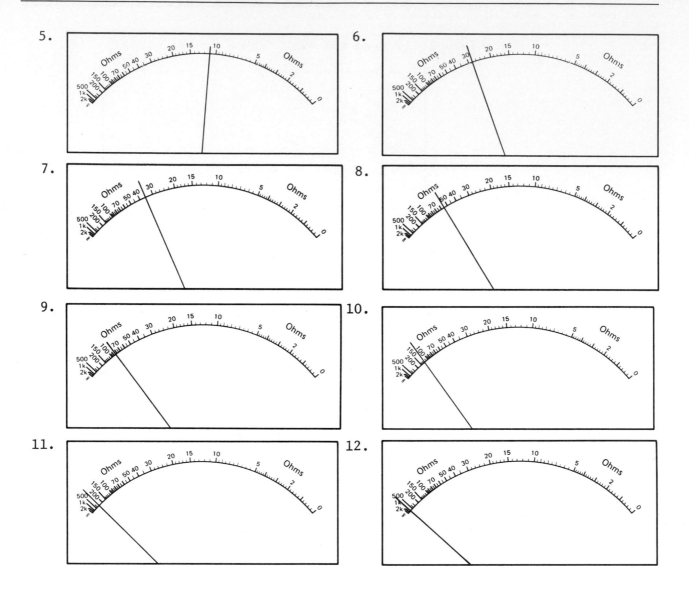

Chapter 5 Review

Give the number of significant digits (the accuracy) of each measurement:

1. 4.06 kg   2. 24,000 mi   3. 36$\overline{0}$0 V   4. 5.60 cm

5. 0.0070 W   6. 0.0651 s   7. 20.00 m   8. 20.050 km

Find (a) the precision and (b) the greatest possible error of each measurement:

9. 6.05 m   10. 15.0 mi   11. 160,500 L

12. 2300 V   13. 17.00 cm   14. 13,0$\overline{0}$0,000 V

15. $1\frac{5}{8}$ in.   16. $10\frac{3}{16}$ mi

Read the measurement shown on the vernier caliper:

17. In metric units   **18.** In English units

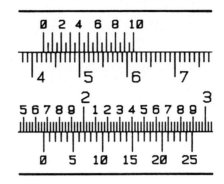

19. Read the measurement shown on the metric micrometer.
20. Read the measurement shown on the English micrometer.

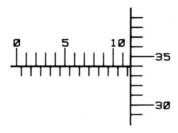   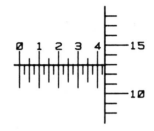

Exercise 19   Exercise 20

21. Find the measurement that is (a) the most accurate, and (b) the most precise.

2500 V; 36,500 V; 60,000 V; 9.6 V; 120 V

22. Find the measurement that is (a) the least accurate, and (b) the least precise.

0.0005 A; 0.0060 A; 0.425 A; 0.0105 A; 0.0055 A

Use the rules for addition of measurements to find the sum of each set of measurements:

23. 18,000 W; 260,000 W; 2300 W; 45,500 W; 398,000 W

24. 16.8 cm; 19.7 m; 0.14 km; 240 m

25. Use the rules for subtraction of measurements to subtract

1,500,000 V
1,125,000 V

Use the rules for multiplication and/or division of measurements to evaluate:

26. 15.6 cm × 18.5 cm × 6.5 cm

27. $\dfrac{98.2 \text{ m}^3}{16.7 \text{ m}}$ 28. $\dfrac{239 \text{ N}}{24.8 \text{ m} \times 6.7 \text{ m}}$ 29. $\dfrac{(220 \text{ V})^2}{365 \text{ }\Omega}$

Find (a) the relative error, and (b) the percent of error (to the nearest hundredth percent) for each measurement:

30. $5\dfrac{7}{16}$ in.   31. 15.60 cm

32. Given a resistor of 2000 Ω with a tolerance of ±10%, find the upper and lower limits.

For each resistor find its resistance and its tolerance written as a percent:

33.

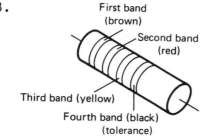

First band (brown)
Second band (red)
Third band (yellow)
Fourth band (black) (tolerance)

34.

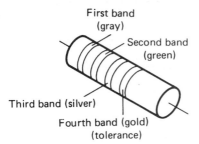

First band (gray)
Second band (green)
Third band (silver)
Fourth band (gold) (tolerance)

Read each scale:

35.

36.

37.

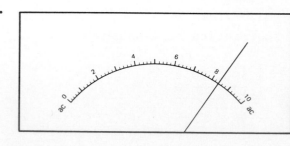

38.

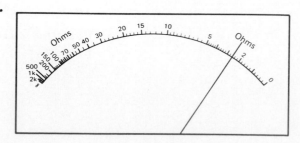

POLYNOMIALS

6.1 Fundamental Operations

Listed below are some basic principles that you will apply in this course. Most of them you probably already know; the rest will be discussed.

Basic Mathematical Principles

1. $a + b = b + a$

2. $ab = ba$

3. $(a + b) + c = a + (b + c)$

4. $(ab)c = a(bc)$

5. $a(b + c) = ab + ac$, or $(b + c)a = ba + ca$

6. $a + 0 = a$

7. $a \cdot 0 = 0$

8. $a + (-a) = 0$

9. $a \cdot 1 = a$

10. $a \cdot \dfrac{1}{a} = 1$ (a cannot be 0)

In mathematics, letters are often used to represent numbers. Thus, it is necessary to know how to indicate arithmetic operations and carry them out using letters.

Addition. $x + y$ means add y to x.

Subtraction. $x - y$ means subtract y from x or add the negative of y to x; that is, $x + (-y)$.

Multiplication. xy or $x \cdot y$ or $(x)(y)$ or $(x)y$ or $x(y)$ means multiply x by y.

Division. $x \div y$ or $\dfrac{x}{y}$ means divide x by y, or find a number z such that $zy = x$.

Exponents. $xxxx$ means use x as a factor 4 times, which is abbreviated by writing x^4. In the expression x^4, x is called the *base*, and 4 is called the *exponent*. For example, 2^4 means $2 \cdot 2 \cdot 2 \cdot 2 = 16$.

Order of Operations

1. Perform all operations inside parentheses first.
2. Evaluate all powers, if any.
   For example, $6 \cdot 2^3 = 6 \cdot 8 = 48$.
3. Perform any multiplications or divisions, in order, from left to right.
4. Do any additions or subtractions, in order, from left to right.
5. If the problem contains a fraction bar, treat the numerator and the denominator separately.

Example 1. Evaluate:

$$4 - 9(6 + 3) \div (-3)$$
$$= 4 - 9(9) \div (-3)$$
$$= 4 - 81 \div (-3)$$
$$= 4 - \quad (-27)$$
$$= 31$$

Example 2. Evaluate:

$$(-6) + 5(-2)^2(-9) - 7(3 - 5)^3$$
$$= (-6) + 5(-2)^2(-9) - 7(-2)^3$$
$$= (-6) + 5(4)(-9) - 7(-8)$$
$$= (-6) - 180 + 56$$
$$= -130$$

To *evaluate an expression*, replace the letters with given numbers; then do the arithmetic in correct order. The result is the value of the expression.

<u>Example 3.</u> Evaluate $\dfrac{x^2 - y + 5}{2x - 2}$, if $x = 4$ and $y = 3$.

Replace x with 4 and y with 3 in the expression.

$$\frac{4^2 - 3 + 5}{2(4) - 2} = \frac{16 - 3 + 5}{8 - 2} = \frac{18}{6} = 3$$

Notice that in a fraction the line between the numerator and denominator serves as parentheses for both. That is, the operations in both numerator and denominator are done first, before the division.

<u>Example 4.</u> Evaluate $\dfrac{ab}{3c} + c$, if $a = 6$, $b = 10$, and $c = -5$.

Replace a with 6, b with 10 and c with -5 in the expression.

$$\frac{6 \cdot 10}{3(-5)} + (-5) = \frac{60}{-15} + (-5) = -4 + (-5) = -9$$

EXERCISES 6.1   *Evaluate each expression:*

1. $3(-5)^2 - 4(-2)$ 2. $(-2)(-3)^2 + 3(-2) \div 6$

3. $4(-3) \div (-6) - (-18) \div 3$

4. $48 \div (-2)(-3) + (-2)^2$

5. $(-72) \div (-3) \div (-6) \div (-2) - (-4)(-2)(-5)$

6. $28 \div (-7)(2)^2 + 3(-4 - 2)^2 - (-3)^2$

7. $[(-2)(-3) + (-24) \div (-2)] \div [-10 + 7(-1)^2]$

8. $(-9)^2 \div 3^3(6) + [3(-2) - 5(-3)]$

9. $[(-2)(-8)^2 \div (-2)^3] - [-4 + (-2)^4]^2$

10. $[(-2)(3) + 5(-2)][5(-4) - 8(-3)]^2$

In Exercises 11—16, let $x = 2$, $y = 3$, and evaluate each expression:

11. $2x - y$ 12. $x - 2y$ 13. $x^2 - y^2$

14. $y^2 - x^2$ 15. $\dfrac{3x + y}{3}$ 16. $\dfrac{2(x + y) - 2x}{2(y - x)}$

In Exercises 17—26, let $x = -1$, $y = 5$, and evaluate each expression:

17. $xy^2 - x$ 18. $4x^3 - y^2$ 19. $\dfrac{2y}{x} - \dfrac{2x}{y}$

20. $3 + 4(x + y)$ 21. $3 - 4(x + y)$ 22. $1.7 - 5(2x - y)$

23. $\dfrac{1}{x} - \dfrac{1}{y} + \dfrac{2}{xy}$

24. $(2.4 - x)(x - xy)$

25. $(y - 2x)(3x - 6xy)$

26. $\dfrac{(y - x)^2 - 4y}{4x^2}$

In Exercises 27—32, let x = -3, y = 4, and z = 6. Evaluate each expression:

27. $(2xy^2z)^2$

28. $(x^2 - y^2)z$

29. $(y^2 - 2x^2)z^2$

30. $\left(\dfrac{x + 3y}{z}\right)^2$

31. $(2x + 3y)(y + z)$

32. $(4 - x)^2(z - y)$

In Exercises 33—38, let x = -1, y = 2, and z = -3. Evaluate each expression:

33. $(2x + 6)(3y - 4)$

34. $z^2 - 5yx^2$

35. $(3x + 5)(2y - 1)(5z + 2)$

36. $(3x - 4z)(2x + 3z)$

37. $(x - xy)^2(z - 2x)$

38. $3x^2(y - 3z)^2 - 6x$

6.2 Simplifying Algebraic Expressions

Parentheses are often used to show the correct order of operations, especially if the order of operations is complicated or may be ambiguous. Sometimes it is easier to simplify such an expression by first removing the parentheses—before doing the indicated operations. Two rules for removing parentheses are:

Removing Parentheses

1. Parentheses preceded by a "+" sign may be removed without changing the signs of the terms within. That is, $3w + (4x + y) = 3w + 4x + y$.
2. Parentheses preceded by a "-" sign may be removed if the signs of <u>all</u> the terms within the, parentheses are changed; then the "-" sign which preceded the parentheses is dropped. That is $3w - (4x - y) = 3w - 4x + y$. (Notice that the sign of the term $4x$ inside the parentheses is not written. It is, therefore, understood to be "+.")

<u>Example 1</u>. Remove the parentheses from the expression $5x - (-3y + 2z)$.

$$5x - (-3y + 2z) = 5x + 3y - 2z$$

<u>Example 2</u>. Remove the parentheses from the expression $7x + (-y + 2z) - (w - 4)$.

$$7x + (-y + 2z) - (w - 4) = 7x - y + 2z - w + 4$$

Terms are the parts of an expression separated by "+" and "-" signs. For example, $3xy + 2y + 8x^2$ is an expression consisting of three terms.

$$\boxed{3xy} + \boxed{2y} + \boxed{8x^2}$$

    ↑   ↑   ↑

  1st   2nd   3rd

  term term term

Terms which consist of the same variables (letters), used as factors the same number of times (same exponents), are called *like terms*.

<u>Example 3</u>. The following chart gives examples of like terms and unlike terms:

Like Terms	*Unlike Terms*
(a) $2x$ and $3x$	(e) $2x^2$ and $3x$
(b) $2ax$ and $5ax$	(f) $2ax$ and $5bx$
(c) $2x^3$ and $18x^3$	(g) $2x^3$ and $18x^2$
(d) $2a^2x^4$, a^2x^4, and $11a^2x^4$	(h) $2a^2x^4$, $3ax^4$, and $11a^2x^3$

Like terms that occur in a single expression can be combined into one term by using Rule 5 from page 235. Thus, $ba + ca = (b + c)a$.

<u>Example 4</u>. Combine like terms $2x + 3x$.

$$2x + 3x = (2 + 3)x = 5x$$

<u>Example 5</u>. Combine like terms $2ax + 3ax$.

$$2ax + 3ax = (2 + 3)ax = 5ax$$

<u>Example 6</u>. Combine like terms $2a^2x^4 + a^2x^4 + 11a^2x^4$.

$$2a^2x^4 + a^2x^4 + 11a^2x^4 = 2a^2x^4 + 1a^2x^4 + 11a^2x^4$$
$$= (2 + 1 + 11)a^2x^4 = 14a^2x^4$$

The number that is a factor of a term is called the *numerical coefficient* of the term. The term $5x$ has 5 as its numerical coefficient. The term ax^2 has the factor 1 (understood) as its numerical coefficient. Like terms may also be defined as: terms which differ only in their numerical coefficients.

Example 7. Combine like terms $3x^2 + 8x^2$.

$$3x^2 + 8x^2 = (3 + 8)x^2 = 11x^2$$

Example 8. Combine like terms $9a^3b^4 + 2a^2b^3 + 7a^3b^4$.
$$9a^3b^4 + 2a^2b^3 + 7a^3b^4 = (9 + 7)a^3b^4 + 2a^2b^3$$
$$= 16a^3b^4 + 2a^2b^3$$

Some expressions contain parentheses that must be removed before combining like terms.

Example 9. Simplify $4x - (x - 2)$.

$$4x - (x - 2) = 4x - x + 2 = 3x + 2$$

Example 10. Simplify $3x - (-2x - 3y) + 2y$.

$$3x - (-2x - 3y) + 2y = 3x + 2x + 3y + 2y = 5x + 5y$$

Example 11. Simplify $(7 - 2x) + (5x + 1)$.

$$(7 - 2x) + (5x + 1) = 7 - 2x + 5x + 1 = 3x + 8$$

See Rule 5 from page 235: $a(b + c) = ab + ac$. This rule is used to remove parentheses when a number, a letter, or some product precedes the parentheses.

Example 12. Simplify $3x + 5(x - 3)$.

$$3x + 5(x - 3) = 3x + 5x - 15 = 8x - 15$$

Example 13. Simplify $4y - 6(-y + 2)$.

$$4y - 6(-y + 2) = 4y + 6y - 12 = 10y - 12$$

EXERCISES 6.2   *Remove the parentheses from each expression:*

1. $a + (b + c)$ 2. $a - (b + c)$

3. $a - (-b - c)$ 4. $a - (-b + c)$

5. $a + (-b - c)$ 6. $x + (y + z + 3)$

7. $x - (-y + z - 3)$ 8. $x - (-y - z + 3)$

9. $x - (y + z + 3)$ 10. $x + (-y - z - 3)$

11. $(2x + 4) + (3y + 4r)$ 12. $(2x + 4) - (3y + 4r)$

13. $(3x - 5y + 8) + (6z - 2w + 3)$

14. $(4x + 6y - 9) + (-2z + 5w + 3)$

15. $(-5x - 3y - 2) - (6z - 3w - 5)$

16. $(-9x + 6) - (3z + 3w - 1)$

17. $(2x + 3y - 5) + (-z - w + 2) - (-3r + 2s + 7)$

18. $(5x - 11y - 2) - (7z + 3) + (3r + 7) - (4s - 2)$

19. $-(2x - 3y) - (z + 4w) - (4r - s)$

20. $-(3x + y) - (2z + 7w) - (3r - 5s + 2)$

Combine like terms:

21. $4b + b$

22. $x^2 + 3x + 2x^2 + 7x$

23. $4h + 6h$

24. $9k + 3k$

25. $5m - 2m$

26. $4x + 6x - 5x$

27. $3a + 5b - 2a + 7b$

28. $11 + 2m - 6 + m$

29. $6a^2 + a + 1 - 2a$

30. $5x^2 + 3x^2 - 8x^2$

31. $2x^2 + 16x + x^2 - 13x$

32. $13x^2 + 14xy + 6y^2 - 3y^2 + x^2$

33. $1.3x + 5.6x - 13.2x + 4.5x$

34. $\frac{1}{2}x - \left(\frac{2}{3}\right)y - \left(\frac{3}{4}\right)x + \left(\frac{5}{6}\right)y$

35. $41a^3 + 7a^2 - 26a^3 - 15 - 5a^2$

36. $2.3x^2 - 4.7x + 0.92x^2 - 2.13x$

37. $4x^2y - 2xy - y^2 - 3x^2 - 2x^2y + 3y^2$

38. $3x^2 - 5x - 2 + 4x^2 + x - 4 + 5x^2 - x + 2$

39. $2x^3 + 4x^2y - 4y^3 + 3x^3 - x^2y + y - y^3$

40. $4x^2 - 5x - 7k^2 - 3x - y^2 + 2x^2 + 3xy - 2y^2$

$-1x^2 - 8x - 3y^2 + 3xy$

Simplify by first removing parentheses and then combining like terms:

41. $y - (y - 1)$

42. $x + (2x + 1)$

43. $4x + (4 - x)$

44. $5x - (2 - 3x)$

45. $10 - (5 + x)$

46. $x - (-x - y) + 2y$

47. $2y - (7 - y)$

48. $-y - (y + 3)$

49. $(5y + 7) - (y + 2)$

50. $(2x + 4) - (x - 7)$

51. $(4 - 3x) + (3x + 1)$

52. $10 - (y + 6) + (3y - 2)$

53. $-5y + 9 - (-5y + 3)$

54. $0.5x + (x - 1) - (0.2x + 8)$

55. $0.2x - (0.2x - 28)$

56. $(0.3x - 0.5) - (-2.3x + 1.4)$

57. $\left(\frac{1}{2}x - \frac{2}{3}\right) - \left(2 - \frac{3}{4}x\right)$

58. $\left(\frac{3}{4}x - 1\right) + \left(-\frac{1}{2}x - \frac{2}{3}\right)$

59. $(5x + 13) - 3(x - 2)$

60. $(7x + 8) - 5(x - 6)$

61. $-9y - 0.5(8 - y)$

62. $12(x + 1) - 3(4x - 2)$

63. $2y - 2(y + 21)$

64. $3x - 3(6 - x)$

65. $6n - (2n - 8)$

66. $14x - 8(2x - 8)$

67. $0.8x - (-x + 7)$

68. $-(x - 3) - 3(4 + x)$

69. $5(3 + x) - 6(x - 3)$

70. $(x + 4) - 2(2x - 7)$

71. $4(2 - 3n) - 2(5 - 3n)$

72. $\left(\dfrac{2}{3}\right)(6x - 9) - \left(\dfrac{3}{4}\right)(12x - 16)$

73. $0.45(x + 3) - 0.75(2x + 13)$

74. $13\left(7x - 2\dfrac{1}{2}\right) - 9\left(8x + 9\dfrac{2}{3}\right)$

6.3   Addition and Subtraction of Polynomials

A *monomial, or term,* is any algebraic expression that contains only products of numbers and variables, which have nonnegative integer exponents. The following expressions are examples of monomials.

$$2x, \qquad 5, \qquad -3b, \qquad \dfrac{3}{4}a^2bw, \qquad \sqrt{315}\ m$$

A *polynomial* is either a monomial or the sum of monomials. There are two special types of polynomials to be considered here. A *binomial* is a polynomial that is the sum of two unlike monomials. A *trinomial* is the sum of three unlike monomials.

Example 1. The following are *binomials*:

$$a + b, \qquad 2a^2 + 3, \qquad 5mn^2 + 7wr^2$$

Example 2. The following are *trinomials*:

$$3 + 5a + 7b, \qquad 2n + 4m + 6p, \qquad 2a^3b + 3a^2b^2 + 4ab^3$$

Expressions that contain variables in the denominator are *not* polynomials. For example, $\dfrac{3}{4x}$, $\dfrac{8x}{3x - 5}$, and $\dfrac{33}{4x^2} + \dfrac{8}{x - 1}$ are *not* polynomials.

The <u>degree of a monomial in one variable</u> is the same as the exponent of the variable.

Example 3. Find the degree of each monomial:

(a) $-7m$ has degree 1

(b) $6x^2$ has degree 2

(c) $5y^3$ has degree 3

(d) 5 has degree 0

The degree of a polynomial in one variable is the same as the highest degree monomial contained in the polynomial.

Example 4. Find the degree of each polynomial:

(a) $5x^4 + x^2$ has degree 4

(b) $6y^3 + 4y^2 - y + 1$ has degree 3

A polynomial is in *decreasing order* if each term is of some degree less than the preceding term.

Example 5. The following polynomial is written in decreasing order:

$$4x^5 - 3x^4 - 4x^2 - x + 5$$
$$\xrightarrow{\text{exponents decrease}}$$

A polynomial is in *increasing order* if each term is of some degree larger than the preceding term.

Example 6. The following polynomial is written in increasing order:

$$5 - x - 4x^2 - 3x^4 + 4x^5$$
$$\xrightarrow{\text{exponents increase}}$$

To add monomials and polynomials, write like terms under each other. Then add by columns, as in the examples below.

Example 7. Add $3x + 5$ and $5x - 7$.

$$\begin{array}{r} 3x + 5 \\ 5x - 7 \\ \hline 8x - 2 \end{array}$$

Example 8. Add $15x$ and $-12x + 3$.

$$
\begin{array}{r}
15x \\
-12x + 3 \\
\hline
3x + 3
\end{array}
$$

Example 9. Add $(2x^2 - 5x) + (3x^2 + 2x - 4) + (-4x^2 + 5)$.

$$
\begin{array}{r}
2x^2 - 5x \\
3x^2 + 2x - 4 \\
-4x^2 \qquad + 5 \\
\hline
x^2 - 3x + 1
\end{array}
$$

Note that the polynomials are usually written in decreasing order. Also note that the like terms are written in the same columns. If the terms are unlike, the addition must be left in the form of an indicated sum, as in Example 10.

Example 10. Add $2x^2 + 3x$.

$$
\begin{array}{r}
2x^2 \\
+ 3x \\
\hline
2x^2 + 3x
\end{array}
$$
is the sum because $2x^2$ and $3x$ are unlike terms

Subtracting Polynomials

To subtract a second polynomial from a first,

1. Write the second polynomial under the first. Place like terms under each other where possible.
2. Change the sign of each term in the second polynomial.
3. Add, as in the previous section.

Example 11. Find the difference: $(3a - 5b) - (2a - 4b)$

$$
\text{Subtract:} \begin{array}{r} 3a - 5b \\ 2a - 4b \\ \hline \end{array} \rightarrow \text{Add:} \begin{array}{r} 3a - 5b \\ -2a + 4b \\ \hline a - b \end{array}
$$

The arrow indicates the change of the subtraction problem to an addition problem. Do this by changing the signs of each of the terms in the second polynomial.

Example 12. Subtract: $\begin{array}{r} 5x^2 - 3x - 4 \\ 2x^2 - 5x + 5 \end{array}$ $\rightarrow$ Add: $\begin{array}{r} 5x^2 - 3x - 4 \\ -2x^2 + 5x - 5 \\ \hline 3x^2 + 2x - 9 \end{array}$

To check subtraction, add the result to the polynomial in the second line of the original subtraction. If the sum is the same as the first polynomial, the result is correct.

Example 13. Check the result from Example 12.

Add: $\begin{array}{r} 2x^2 - 5x + 5 \\ 3x^2 + 2x - 9 \\ \hline 5x^2 - 3x - 4 \end{array}$

This result is the same as the upper polynomial in Example 12; therefore, the answer for Example 12 is correct.

A second method shown in Example 14 uses the rules for removing parentheses given in Sec. 6.2.

Example 14. Subtract $x^2 - 2x$ from $3x^2 + 4x$.

Write using parentheses:

$$(3x^2 + 4x) - (x^2 - 2x)$$

Remove parentheses and combine like terms:

$$3x^2 + 4x - x^2 + 2x = 2x^2 + 6x$$

EXERCISES 6.3   *Classify each expression as a monomial, a binomial, or a trinomial:*

1. $3m + 27$ 2. $4a^2bc^3$

3. $-5x - 7y$ 4. $2x^2 + 7y + 3z^2$

5. $-5xy$ 6. $a + b + c$

7. $2x + 3y - 5z$ 8. $2a - 3b^3$

9. $-42x^3 - y^4$ 10. $15x^{14} - 3x^2 + 5x$

Rearrange each polynomial in decreasing order and state its degree:

11. $1 - x + x^2$ 12. $2x^3 - 3x^4 + 2x$

13. $x + x^2 - 1$ 14. $y^3 - 1 + y^2$

15. $-4x^2 + 5x^3 - 2$ 16. $3x^3 + 6 - 2x + 4x^5$

17. $7 - 3y + 4y^3 - 6y^2$

18. $x^3 - 4x^4 + 2x^2 - 7x^5 + 5x - 3$

19. $1 - x^5$

20. $360x^2 - 720x - 120x^3 + 30x^4 + 1 - 6x^5 + x^6$

Add:

21. $\begin{array}{r} -5a \\ 7a \\ \hline \end{array}$

22. $\begin{array}{r} 3a - 5b + c \\ 4a + 6b - 2c \\ \hline \end{array}$

23. $\begin{array}{r} 4x^2 - 7x \\ -2x^2 + 5x \\ \hline \end{array}$

24. $\begin{array}{r} 5y^3 - 4y^2 - 2y \\ -7y^3 - 3y^2 - y \\ 2y^3 - y^2 + y \\ \hline \end{array}$

25. $\begin{array}{r} 4y^2 - 3y - 15 \\ 7y^2 - 6y + 8 \\ -3y^2 + 4y + 13 \\ \hline \end{array}$

26. $\begin{array}{r} 129a - 13b - 56c \\ -13a - 52b + 21c \\ 44a \quad\quad + 11c \\ \hline \end{array}$

27. $\begin{array}{r} 3a^3 + 2a^2 \quad\quad + 5 \\ a^3 \quad\quad - 7a - 2 \\ - 5a^2 + 4a \\ - 2a - 3 \\ \hline \end{array}$

28. $\begin{array}{r} 3a - 15a \end{array}$

29. $(5a^2 - 7a + 5) + (2a^2 - 3a - 4)$

30. $(2b - 5) + 3b + (-4b - 7)$

31. $(6x^2 - 7x + 5) + (3x^2 + 2x - 5)$

32. $(4x - 7y - z) + (2x - 5y - 3z) + (-3x - 6y - 4z)$

33. $(2a^3 - a) + (4a^2 + 7a) + (7a^3 - a - 5)$

34. $(5y - 7x + 4z) + (3z - 6y + 2x) + (13y + 7z - 6x)$

35. $(3x^2 + 4x - 5) + (-x^2 - 2x + 2) + (-2x^2 + 2x + 7)$

36. $(-x^2 + 6x - 8) + (10x^2 - 13x + 3) + (-12x^2 - 14x + 3)$

37. $(3x^2 + 7) + (6x - 7) + (2x^2 + 5x - 13) + (7x - 9)$

38. $(5x + 3y) + (-3x - 3y) + (-x - 6y) + (3x - 4y)$

39. $(5x^3 - 11x - 1) + (11x^2 + 3) + (3x + 7) + (2x^2 - 2)$

40. $(3x^4 - 5x^2 + 4) + (6x^4 - 6x^2 + 1) + (2x^4 - 7x^2)$

Find each difference and check your answer:

41. $3a - 4b$
    $2a - 5b$

42. $4x^2 - 3x - 5$
    $2x^2 - 7x$

43. $57m - 16n + 32$
    $41m + 42n - 13$

44. $6a - 3b - 4c$
    $-2a - b + 15c$

45. $-12a^2 - 13a - 5$
    $2a^2 - 2a - 2$

46. $-5a^2 - 3a$
    $2a^2 - 4a - 5$

47. $4y^2 + 21y - 7$
    $-2y^2 - 27y + 8$

48. $(5y^2 - 5y - 4) - (4y^2 - 2y + 4)$

49. $(3x^2 + 4x + 7) - (x^2 - 2x + 5)$

50. $(2x^2 + 5x - 9) - (3x^2 - 4x + 7)$

51. $(3x^2 - 5x + 4) - (6x^2 - 7x + 2)$

52. $(1 - 3x - 2x^2) - (-1 - 5x + x^2)$

53. $2x^3 + 4x - 1$
    $x^3 + x + 2$

54. $7x^4 + 3x^3 + 5x$
    $-2x^4 + x^3 - 6x + 6$

55. $12x^5 - 13x^4 + 7x^2$
    $4x^5 + 5x^4 + 2x^2 - 1$

56. $8x^3 + 6x^2 - 15x + 7$
    $14x^3 + 2x^2 + 9x - 1$

Find each difference:

57. $(3a - 4b) - (2a - 7b)$

58. $(-13x^2 - 3y^2 - 4y) - (-5x - 4y + 5y^2)$

59. $(7a - 4b) - (3x - 4y)$

60. $(-16y^3 - 42y^2 - 3y - 5) - (12y^2 - 4y + 7)$

61. $(12x^2 - 3x - 2) - (11x^2 - 7)$

62. $(14z^3 - 6y^3) - (2y^2 + 4z^3)$

63. $(20w^2 - 17w - 6) - (13w^2 + 7w)$

64. $(y^2 - 2y + 1) - (2y^2 + 3y + 5)$

65. $(2x^2 - 5x - 2) - (x^2 - x + 8)$

66. $(3 - 5z + 3z^2) - (14 + z - 2z^2)$

67. Subtract $4x^2 + 2x - 7$ from $8x^2 - 2x + 5$

68. Subtract $-6x^2 - 3x + 4$ from $2x^2 - 6x - 2$

69. Subtract $9x^2 + 6$ from $3x^2 + 2x - 4$

70. Subtract $-4x^2 - 6x + 2$ from $4x^2 + 6$

Multiplication of Monomials

To multiply two monomials, multiply their numerical coefficients and combine their variable factors according to the following rule for exponents.

Rule 1 for Exponents: Multiplying Powers

$$x^a \cdot x^b = x^{a+b}$$

That is, to multiply powers of the same base, add the exponents.

Example 1. Multiply $(2x^3)(5x^4)$.

$$(2x^3)(5x^4) = 2 \cdot 5 \cdot x^3 \cdot x^4 = 10x^{3+4} = 10x^7$$

Example 2. Multiply $(3a)(-15a^2)(4a^4b^2)$.

$$(3a)(-15a^2)(4a^4b^2) = (3)(-15)(4)(a)(a^2)(a^4)(b^2)$$
$$= -180a^7b^2$$

A special note about the meaning of $-x^2$ is needed here. Note that only x is squared. If you write this expression as a product,

$$-x^2 = -x \cdot x.$$

The expression $(-x)^2 = (-x)(-x) = x^2$.

Rule 2 for Exponents: Raising a Power to a Power

$$(x^a)^b = x^{ab}$$

That is, to raise a power to a power, multiply the exponents.

Example 3. Find $(x^3)^5$.

$$(x^3)^5 = x^3 \cdot x^3 \cdot x^3 \cdot x^3 \cdot x^3 = x^{15} \text{ by Rule 1, or}$$
$$(x^3)^5 = x^{3 \cdot 5} = x^{15} \text{ by Rule 2.}$$

Example 4. Find $(x^5)^9$.

$$(x^5)^9 = x^{45}$$

Rule 3 for Exponents: Raising a Product to a Power

$$(xy)^a = x^a y^a$$

That is, to raise a product to a power, raise each factor to this same power.

Example 5. Find $(xy)^3$.

$$(xy)^3 = x^3 y^3$$

Example 6. Find $(2x^3)^2$.

$$(2x^3)^2 = 2^2(x^3)^2 = 4x^6$$

Example 7. Find $(-3x^4)^5$.

$$(-3x^4)^5 = (-3)^5(x^4)^5 = -243x^{20}$$

Example 8. Find $(ab^2c^3)^4$.

$$(ab^2c^3)^4 = a^4(b^2)^4(c^3)^4 = a^4 b^8 c^{12}$$

Example 9. Find $(2a^2bc^3)^2$.

$$(2a^2bc^3)^2 = 2^2(a^2)^2(b)^2(c^3)^2 = 4a^4 b^2 c^6$$

Example 10. Evaluate $(2a^2)(-3ab^2)$ when $a = 2$ and $b = 3$.

$$(2a^2)(-3ab^2) = 2(-3)(a^2)(a)(b^2)$$
$$= -6a^3 b^2$$
$$= -6(2)^3(3)^2$$
$$= -6(8)(9)$$
$$= -432$$

EXERCISES 6.4   *Find each product:*

1. $(3a)(-5)$ 2. $(7x)(2x)$ 3. $(4a^2)(7a)$

4. $(4x)(6x^2)$ 5. $(-4m^2)(-7m^2)$ 6. $(5x^2)(-8x^3)$

7. $(8a^6)(4a^2)$ 8. $(-4y^4)(-9y^3)$ 9. $(13p)(-2pq)$

10. $(4ab)(10a)$ 11. $(6n)(5n^2m)$ 12. $(-9ab^2)(6a^2b^3)$

13. $(-42a)\left(-\frac{1}{2}a^3b\right)$ 14. $(28m^3)\left(\frac{1}{4}m^2\right)$ 15. $\left(\frac{2}{3}x^2y^2\right)\left(\frac{9}{16}xy^2\right)$

16. $\left(-\frac{5}{6}a^2b^6\right)\left(\frac{9}{20}a^5b^4\right)$ 17. $(8a^2bc)(3ab^3c^2)$

18. $(-4xy^2z^3)(4x^5z^3)$ 19. $\left(\frac{2}{3}x^2y\right)\left(\frac{9}{32}xy^4z^3\right)$

20. $\left(\frac{3}{5}m^4n^7\right)\left(\frac{20}{9}m^2nq^3\right)$ 21. $(32.6mnp^2)(-11.4m^2n)$

22. $(5.6a^2b^3c)(6.5a^4b^5)$ 23. $(5a)(-17a^2)(3a^3b)$

24. $(-4a^2b)(-5ab^3)(-2a^4)$

Use the rules for exponents to simplify:

25. $(x^3)^2$ 26. $(xy)^4$ 27. $(x^4)^5$

28. $(2x^2)^5$ 29. $(-3x^4)^2$ 30. $(5x^2)^3$

31. $(-x^3)^3$ 32. $(-x^2)^4$ 33. $(x^2 \cdot x^3)^2$

34. $(3x)^4$ 35. $(x^5)^6$ 36. $(-3xy)^3$

37. $(-5x^3y^2)^2$ 38. $(-x^2y^4)^5$ 39. $(15m^2)^2$

40. $(-7w^2)^3$ 41. $(25n^4)^3$ 42. $(36a^5)^3$

43. $(3x^2 \cdot x^4)^2$ 44. $(16x^4 \cdot x^5)^3$ 45. $(2x^3y^4z)^3$

46. $(-4a^2b^3c^4)^4$ 47. $(-2h^3k^6m^2)^5$ 48. $(4p^5q^7r)^3$

Evaluate each expression when $a = 2$ *and* $b = -3$:

49. $(4a)(17b)$ 50. $(3a)(-5b^2)$ 51. $(9a^2)(-2a)$

52. $(a^2)(ab)$ 53. $(41a^3)(-2b^3)$ 54. $(ab)^2$

55. $(a^2)^2$ 56. $(3a)^2$ 57. $(4b)^3$

58. $(-2ab^2)^2$ 59. $(5a^2b^3)^2$ 60. $(-7ab)^2$

61. $(5ab)(a^2b^2)$ 62. $(-3ab)^3$ 63. $(9a)(ab^2)$

64. $(-2a^2)(6ab)$ 65. $-a^2b^4$ 66. $(-ab^2)^2$

67. $(-a)^4$ 68. $-b^4$

6.5 Multiplication of Polynomials

To multiply a polynomial by a monomial, multiply each term of the polynomial by the monomial, and then add the products as shown in the following examples.

Example 1. Multiply $3a(a^2 - 2a + 1)$.

$$3a(a^2 - 2a + 1) = 3a(a^2) + 3a(-2a) + 3a(1)$$
$$= 3a^3 - 6a^2 + 3a$$

Example 2. Multiply $(-5a^3b)(3a^2 - 4ab + 5b^3)$.

$$(-5a^3b)(3a^2 - 4ab + 5b^3)$$
$$= (-5a^3b)(3a^2) + (-5a^3b)(-4ab) + (-5a^3b)(5b^3)$$
$$= -15a^5b + 20a^4b^2 - 25a^3b^4$$

To multiply a polynomial by another polynomial, multiply each term of the first polynomial by each term of the second polynomial. Then add the products. Arrange the work as shown in the example below.

Example 3. Multiply $(5x - 3)(2x + 4)$.

Step 1:
$$\begin{array}{c} 5x \;-\; 3 \\ 2x \;+\; 4 \end{array}$$
Write each polynomial in decreasing order, one under the other.

Step 2: $10x^2 - 6x$ Multiply each term of the upper polynomial by the first term in the lower one.

Step 3: $\underline{\qquad\quad 20x - 12}$ Multiply each term of the upper polynomial by the second term in the lower one. Place like terms in the same columns.

Step 4: $10x^2 + 14x - 12$ Add the like terms.

Example 4. Multiply $(x + 3)(x^2 + 2x - 4)$.

Step 1:
$$\begin{array}{r} x^2 + 2x - 4 \\ x + 3 \end{array}$$

Step 2: $x^3 + 2x^2 - 4x$

Step 3: $\underline{\qquad\qquad 3x^2 + 6x - 12}$

Step 4: $x^3 + 5x^2 + 2x - 12$

Example 5. Multiply $(a + b)(c + 2d)$.

Step 1: $a + b$
          $c + 2d$

Step 2: $ac + bc$

Step 3: $2ad + 2bd$

Step 4: $ac + bc + 2ad + 2bd$

Note that in Step 3, there are no terms which are similar to any terms from Step 2, so we place them to one side of the terms found in Step 2.

EXERCISES 6.5   *Find each product:*

1. $4(a + 6)$ 2. $3(a^2 - 5)$ 3. $-6(3x^2 + 2y)$

4. $-5(8x - 4y^2)$ 5. $a(4x^2 - 6y + 1)$ 6. $c(2a + b + 3c)$

7. $x(3x^2 - 2x + 5)$ 8. $y(3x + 2y^2 + 4y)$

9. $2a(3a^2 + 6a - 10)$ 10. $5x(8x^2 - x + 5)$

11. $-3x(4x^2 - 7x - 2)$ 12. $-6x(8x^2 + 5x - 9)$

13. $4x(-7x^2 - 3y + 2xy)$ 14. $7a(2a + 3b - 4ab)$

15. $3xy(x^2y - xy^2 + 4xy)$ 16. $-2ab(3a^2 + 4ab - 2b^2)$

17. $-6x^3(1 - 6x^2 + 9x^4)$ 18. $5x^4(2x^3 + 8x^2 - 1)$

19. $5ab^2(a^3 - b^3 - ab)$ 20. $7w^2y(w^2 - 4y^2 + 6w^2y^3)$

21. $\frac{2}{3}m(14n - 12m)$ 22. $\frac{1}{2}a^2b(8ab^2 - 2a^2b)$

23. $\frac{4}{7}yz^3\left(28y - \frac{2}{5}z\right)$ 24. $-\frac{1}{8}rs(s - t)$

25. $-4a(1.3a^5 + 2.5a^2 + 1)$ 26. $1.28m(2.3m^2 + 4.7n^2)$

27. $417a(3.2a^2 + 4a)$

28. $1.2m^2n^3(9.7m + 6.5mn - 13n^2)$

29. $4x^2y(6x^2 - 4xy + 5y^2)$

30. $x^2y^3z(x^4 - 3x^2y - 3yz + 4z^2)$

31. $\frac{2}{3}ab^3\left(\frac{3}{4}a^2 - \frac{1}{2}ab^2 + \frac{5}{6}b^3\right)$

32. $-\frac{5}{9}a^2b^4\left(\frac{3}{7}a^3b^2 - \frac{3}{5}ab - \frac{15}{16}b^4\right)$

33. $3x(x - 4) + 2x(1 - 5x) - 6x(2x - 3)$

34. $x(x - 2) - 3x(x + 8) - 2(x^2 + 3x - 5)$

35. $xy(3x + 2xy - y^2) - 2xy^2(2x - xy + 3y)$

36. $ab^2(2a - 3a^2b + b) - a^2b(1 + 2ab^2 - 4b)$

37. $(x + 1)(x + 6)$ 38. $(x + 10)(x - 3)$

39. $(x + 7)(x - 2)$ 40. $(x - 3)(x - 7)$

41. $(x - 5)(x - 8)$ 42. $(x + 9)(x + 4)$

43. $(3a - 5)(a - 4)$ 44. $(5x - 2)(3x - 4)$

45. $(6a + 4)(2a - 3)$ 46. $(3x + 5)(6x - 7)$

47. $(4a + 8)(6a + 9)$ 48. $(5x - 4)(5x - 4)$

49. $(3x - 2y)(5x + 2y)$ 50. $(4x - 6y)(6x + 9y)$

51. $(2x - 3)(2x - 3)$ 52. $(5m - 9)(5m + 9)$

53. $(2c - 5d)(2c + 5d)$ 54. $(3a + 2b)(2a - 3b)$

55. $(-7m - 3)(-13m + 1)$ 56. $(w - r)(w - s)$

57. $(x^5 - x^2)(x^3 - 1)$ 58. $(7w^4 - 6r^2)(7w^4 + 5r^2)$

59. $(2y^2 - 4y - 8)(5y - 2)$ 60. $(m^2 + 2m + 4)(m - 2)$

61. $(4x - 2y - 13)(6x + 3y)$ 62. $(4y - 3z)(2y^2 - 5yz + 6z^2)$

63. $(g + h - 6)(g - h + 3)$ 64. $(2x - 3y + 4)(4x - 5y - 2)$

65. $(8x - x^3 + 2x^4 - 1)(x^2 + 2 + 5x^3)$

66. $(y^5 - y^4 + y^3 - y^2 + y - 1)(y + 1)$

6.6 Division by a Monomial

To divide a monomial by a monomial, first write the quotient in fraction form. Then, factor both numerator and denominator into prime factors. Reduce to lowest terms by dividing both the numerator and the denominator by their common factors. The remaining factors in the numerator and the denominator give the quotient.

Example 1. Divide: $6a^2 \div 2a$

$$6a^2 \div 2a = \frac{6a^2}{2a}$$

$$= \frac{\overset{1}{\cancel{2}} \cdot 3 \cdot \overset{1}{\cancel{a}} \cdot a}{\underset{1}{\cancel{2}} \cdot \underset{1}{\cancel{a}}}$$

$$= \frac{3a}{1}$$

$$= 3a$$

Example 2. Divide: $4a \div 28a^3$

$$4a \div 28a^3 = \frac{4a}{28a^3}$$

$$= \frac{\overset{1}{\cancel{2}} \cdot \overset{1}{\cancel{2}} \cdot \overset{1}{\cancel{a}}}{\underset{1}{\cancel{2}} \cdot \underset{1}{\cancel{2}} \cdot 7 \cdot \underset{1}{\cancel{a}} \cdot a \cdot a}$$

$$= \frac{1}{7a^2}$$

Example 3. Divide: $\dfrac{6a^2bc^3}{10ab^2c}$.

$$\frac{6a^2bc^3}{10ab^2c} = \frac{\cancel{2} \cdot 3 \cdot \cancel{a} \cdot a \cdot \cancel{b} \cdot \cancel{c} \cdot c \cdot c}{\cancel{2} \cdot 5 \cdot \cancel{a} \cdot \cancel{b} \cdot b \cdot \cancel{c}}$$

$$= \frac{3ac^2}{5b}$$

Note: Since division by zero is undefined, we will assume that there are no zero denominators here and for the remainder of this chapter.

To divide a polynomial by a monomial, divide each term of the polynomial by the monomial.

Example 4. Divide: $\dfrac{15x^3 - 3x^2 + 15x}{3x}$

$$\frac{15x^3 - 3x^2 + 15x}{3x} = \frac{15x^3}{3x} - \frac{3x^2}{3x} + \frac{15x}{3x}$$

$$= \frac{\cancel{3} \cdot 5 \cdot \cancel{x} \cdot x \cdot x}{\cancel{3} \cdot \cancel{x}} - \frac{\cancel{3} \cdot \cancel{x} \cdot x}{\cancel{3} \cdot \cancel{x}} + \frac{\cancel{3} \cdot 5 \cdot \cancel{x}}{\cancel{3} \cdot \cancel{x}}$$

$$= 5x^2 - x + 5$$

Example 5. Divide: $(30d^4y - 28d^2y^2 + 12dy^2) \div (-6dy^2)$

$$(30d^4y - 28d^2y^2 + 12dy^2) \div (-6dy^2)$$

$$= \frac{30d^4y}{-6dy^2} + \frac{-28d^2y^2}{-6dy^2} + \frac{12dy^2}{-6dy^2}$$

$$= \frac{\cancel{6} \cdot 5 \cdot \cancel{d} \cdot d \cdot d \cdot d \cdot \cancel{y}}{-6 \cdot \cancel{d} \cdot \cancel{y} \cdot y} + \frac{-\cancel{2} \cdot 14 \cdot \cancel{d} \cdot d \cdot \cancel{y} \cdot \cancel{y}}{-\cancel{2} \cdot 3 \cdot \cancel{d} \cdot \cancel{y} \cdot \cancel{y}}$$

$$+ \frac{\cancel{6} \cdot 2 \cdot \cancel{d} \cdot y \cdot \cancel{y}}{-\cancel{6} \cdot \cancel{d} \cdot \cancel{y} \cdot \cancel{y}}$$

$$= \frac{5d^3}{-y} + \frac{14d}{3} + \frac{2}{-1} = -\frac{5d^3}{y} + \frac{14d}{3} - 2$$

EXERCISES 6.6   *Divide:*

1. $\dfrac{9x^5}{3x^3}$ 2. $\dfrac{15x^6}{5x^4}$ 3. $\dfrac{18x^{12}}{12x^4}$ 4. $\dfrac{20x^7}{4x^5}$ 5. $\dfrac{18x^3}{3x^5}$

6. $\dfrac{4x^2}{12x^6}$ 7. $\dfrac{8x^2}{12x}$ 8. $\dfrac{-6x^3}{2x}$ 9. $\dfrac{x^2y}{xy}$ 10. $\dfrac{xy^2}{x^2y}$

11. $(17x) \div (3x)$ 12. $(14x^2) \div (2x^3)$

13. $(15a^3b) \div (3ab^2)$ 14. $(-13a^5) \div (7a)$

15. $(16m^2n) \div (2m^3n^2)$ 16. $(-108m^4n) \div (27m^3n)$

17. $0 \div (113w^2r^3)$ 18. $(-148wr^3) \div (148wr^3)$

19. $(207p^3) \div (9p)$ 20. $(42x^2y^3) \div (-14x^2y^4)$

21. $\dfrac{92mn}{-46mn}$ 22. $\dfrac{-132rs^3}{-33r^2s^2}$ 23. $\dfrac{252}{7r^2}$ 24. $\dfrac{118a^3}{-2a^4}$

25. $\dfrac{92x^3y}{-7xy^3}$ 26. $\dfrac{45x^6}{-72x^3y^2}$

27. $(-16a^5b^2) \div (-14a^8b^4)$ 28. $(35a^3b^4c^6) \div (63a^3b^2c^8)$

29. $(-72x^3yz^4) \div (-162xy^2)$ 30. $(-144x^2z^3) \div (216x^5y^2z)$

31. $(4x^2 - 8x + 6) \div 2$ 32. $(18y^3 + 12y^2 + 6y) \div 6$

33. $(x^4 + x^3 + x^2) \div x^2$ 34. $(20r^2 - 16r - 12) \div (-4)$

35. $(ax - ay - az) \div a$ 36. $(14c^3 - 28c^2 - 2c) \div (-2c)$

37. $\dfrac{24a^4 - 16a^2 - 8a}{8}$ 38. $\dfrac{88x^5 - 110x^4 + 11x^3}{11x^3}$

39. $\dfrac{b^{12} - b^9 - b^6}{b^3}$ 40. $\dfrac{27a^3 - 18a^2 + 9a}{-9a}$

41. $(bx^4 - bx^3 + bx^2 - bx) \div (-bx)$

42. $(3.5ax^2 - 0.42a^2x) \div 0.07ax$

43. $(24x^2y^3 + 12x^3y^4 - 6xy^3) \div 2xy^3$

44. $(4a^5 - 32a^4 + 8a^3 - 4a^2) \div (-4a^2)$

45. $\dfrac{224x^4y^2z^3 - 168x^3y^3z^4 - 112xy^4z^2}{28xy^2z^2}$

46. $(55w^2 - 11w - 33) \div 11w$ 47. $(24y^5 - 18y^3 - 12y) \div 6y^2$

48. $(3a^2b + 4a^2b^2) \div 2ab^2$ 49. $(1 - 6x^2 - 4x^4) \div 2x^2$

50. $(18w^4r^4 + 27w^2r^2) \div (-9w^3r^3)$

6.7 Division by a Polynomial

Dividing a polynomial by a polynomial of more than one term is very similar to the long-division procedure in arithmetic. We use the same names, as shown below.

$$
\begin{array}{r}
25 \leftarrow \text{quotient} \\
\text{divisor} \rightarrow 25\,\overline{\smash{)}\,643} \leftarrow \text{dividend} \\
50 \\
\hline
143 \\
125 \\
\hline
18 \leftarrow \text{remainder}
\end{array}
$$

$$
\begin{array}{r}
x + 3 \leftarrow \text{quotient} \\
\text{divisor} \rightarrow x + 1\,\overline{\smash{)}\,x^2 + 4x + 4} \leftarrow \text{dividend} \\
x^2 + x \\
\hline
3x + 4 \\
3x + 3 \\
\hline
1 \leftarrow \text{remainder}
\end{array}
$$

As a check, you may use the relationship *dividend = divisor × quotient + remainder*. A similar procedure is followed in both cases. Compare the solutions of the two problems in Example 5.

Example 5. Divide:

Arithmetic

$$
\begin{array}{r}
21 \text{ r } 1 \\
16\overline{)337} \\
\underline{32} \\
17 \\
\underline{16} \\
1
\end{array}
$$

(1)
(2)
(3)
(4)
(5)
(6)

1. Divide 16 into 33. It will go at most 2 times, so write the 2 above the line over the dividend (337).

2. Multiply the divisor (16) by 2. Write the result (32) under the first 2 digits of 337.

3. Subtract 32 from 33, leaving 1. Bring down the 7, giving 17.

4. Divide 16 into 17. It will go at most 1 time, so write the 1 to the right of the 2 above the dividend.

5. Multiply the divisor (16) by 1. Write the result (16) under 17.

6. Subtract 16 from 17, leaving 1. The remainder, 1, is less than 16, so the problem is finished. The quotient is 21 with remainder 1.

Algebra

$$
\begin{array}{r}
2x - 5 \text{ r } 1 \\
x + 3\overline{)2x^2 + x - 14} \\
\underline{2x^2 + 6x} \\
-5x - 14 \\
\underline{-5x - 15} \\
1
\end{array}
$$

1. Divide x into $2x^2$, the first term in the dividend. It will go exactly $2x$ times, so write the $2x$ on the line above the dividend $(2x^2 + x - 14)$.

2. Multiply the divisor $(x + 3)$ by $2x$. Write the result $(2x^2 + 6x)$ under the first 2 terms of the dividend.

3. Subtract this result $(2x^2 + 6x)$ from the first 2 terms of the dividend, leaving $-5x$. Bring down the last term of the dividend (-14).

4. Divide x into $-5x$ (the first term in line 4). It goes exactly -5 times, so write -5 to the right of $2x$ above the dividend.

5. Multiply the divisor $(x + 3)$ by the -5. Write the result $(-5x - 15)$ on line 5.

6. Subtract line 5 from line 4, leaving 1. The remainder, 1, is of a lower degree than the divisor $(x + 3)$, so the problem is finished. The quotient is $2x - 5$ with remainder 1.

Example 6. Divide: $(x^3 - 1) \div (x - 1)$

The dividend should always be arranged in decreasing order of degree. Any missing powers of x should be filled in by using zeros as coefficients. For example, $x^3 - 1 = x^3 + 0x^2 + 0x - 1$ and we write:

$$
\begin{array}{r}
x^2 + x + 1 \\
x - 1 \overline{\smash{\big)}\ x^3 + 0x^2 + 0x - 1} \\
\underline{x^3 - x^2} \\
x^2 + 0x \\
\underline{x^2 - x} \\
x - 1 \\
\underline{x - 1} \\
0
\end{array}
$$

The remainder is 0, so the quotient is $x^2 + x + 1$.

Check:
$$
\begin{array}{l}
x^2 + x + 1 \quad\quad \text{quotient} \\
\underline{\quad\quad x - 1} \quad\quad \text{divisor} \\
x^3 + x^2 + x \\
\underline{\quad\ - x^2 - x - 1} \\
x^3 \quad\quad\quad\ - 1 \quad \text{dividend}
\end{array}
$$

When you subtract, be especially careful to change all of the signs of each expression being subtracted. Then follow the rules for addition.

EXERCISES 6.7   *Find each quotient and check:*

1. $(x^2 + 3x + 2) \div (x + 1)$

2. $(y^2 - 5y + 6) \div (y - 2)$

3. $(6a^2 - 3a + 2) \div (2a - 3)$

4. $(21y^2 + 2y - 10) \div (3y + 2)$

5. $(12x^2 - x - 9) \div (3x + 2)$

6. $(20x^2 + 57x + 30) \div (4x + 9)$

7. $\dfrac{2y^2 + 3y - 5}{2y - 1}$ 8. $\dfrac{3x^2 - 5x - 10}{3x - 8}$

9. $\dfrac{6b^2 + 13b - 28}{2b + 7}$ 10. $\dfrac{8x^2 + 13x - 27}{x + 3}$

11. $(6x^3 + 13x^2 + x - 2) \div (x + 2)$

12. $(8x^3 - 18x^2 + 7x + 3) \div (x - 1)$

13. $(8x^3 - 14x^2 - 79x + 110) \div (2x - 7)$

14. $(3x^3 - 17x^2 + 18x + 10) \div (3x + 1)$

15. $\dfrac{2x^3 - 14x - 12}{x + 1}$

16. $\dfrac{x^3 + 7x^2 - 36}{x + 3}$

17. $\dfrac{4x^3 - 24x^2 + 128}{2x + 4}$

18. $\dfrac{72x^3 + 22x + 4}{6x - 1}$

19. $\dfrac{3x^3 + 4x^2 - 6}{x - 2}$

20. $\dfrac{2x^3 + 3x^2 - 9x + 5}{x + 3}$

21. $\dfrac{4x^3 + 2x^2 + 30x + 20}{2x - 5}$

22. $\dfrac{18x^3 + 6x^2 + 4x}{6x - 2}$

23. $\dfrac{8x^4 - 10x^3 + 16x^2 + 4x - 30}{4x - 5}$

24. $\dfrac{9x^4 + 12x^3 - 6x^2 + 10x + 24}{3x + 4}$

25. $\dfrac{8x^3 - 1}{2x + 1}$

26. $\dfrac{x^3 + 1}{x + 1}$

27. $\dfrac{x^4 - 16}{x + 2}$

28. $\dfrac{16x^4 + 1}{2x - 1}$

29. $\dfrac{3x^4 + 5x^3 - 17x^2 + 11x - 2}{x^2 + 3x - 2}$

30. $\dfrac{6x^4 + 5x^3 - 11x^2 + 9x - 5}{2x^2 - x + 1}$

Chapter 6 Review

won't best Test

1. For any number a, $a \cdot 1 = ?$
2. For any number a, $a \cdot 0 = ?$
3. For any number a except 0, $a \cdot \dfrac{1}{a} = ?$

In Exercises 4—9, let $x = 3$ and $y = -2$ and evaluate each expression:

4. $x + y$

5. $x - 3y$

6. xy

7. $\dfrac{x^2}{y}$

8. y^3

9. $\dfrac{2x^3 - 3y}{xy^2}$

In Exercises 10—13, simplify by removing parentheses and combining like terms:

10. $(5y - 3) - (2 - y)$

11. $(7 - 3x) - (5x + 1)$

12. $11(2x + 1) - 4(3x - 4)$

13. $(x^3 + 2x^2y) - (3y^3 - 2x^3 + x^2y + y)$

14. Is $1 - 8x^2$ a monomial, a binomial, or a trinomial?

15. What is the degree of the polynomial $x^4 + 2x^3 - 6$?

Perform the indicated operations:

16. $(3a^2 + 7a - 2) + (5a^2 - 2a + 4)$

17. $(6x^3 + 3x^2 + 1) - (-3x^3 - x^2 - x - 1)$

18. $(3x^2 + 5x + 2) + (9x^2 - 6x - 2) - (2x^2 + 6x - 4)$

19. $(6x^2)(4x^3)$ 20. $(-7x^2y)(8x^3y^2)$

21. $(3x^2)^3$ 22. $5a(3a + 4b)$

23. $-4x^2(8 - 2x + 3x^2)$ 24. $(5x + 3)(3x - 4)$

25. $(3x^2 - 6x + 1)(2x - 4)$ 26. $(49x^2) \div (7x^3)$

27. $\dfrac{15x^3y}{3xy}$ = $5x^2y$ 28. $\dfrac{36a^3 - 27a^2 + 9a}{9a}$ $4a^2 - 3a + a$

29. $\dfrac{6x^2 + x - 12}{2x + 3}$ 30. $\dfrac{3x^3 + 2x^2 - 6x + 4}{x + 2}$

$3x +$

7

EQUATIONS AND FORMULAS

7.1 Equations

In technical work, success with the use of equations and formulas is indispensable. An *equation* is a statement that two quantities are equal. The symbol "=" is read "equals" and separates an equation into two parts, the left member and the right member. For example, in the equation

$$2x + 3 = 11$$

the left member is $2x + 3$, and the right member is 11.

To solve an equation means to find what number or numbers can replace the variable to make the equation a true statement. The variable (the unknown) is usually indicated by a letter such as x. In the equation $2x + 3 = 11$, the solution is 4. That is, when x is replaced by 4, the resulting equation is a true statement.

$$2x + 3 = 11$$

For $x = 4$, $\qquad 2(4) + 3 = 11 \qquad ?$

$$8 + 3 = 11 \qquad \text{True}$$

A replacement number (or numbers) that produces a true statement is called a *solution* or a *root* of the equation.

Note that replacing x by any other number, such as $x = 5$, results in a false statement.

$$2x + 3 = 11$$
$$2(5) + 3 = 11 \quad ?$$
$$10 + 3 = 11 \quad \text{False}$$

One method of solving equations is by trial and error—simply "guesstimate" a number. Replace the variable with this guessed number and see if the resulting statement is true. If the statement is true, the number chosen is a root.

<u>Example 1</u>. Solve $2x + 4 = 8$.

<u>*"Guesstimate"*: 3</u>	$2(3) + 4 = 8 \quad ?$ $6 + 4 = 8 \quad \text{False}$
<u>*"Guesstimate"*: 2</u>	$2(2) + 4 = 8 \quad ?$ $4 + 4 = 8 \quad \text{True}$

Therefore, 2 is a root of $2x + 4 = 8$.

<u>Example 2</u>. Solve $3x - 5 = 10$.

<u>*"Guesstimate"*: 6</u>	$3(6) - 5 = 10 \quad ?$ $18 - 5 = 10 \quad \text{False}$
<u>*"Guesstimate"*: 4</u>	$3(4) - 5 = 10 \quad ?$ $12 - 5 = 10 \quad \text{False}$
<u>*"Guesstimate"*: 5</u>	$3(5) - 5 = 10 \quad ?$ $15 - 5 = 10 \quad \text{True}$

Therefore, 5 is a root of $3x - 5 = 10$.

A more systematic method of solving equations involves changing the given equation to an *equivalent equation*. This is done by performing the same arithmetic operation on both sides of the equation. The basic arithmetic operations used are addition, subtraction, multiplication, and division.

Equivalent equations are equations that have the same roots. For example, $3x = 6$ and $x = 2$ are equivalent equations, since 2 is the root of each. In solving an equation by this method, you continue to change the given equation to an equivalent equation until you find an equation whose root is obvious.

Four Basic Rules Used to Solve Equations

1. If the same quantity is added to both sides of an equation, the resulting equation is equivalent to the original equation.

Example 3. Solve $x - 2 = 8$.

$$x - 2 + 2 = 8 + 2$$
$$x = 10$$

2. If the same quantity is subtracted from both sides of an equation, the resulting equation is equivalent to the original equation.

Example 4. Solve $x + 5 = 2$.

$$x + 5 - 5 = 2 - 5$$
$$x = -3$$

3. If both sides of an equation are multiplied by the same (nonzero) quantity, the resulting equation is equivalent to the original equation.

Example 5. Solve $\frac{x}{9} = 4$.

$$9\left(\frac{x}{9}\right) = (4)9$$

$$x = 36$$

4. If both sides of an equation are divided by the same (nonzero) quantity, the resulting equation is equivalent to the original equation.

Example 6. Solve $4x = 20$.

$$\frac{4x}{4} = \frac{20}{4}$$

$$x = 5$$

Basically, to solve an equation, use one of the rules and use a number that will *undo* what has been done to the variable.

Example 7. Solve $x + 3 = 8$.

Since 3 has been added to the variable, use Rule 2 and subtract 3 from both sides of the equation.

$$x + 3 = 8$$
$$x + 3 - 3 = 8 - 3$$
$$x = 5$$

A check is necessary, since an error could have been made in doing the arithmetic in the equation. To check, replace the variable in the original equation by 5, the apparent root, to make sure the resulting statement is true.

Check: $x + 3 = 8$
              $5 + 3 = 8$ True

Thus, the root is 5.

Example 8. Solve $x - 4 = 7$.

Since 4 has been subtracted from the variable, use Rule 1 and add 4 to both sides of the equation.

$$x - 4 = 7$$
$$x - 4 + 4 = 7 + 4$$
$$x = 11$$

The apparent root is 11.

Check: $x - 4 = 7$
              $11 - 4 = 7$ True

Thus, the root is 11.

Example 9. Solve $2x = 9$.

Since the variable has been multiplied by 2, use Rule 4 and divide both sides of the equation by 2.

$$2x = 9$$
$$\frac{2x}{2} = \frac{9}{2}$$
$$x = \frac{9}{2}$$

Note: Each solution should be checked by substituting it into the original equation. When a check is not provided in the text, the check is left to the student.

<u>Example 10</u>. Solve $\frac{x}{3} = 9$.

Since the variable has been divided by 3, use Rule 3 and multiply both sides of the equation by 3.

$$\frac{x}{3} = 9$$

$$3\left(\frac{x}{3}\right) = (9)3$$

$$x = 27$$

<u>Example 11</u>. Solve $-4 = -6x$.

Since the variable has been multiplied by -6, use Rule 4 and divide both sides of the equation by -6.

$$-4 = -6x$$

$$\frac{-4}{-6} = \frac{-6x}{-6}$$

$$\frac{2}{3} = x$$

The apparent root is $\frac{2}{3}$.

Check:

$$-4 = -6x$$

$$-4 = -6\left(\frac{2}{3}\right) \quad ?$$

$$-4 = -4 \qquad \text{True}$$

Thus, the root is $\frac{2}{3}$.

    Some equations have more than one operation indicated on the variable. For example, the equation $2x + 5 = 6$ has both addition of 5 and multiplication by 2 indicated. Use the following rule to solve equations like these.

When more than one operation is indicated on the variable, usually undo the additions and subtractions first, then undo the multiplications and divisions.

Example 12. Solve $2x + 5 = 6$.

$$2x + 5 - 5 = 6 - 5 \qquad \text{Subtract 5 from both sides.}$$

$$2x = 1$$

$$\frac{2x}{2} = \frac{1}{2} \qquad \text{Divide both sides by 2.}$$

The apparent root is $\frac{1}{2}$.

Check: $\qquad 2x + 5 = 6$

$$2\left(\frac{1}{2}\right) + 5 = 6 \qquad ?$$

$$1 + 5 = 6 \qquad \text{True}$$

Thus, the root is $\frac{1}{2}$.

Example 13. Solve $\frac{x}{3} - 6 = 9$.

$$\frac{x}{3} - 6 + 6 = 9 + 6 \qquad \text{Add 6 to both sides.}$$

$$\frac{x}{3} = 15$$

$$3\left(\frac{x}{3}\right) = (15)3 \qquad \text{Multiply both sides by 3.}$$

$$x = 45$$

Example 14. Solve $118 - 22m = 30$.

Think of $118 - 22m$ as $118 + (-22m)$.

$$118 - 22m - 118 = 30 - 118 \qquad \text{Subtract 118 from both sides.}$$

$$-22m = -88$$

$$\frac{-22m}{-22} = \frac{-88}{-22} \qquad \text{Divide both sides by } -22.$$

$$m = 4$$

The apparent root is 4.

Check: $\quad 118 - 22m = 30$

$$118 - 22(4) = 30 \qquad ?$$

$$118 - 88 = 30 \qquad ?$$

$$30 = 30 \qquad \text{True}$$

Thus, the root is 4.

Here is another approach to solving this equation.

$$118 - 22m = 30$$

$$118 - 22m + 22m = 30 + 22m \qquad \text{Add } 22m \text{ to both sides.}$$

$$118 = 30 + 22m$$

$$118 - 30 = 30 + 22m - 30 \qquad \text{Subtract 30 from both sides.}$$

$$88 = 22m$$

$$\frac{88}{22} = \frac{22m}{22} \qquad \text{Divide both sides by 22.}$$

$$4 = m$$

EXERCISES 7.1   *Solve each equation and check:*   2-60

1. $x + 2 = 8$ 2. $3a = 7$ 3. $y - 5 = 12$

4. $\frac{2}{3}n = 6$ 5. $w - 7\frac{1}{2} = 3$ 6. $2m = 28.4$

7. $\frac{x}{13} = 1.5$ 8. $n + 12 = -5$ 9. $3b = 15.6$

10. $y - 17 = 25$ 11. $17x = 5117$ 12. $28 + m = 3$

13. $2 = x - 5$ 14. $-29 = -4y$ 15. $17 = -3 + w$

16. $49 = 32 + w$ 17. $14b = 57$ 18. $y + 28 = 13$

19. $5m = 0$ 20. $28 + m = 28$ 21. $x + 5 = 5$

22. $y + 7 = -7$ 23. $4x = 64$ 24. $\frac{5x}{5} - \frac{125}{5} = \frac{0}{5}$

25. $\frac{x}{7} = 56$ 26. $\frac{y}{5} = 35$ $x=25$ 27. $-48 = 12y$

28. $13x = -78$ 29. $-x = 2$ 30. $-y = 7$

31. $5y + 3 = 13$ 32. $4x - 2 = 18$ 33. $10 - 3x = 16$

34. $8 - 2y = 4$ 35. $\frac{x}{4} - 5 = 3$ 36. $\frac{x}{5} + 4 = 9$

37. $2 - x = 6$ 38. $8 - y = 3$ 39. $\frac{2}{3}y - 4 = 8$

40. $5 - \frac{1}{4}x = 7$ 41. $3x - 5 = 12$ 42. $5y + 7 = 28$

43. $\frac{m}{3} - 6 = 8$ 44. $\frac{w}{5} + 7 = 13$ 45. $\frac{2x}{3} = 7$

46. $\frac{4b}{5} = 15$ 47. $-3y - 7 = -6$ 48. $28 = -7 - 3r$

49. $5 - x = 6$ 50. $17 - 5w = -68$ 51. $54y - 13 = 17.8$

52. $37a - 7 = 67$

53. $28w - 56 = -8$

54. $52 - 4x = -8$

55. $29r - 13 = 57$

56. $15x - 32 = 18$

57. $31 - 3y = 41$

58. $62 = 13y - 3$

59. $-83 = 17 - 4x$

60. $58 = 5m + 52$

7.2 Applications Involving Simple Equations*

A technical problem can often be expressed mathematically as a simple equation. The problem can be solved, then, by solving the equation. To solve such an application problem, we suggest the following:

Solving Application Problems

Step 1: Read the problem carefully at least twice.

Step 2: If possible, draw a diagram. This will often help you to visualize the mathematical relationship needed to write the equation.

Step 3: Choose a symbol to represent the unknown quantity in the problem and write what it represents.

Step 4: Write an equation which expresses the information given in the problem and which involves the unknown.

Step 5: Solve the equation from Step 4.

Step 6: Check your solution both in the equation from Step 4 and in the original problem itself.

Example 1. A person decides to plant 15 poplar trees across the back of his 175-ft lot. How far apart should he plant them if he wants them evenly spaced? (He wants one tree planted in each corner.)

*Do not use the rules for calculating with measurements in this chapter.

Note that if there are 15 trees, then there are 14 spaces between the trees (Fig. 7.1).

Let x = the distance between each tree, and $14x$ = the distance of 14 spaces.

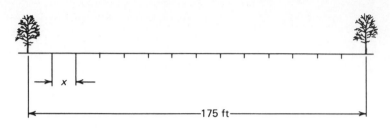

Figure 7.1

$$14x = 175$$

$$\frac{14x}{14} = \frac{175}{14} \qquad \text{Divide both sides by 14.}$$

$$x = 12.5 \text{ ft}$$

Check:
$$14x = 175$$
$$14(12.5) = 175 \qquad ?$$
$$175 = 175 \qquad \text{True}$$

Example 2. An interior wall measures 30 ft 4 in. long. It is to be divided by 10 evenly spaced posts; each post is 4 in. by 4 in. (Posts are to be located in the corners.) What is the distance between each post? Note that there are 9 spaces between posts (Fig. 7.2).

Let x = the distance between each post and $9x$ = the distance of 9 spaces. Then $(10)(4 \text{ in.})$ = the distance used up by ten 4-in. posts.

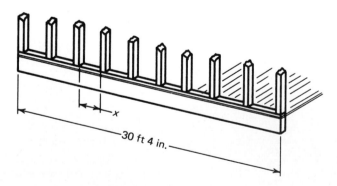

Figure 7.2

$$9x + (10)(4 \text{ in.}) = 364 \text{ in.}$$

30 ft 4 in.
= 364 in.

$$9x + 40 = 364$$

$$9x + 40 - 40 = 364 - 40$$

Subtract 40 from both sides.

$$9x = 324$$

$$\frac{9x}{9} = \frac{324}{9}$$

Divide both sides by 9.

$$x = 36 \text{ in.}$$

Check:
$$9x + 40 = 364$$
$$9(36) + 40 = 364 \quad ?$$
$$324 + 40 = 364 \quad ?$$
$$364 = 364 \quad \text{True}$$

Checking in the original problem:

$$9 \text{ spaces of } 36 \text{ in.} = 324 \text{ in.}$$
$$\text{ten 4-in. posts} = \underline{\ 40 \text{ in.}}$$
$$\text{total} = 364 \text{ in.} = 30 \text{ ft } 4 \text{ in.}$$

Example 3. A farmer orders 40,000 lb of feed for 120 steers. After 18 days, he has 7600 lb of feed left. How much feed has each steer consumed per day (on the average)?

Let x = number of pounds of feed consumed per steer per day

$18x$ = number of pounds of feed consumed per steer during the 18-day period.

$$(120)(18x) + 7600 = 40,000$$

$$2160x + 7600 = 40,000$$

$$2160x + 7600 - 7600 = 40,000 - 7600$$

Subtract 7600 from both sides.

$$2160x = 32,400$$

$$\frac{2160x}{2160} = \frac{32,400}{2160}$$

Divide both sides by 2160.

$$x = 15 \text{ lb}$$

Check:

$$(120)(18x) + 7600 = 40,000$$
$$(120)(18)(15) + 7600 = 40,000 \quad ?$$
$$32,400 + 7600 = 40,000 \quad ?$$
$$40,000 = 40,000 \quad \text{True}$$

EXERCISES 7.2

1. An apartment building owner finds that his monthly profit is $490 on seven apartments. His total expenses are $1400 per month. How much does each apartment rent for? (Assume each apartment rents for the same amount.)

2. A furniture store has a "six-months-same-as-cash" policy. This states that the customer may spread his payments over six months without paying an interest charge. You buy $2950 worth of furniture. What will each of your six monthly payments be after putting $700 down?

3. A camping-supplies store sold nine cookstoves for a total of $405 in June. The cost to the owner was $25 per stove. How much did he make on each stove?

4. An appliance dealer buys 22 washing machines. He finds his total bill comes to $8300. How much did he pay per washer if the bill included $1150 in transportation charges?

5. A plumber installs a pump to empty a basement flooded to a depth of 29 in. He checks six hours later and finds the water level down to 11 in. At how many inches per hour is the pump emptying the basement?

6. A bricklayer starts out with a stack of 300 bricks. He lays six equal rows of bricks and then finds that he has 48 bricks left. How many bricks are in each row?

7. A set of eight built-in bookshelves is to be constructed in a room. The floor-to-ceiling clearance is 8 ft 2 in. Each shelf is 1 in. thick. An equal space is to be left between shelves. What space should there be between each shelf and the next? (There is no shelf against the ceiling and no shelf on the floor.)

8. A piece of lumber 5 ft 4 in. long is to be cut into nine equal pieces. The loss per cut is one-eighth inch. How long should each piece be? (Eight cuts will be needed.)

9. A farmer has a total crop of 81,300 bu of corn. Of these, 19,550 bu came from a 170-acre farm. The rest came from a 475-acre farm. What was the yield per acre on the second farm?

10. Corn is worth $3.15 bu. Hogs are worth 56¢/lb live weight. How many bushels of corn are needed to equal the value of 450 lb of live-weight hogs?

11. A farmer ordered 7500 lb of feed. Three months later he finds that he has 2250 lb of feed left. How much feed did he use per month on the average?

12. A farmer has a total crop of 109,100 bu of corn. To cushion market changes, he decides to sell all but 35,000 bu of his crop steadily, over a six-month period. How many bushels should he sell per month to do this?

13. An electrician contracts to do a certain job for $82. His materials cost $23.50. The job takes 3 h. What is his hourly rate of pay?

14. The total current through a parallel circuit with five branches is 1.5 A. Two of the branches have a combined current of 0.45 A. The currents through each of the other three branches are known to be equal. What are they? <u>Note</u>. The sum of the currents through all of the branches must equal the total current.

 15. Two resistors are connected in series to a 30-V power source. The voltage drop across the first resistor measures 18 V. What is the current in amperes through the other resistor, which has a resistance of 60 Ω? (The voltage drop across a resistor in volts equals the product of the resistance in ohms and the current through it in amperes.)

 16. Ten resistors are connected in series to a 117-V power source. One resistor has a voltage drop of 36 V across it. All the other resistors are of equal resistance, and thus have the same voltage drop across each one. What is the voltage drop across each of the nine other resistors?

 17. In testing the performance of a fuel pump, a mechanic starts with 300 mL of fuel. After 450 strokes, he finds that 120 mL of fuel remain. How much fuel is being delivered with each stroke of the pump?

 18. The charge on a battery is 44 ampere-hours (Ah). After being connected to a charger for 8 h, the charge is found to be 140 Ah. At what average hourly rate was the battery being charged?

 19. A mechanic in charge of maintaining a fleet of trucks begins the year with a stock of 1200 spark plugs. Nine weeks later an inventory reveals 606 spark plugs still on hand. At what rate per week is the fleet using spark plugs?

 20. A mechanic needs to insert shims to fill a space of total thickness 0.070 in. There already is an 0.018-in. shim in place. How many 0.004-in. shims must be inserted?

 21. A dispensary has a stock of 600 disposable syringes. At the end of a two-week (14 days) period, an inventory shows that 236 syringes remain. What is the average daily rate of use of the syringes?

 22. A 1000-cm^3 I.V. (intravenous) solution has been administered for 4 h, at a rate of 130 cm^3/h. The remaining solution is to be given over a 6-h period. At what rate should the remaining solution be administered?

 23. On July 1 a hospital has 300 vials of a certain medication. A week later, an inventory reveals 181 vials still on hand. On the average, how many vials per day did the hospital use?

 24. A nurse is monitoring the administration of 1000 cm^3 of an I.V. solution. After 4 h, he finds that 320 cm^3 of solution remain. At what hourly rate is the solution being administered?

7.3 Equations with Variables in Both Members

To solve equations with variables in both members (both sides), such as $3x + 4 = 5x - 12$, do the following:

1. Add or subtract either variable term from both sides of the equation.

$$3x + 4 = 5x - 12$$
$$3x + 4 - 3x = 5x - 12 - 3x$$
$$4 = 2x - 12$$

2. Take the constant term (which now appears on the same side of the equation with the variable term) and add it to, or subtract it from, both sides. Then solve the resulting equation.

$$4 + 12 = 2x - 12 + 12$$

$$16 = 2x$$

$$\frac{16}{2} = \frac{2x}{2} \qquad \text{Divide both sides by 2.}$$

$$8 = x$$

This equation could also have been solved as follows:

$$3x + 4 = 5x - 12$$

$$3x + 4 - 5x = 5x - 12 - 5x \qquad \text{Subtract } 5x \text{ from both sides.}$$

$$-2x + 4 = -12$$

$$-2x + 4 - 4 = -12 - 4 \qquad \text{Subtract 4 from both sides.}$$

$$-2x = -16$$

$$\frac{-2x}{-2} = \frac{-16}{-2} \qquad \text{Divide both sides by } -2.$$

$$x = 8$$

<u>Example 1.</u> Solve $5x - 4 = 8x - 13$.

$$5x - 4 - 8x = 8x - 13 - 8x \qquad \text{Subtract } 8x \text{ from both sides.}$$

$$-3x - 4 = -13$$

$$-3x - 4 + 4 = -13 + 4 \qquad \text{Add 4 to both sides.}$$

$$-3x = -9$$

$$\frac{-3x}{-3} = \frac{-9}{-3} \qquad \text{Divide both sides by } -3.$$

$$x = 3$$

Check:
$$5x - 4 = 8x - 13$$
$$5(3) - 4 = 8(3) - 13 \quad ?$$
$$15 - 4 = 24 - 13 \quad ?$$
$$11 = 11 \qquad \text{True}$$

Example 2. Solve $-2x + 5 = 6x - 11$.

$$-2x + 5 + 2x = 6x - 11 + 2x \qquad \text{Add } 2x \text{ to both sides.}$$

$$5 = 8x - 11$$

$$5 + 11 = 8x - 11 + 11 \qquad \text{Add 11 to both sides.}$$

$$16 = 8x$$

$$\frac{16}{8} = \frac{8x}{8} \qquad \text{Divide both sides by 8.}$$

$$2 = x$$

Check:
$$-2x + 5 = 6x - 11$$
$$-2(2) + 5 = 6(2) - 11 \quad ?$$
$$-4 + 5 = 12 - 11 \quad ?$$
$$1 = 1 \qquad \text{True}$$

Example 3. Solve $5x + 7 = 2x - 14$.

$$5x + 7 - 2x = 2x - 14 - 2x \qquad \text{Subtract } 2x \text{ from both sides.}$$

$$3x + 7 = -14$$

$$3x + 7 - 7 = -14 - 7 \qquad \text{Subtract 7 from both sides.}$$

$$3x = -21$$

$$\frac{3x}{3} = \frac{-21}{3} \qquad \text{Divide both sides by 3.}$$

$$x = -7$$

Check:
$$5x + 7 = 2x - 14$$
$$5(-7) + 7 = 2(-7) - 14 \quad ?$$
$$-35 + 7 = -14 - 14 \quad ?$$
$$-28 = -28 \qquad \text{True}$$

Example 4. Solve $4 - 5x = 28 + x$.

$$4 - 5x + 5x = 28 + x + 5x \qquad \text{Add } 5x \text{ to both sides.}$$

$$4 = 28 + 6x$$

$$4 - 28 = 28 + 6x - 28 \qquad \text{Subtract 28 from both sides.}$$

$$-24 = 6x$$

$$\frac{-24}{6} = \frac{6x}{6} \qquad \text{Divide both sides by 6.}$$

$$-4 = x$$

EXERCISES 7.3   *Solve each equation and check:*

1. $4y + 9 = 7y - 15$

2. $2y - 45 = -y$

3. $5x + 3 = 7x - 5$

4. $2x - 3 = 3x - 13$

5. $-2x + 7 = 5x - 21$

6. $5x + 3 = 2x - 15$

7. $3y + 5 = 5y - 1$

8. $3x - 4 = 7x - 32$

9. $-3x + 17 = 6x - 37$

10. $3x + 13 = 2x - 12$

11. $7x + 9 = 9x - 3$

12. $-5y + 12 = 12y - 5$

13. $3x - 2 = 5x + 8$

14. $13y + 2 = 20y - 5$

15. $-4x + 25 = 6x - 45$

16. $5x - 7 = 6x - 5$

17. $5x + 4 = 10x - 7$

18. $3x - 2 = 5x - 20$

19. $27 + 5x = 9 + 3x$

20. $2y + 8 = 5y - 1$

21. $-7x + 18 = 11x - 36$

22. $4x + 5 = 2x - 7$

23. $4y + 11 = 7y - 28$

24. $4x = 2x - 12$

25. $-4x + 2 = 8x - 7$

26. $6x - 1 = 9x - 9$

27. $13x + 6 = 6x - 1$

28. $6y + 7 = 18y - 1$

29. $3x + 1 = 17 - x$

30. $17 - 4y = 14 - y$

7.4 Equations with Parentheses

To solve an equation having parentheses in one or both members, always remove parentheses first. Then combine like terms, before using the previous methods to solve the resulting equation.

<u>Example 1</u>. Solve $5 - (2x - 3) = 7$.

$5 - 2x + 3 = 7$	Remove parentheses.
$8 - 2x = 7$	Combine like terms.
$8 - 2x - 8 = 7 - 8$	Subtract 8 from both sides.
$-2x = -1$	
$\dfrac{-2x}{-2} = \dfrac{-1}{-2}$	Divide both sides by -2.
$x = \dfrac{1}{2}$	

<u>Check</u>: $5 - (2x - 3) = 7$

$5 - \left[2\left(\dfrac{1}{2}\right) - 3\right] = 7$?

$5 - (1 - 3) = 7$?

$5 - (-2) = 7$ True

Therefore, $\dfrac{1}{2}$ is the root.

Example 2. Solve $7x - 6(5 - x) = 9$.

$$7x - 30 + 6x = 9 \qquad \text{Remove parentheses.}$$

$$13x - 30 = 9 \qquad \text{Combine like terms.}$$

$$13x - 30 + 30 = 9 + 30 \qquad \text{Add 30 to both sides.}$$

$$13x = 39$$

$$\frac{13x}{13} = \frac{39}{13} \qquad \text{Divide both sides by 13.}$$

$$x = 3$$

In the following examples we have parentheses as well as the variable in both members.

Example 3. Solve $3(x - 5) = 2(4 - x)$.

$$3x - 15 = 8 - 2x \qquad \text{Remove parentheses.}$$

$$3x - 15 + 2x = 8 - 2x + 2x \qquad \text{Add } 2x \text{ to both sides.}$$

$$5x - 15 = 8 \qquad \text{Combine like terms.}$$

$$5x - 15 + 15 = 8 + 15 \qquad \text{Add 15 to both sides.}$$

$$5x = 23$$

$$\frac{5x}{5} = \frac{23}{5} \qquad \text{Divide both sides by 5.}$$

$$x = \frac{23}{5}$$

Check: $\qquad 3(x - 5) = 2(4 - x)$

$$3\left(\frac{23}{5} - 5\right) = 2\left(4 - \frac{23}{5}\right) \qquad ?$$

$$3\left(-\frac{2}{5}\right) = 2\left(-\frac{3}{5}\right) \qquad ?$$

$$-\frac{6}{5} = -\frac{6}{5} \qquad \text{True}$$

Therefore, $\frac{23}{5}$ is the root.

Example 4. Solve $8x - 4(x + 2) = 12(x + 1) - 14$.

$$8x - 4x - 8 = 12x + 12 - 14 \qquad \text{Remove parentheses.}$$

$$4x - 8 = 12x - 2 \qquad \text{Combine like terms.}$$

$$4x - 8 - 12x = 12x - 2 - 12x \qquad \text{Subtract } 12x \text{ from both sides.}$$

$$-8x - 8 = -2$$

$$-8x - 8 + 8 = -2 + 8 \qquad \text{Add 8 to both sides.}$$

$$-8x = 6$$

$$\frac{-8x}{-8} = \frac{6}{-8} \qquad \text{Divide both sides by } -8.$$

$$x = -\frac{3}{4}$$

Check:

$$8x - 4(x + 2) = 12(x + 1) - 14$$

$$8\left(-\frac{3}{4}\right) - 4\left(-\frac{3}{4} + 2\right) = 12\left(-\frac{3}{4} + 1\right) - 14 \qquad ?$$

$$-6 - 4\left(\frac{5}{4}\right) = 12\left(\frac{1}{4}\right) - 14 \qquad ?$$

$$-6 - 5 = 3 - 14 \qquad ?$$

$$-11 = -11 \qquad \text{True}$$

Therefore, $-\frac{3}{4}$ is the root.

EXERCISES 7.4   *Solve each equation and check:*

1. $2(x + 3) - 6 = 10$

2. $-3x + 5(x - 6) = 32$

3. $3n + (2n + 4) = 6$

4. $5m - (2m - 7) = -5$

5. $16 = -3(x - 4)$

6. $5y + 6(y - 3) = 15$

7. $5a - (3a + 4) = 8$

8. $2(b + 4) - 3 = 15$

9. $5a - 4(a - 3) = 7$

10. $29 = 4 + (2m + 1)$

11. $5(x - 3) = 21$

12. $27 - 8(2 - y) = -13$

13. $2a - (5a - 7) = 22$

14. $2(5m - 6) - 13 = -1$

15. $2(w - 3) + 6 = 0$

16. $6r - (2r - 3) + 5 = 0$

17. $3x - 7 + 17(1 - x) = -6$

18. $4y - 6(2 - y) = 8$

19. $6b = 27 + 3b$

20. $2a + 4 = a - 3$

21. $4(25 - x) = 3x + 2$

22. $4x - 2 = 3(25 - x)$

23. $x + 3 = 4(57 - x)$

24. $2(y + 1) = y - 7$

25. $6x + 2 = 2(17 - x)$

26. $6(17 - x) = 2 - 4x$

27. $5(x - 8) - 3x - 4 = 0$

28. $5(28 - 2x) - 7x = 4$

29. $3(x + 4) + 3x = 6$

30. $8x - 4(x + 2) + 11 = 0$

31. $y - 4 = 2(y - 7)$

32. $7(w - 4) = w + 2$

33. $9m - 3(m - 5) = 7m - 3$

34. $4(x + 18) = 2(4x + 18)$

35. $3(2x + 7) = 13 + 2(4x + 2)$

36. $5y - 3(y - 2) = 6(y + 1)$

37. $8(x - 5) = 13x$

38. $4(x + 2) = 30 - (x - 3)$

39. $5 + 3(x + 7) = 26 - 6(5x + 11)$

40. $2(y - 3) = 4 + (y - 14)$

41. $5(2y - 3) = 3(7y - 6) + 19(y + 1) + 14$

42. $3(7y - 6) - 11(y + 1) = 58 - 7(9y + 4)$

43. $16(x + 3) = 7(x - 5) - 9(x + 4) - 7$

44. $31 - 2(x - 5) = -3(x + 4)$

45. $4(y + 2) = 8(y - 4) + 7$

46. $12x - 13(x + 4) = 4x - 6$

47. $4(5y - 2) + 3(2y + 6) = 25(3y + 2) - 19y$

48. $6(3x + 1) = 5x - (2x + 2)$

49. $12 + 8(2y + 3) = (y + 7) - 16$

50. $-2x + 6(2 - x) - 4 = 3(x + 1) - 6$

51. $5x - 10(3x - 6) = 3(24 - 9x)$

52. $4y + 7 - 3(2y + 3) = 4(3y - 4) -7y + 7$

53. $6(y - 4) - 4(5y + 1) = 3(y - 2) - 4(2y + 1)$

54. $2(5y + 1) + 16 = 4 + 3(y - 7)$

55. $-6(x - 5) + 3x = 6x - 10(-3 + x)$

56. $14x + 14(3 - 2x) + 7 = 4 - x + 5(2 - 3x)$

57. $2.3x - 4.7 + 0.6(3x + 5) = 0.7(3 - x)$

58. $5.2(x + 3) + 3.7(2 - x) = 3$

59. $0.089x - 0.32 + 0.001(5 - x) = 0.231$

60. $5x - 2.5(7 - 4x) = x - 7(4 + x)$

7.5 Equations with Fractions

To Solve an Equation with Fractions

1. Find the least common denominator (LCD) of all the fractional terms on both sides of the equation.
2. Multiply both sides of the equation by the LCD. (If this step has been done correctly, no fractions should now appear in the resulting equation.)
3. Solve the resulting equation from Step 2 using earlier methods in this chapter.

__Example 1.__ Solve $\dfrac{3x}{4} = \dfrac{45}{20}$.

The LCD of 4 and 20 is 20; therefore, multiply both sides of the equation by 20.

$$\frac{3x}{4} = \frac{45}{20}$$

$$20\left(\frac{3x}{4}\right) = \left(\frac{45}{20}\right)20$$

$$15x = 45$$

$$\frac{15x}{15} = \frac{45}{15} \qquad \text{Divide both sides by 15.}$$

$$x = 3$$

__Example 2.__ Solve $\dfrac{3}{4} + \dfrac{x}{6} = \dfrac{13}{12}$.

The LCD of 4, 6, and 12 is 12; therefore, multiply both sides of the equation by 12.

$$\frac{3}{4} + \frac{x}{6} = \frac{13}{12}$$

$$12\left(\frac{3}{4} + \frac{x}{6}\right) = \left(\frac{13}{12}\right)12$$

$$12\left(\frac{3}{4}\right) + 12\left(\frac{x}{6}\right) = \left(\frac{13}{12}\right)12$$

$$9 + 2x = 13$$

$$9 + 2x - 9 = 13 - 9 \qquad \text{Subtract 9 from both sides.}$$

$$2x = 4$$

$$\frac{2x}{2} = \frac{4}{2} \qquad \text{Divide both sides by 2.}$$

$$x = 2$$

Check:
$$\frac{3}{4} + \frac{x}{6} = \frac{13}{12}$$

$$\frac{3}{4} + \frac{2}{6} = \frac{13}{12} \quad ?$$

$$\frac{9}{12} + \frac{4}{12} = \frac{13}{12} \quad \text{True}$$

Example 3. Solve $\frac{2x}{9} - 4 = \frac{x}{6}$.

The LCD of 9 and 6 is 18; therefore, multiply both sides of the equation by 18.

$$\frac{2x}{9} - 4 = \frac{x}{6}$$

$$18\left(\frac{2x}{9} - 4\right) = \left(\frac{x}{6}\right)18$$

$$18\left(\frac{2x}{9}\right) - 18(4) = \left(\frac{x}{6}\right)18$$

$$4x - 72 = 3x$$

$$4x - 72 - 4x = 3x - 4x \qquad \text{Subtract } 4x \text{ from}$$
$$-72 = -x \qquad \qquad \text{both sides.}$$

$$\frac{-72}{-1} = \frac{-x}{-1} \qquad \text{Divide both sides}$$
$$\qquad \qquad \qquad \text{by } -1.$$

$$72 = x$$

Check:
$$\frac{2x}{9} - 4 = \frac{x}{6}$$

$$\frac{2(72)}{9} - 4 = \frac{72}{6} \quad ?$$

$$16 - 4 = 12 \quad \text{True}$$

Example 4. Solve $\frac{2}{3}x + \frac{3}{4}(36 - 2x) = 32$.

The LCD of 3 and 4 is 12; therefore, multiply both sides of the equation by 12.

$$\frac{2}{3}x + \frac{3}{4}(36 - 2x) = 32$$

$$12\left[\frac{2}{3}x + \frac{3}{4}(36 - 2x)\right] = (32)12$$

$$12\left(\frac{2}{3}x\right) + 12\left(\frac{3}{4}\right)(36 - 2x) = (32)12 \qquad \text{Remove brackets.}$$

$$8x + 9(36 - 2x) = 384$$

$$8x + 324 - 18x = 384 \qquad \text{Remove parentheses.}$$

$$-10x + 324 = 384 \qquad \text{Combine like terms.}$$

$$-10x + 324 - 324 = 384 - 324 \qquad \text{Subtract 324}$$
$$-10x = 60 \qquad \qquad \text{from both sides.}$$

$$\frac{-10x}{-10} = \frac{60}{-10} \qquad \text{Divide both sides}$$
$$\qquad \qquad \qquad \text{by } -10.$$

$$x = -6$$

Check: $\frac{2}{3}x + \frac{3}{4}(36 - 2x) = 32$

$\frac{2}{3}(-6) + \frac{3}{4}[36 - 2(-6)] = 32$?

$-4 + \frac{3}{4}(36 + 12) = 32$?

$-4 + \frac{3}{4}(48) = 32$?

$-4 + 36 = 32$ True

Example 5. Solve $\frac{2x + 1}{3} - \frac{x - 6}{4} = \frac{2x + 4}{8} + 2.$

The LCD of 3, 4, and 8 is 24; therefore, multiply both sides of the equation by 24.

$$\frac{2x + 1}{3} - \frac{x - 6}{4} = \frac{2x + 4}{8} + 2$$

$$24\left(\frac{2x + 1}{3} - \frac{x - 6}{4}\right) = \left(\frac{2x + 4}{8} + 2\right)24$$

$$24\left(\frac{2x + 1}{3}\right) - 24\left(\frac{x - 6}{4}\right) = \left(\frac{2x + 4}{8}\right)24 + (2)24$$

$$8(2x + 1) - 6(x - 6) = 3(2x + 4) + 48$$

$16x + 8 - 6x + 36 = 6x + 12 + 48$ Remove parentheses.

$10x + 44 = 6x + 60$ Combine like terms.

$10x + 44 - 6x = 6x + 60 - 6x$ Subtract $6x$ from both sides.

$4x + 44 = 60$

$4x + 44 - 44 = 60 - 44$ Subtract 44 from both sides.

$4x = 16$

$\frac{4x}{4} = \frac{16}{4}$ Divide both sides by 4.

$x = 4$

Check: $\frac{2x + 1}{3} - \frac{x - 6}{4} = \frac{2x + 4}{8} + 2$

$\frac{2(4) + 1}{3} - \frac{4 - 6}{4} = \frac{2(4) + 4}{8} + 2$?

$\frac{9}{3} - \frac{-2}{4} = \frac{12}{8} + 2$?

$3 + \frac{1}{2} = \frac{3}{2} + 2$?

$\frac{7}{2} = \frac{7}{2}$ True

When the variable appears in the denominator of a fraction in an equation, multiply both members by the LCD. Be careful that the replacement for the variable does not make the denominator zero.

Example 6. Solve: $\dfrac{3}{x} = 2$

$$x\left(\dfrac{3}{x}\right) = (2)x \qquad \text{Multiply both sides by the LCD, } x.$$

$$3 = 2x$$

$$\dfrac{3}{2} = \dfrac{2x}{2} \qquad \text{Divide both sides by 2.}$$

$$\dfrac{3}{2} = x$$

Check: $\dfrac{3}{x} = 2$

$$\dfrac{3}{\frac{3}{2}} = 2 \qquad ?$$

$$3 \div \dfrac{3}{2} = 2 \qquad ?$$

$$3 \cdot \dfrac{2}{3} = 2 \qquad ?$$

$$2 = 2 \qquad \text{True}$$

Thus, the root is $\dfrac{3}{2}$.

Example 7. Solve: $\dfrac{5}{x} - 2 = 3$

$$\dfrac{5}{x} - 2 + 2 = 3 + 2 \qquad \text{Add 2 to both sides.}$$

$$\dfrac{5}{x} = 5$$

$$x\left(\dfrac{5}{x}\right) = (5)x \qquad \text{Multiply both sides by } x.$$

$$5 = 5x$$

$$\dfrac{5}{5} = \dfrac{5x}{5} \qquad \text{Divide both sides by 5.}$$

$$1 = x$$

EXERCISES 7.5   *Solve each equation and check:*

1. $\dfrac{2x}{3} = \dfrac{32}{6}$

2. $\dfrac{5x}{7} = \dfrac{20}{14}$

3. $\dfrac{3}{8}y = 1\dfrac{14}{16}$

4. $\dfrac{5}{3}x = -13\dfrac{1}{3}$

5. $2\dfrac{1}{2}x = 7\dfrac{1}{2}$

6. $\dfrac{1}{2} + \dfrac{x}{3} = \dfrac{5}{2}$

7. $\dfrac{2}{3} + \dfrac{x}{4} = \dfrac{28}{6}$

8. $\dfrac{x}{7} - \dfrac{1}{14} = \dfrac{70}{28}$

9. $\dfrac{3}{4} - \dfrac{x}{3} = \dfrac{5}{12}$

10. $1\dfrac{1}{4} + \dfrac{x}{3} = \dfrac{7}{12}$

11. $\dfrac{3}{5}x - 25 = \dfrac{x}{10}$

12. $\dfrac{2x}{3} - 7 = -\dfrac{x}{2}$

13. $\dfrac{y}{3} - 1 = \dfrac{y}{6}$

14. $\dfrac{1}{2}x - 3 = \dfrac{x}{5}$

15. $\dfrac{3x}{4} - \dfrac{7}{20} = \dfrac{2}{5}x$

16. $\dfrac{5x}{6} + \dfrac{1}{3}(6 + x) = 37$

17. $\dfrac{x}{2} + \dfrac{2}{3}(2x + 3) = 46$

18. $\dfrac{1}{6}x - \dfrac{1}{9}(2x - 3) = 1$

19. $0.96 = 0.06(12 + x)$

20. $\dfrac{1}{2}(x + 2) + \dfrac{3}{8}(28 - x) = 11$

21. $\dfrac{3x - 24}{16} - \dfrac{3x - 12}{12} = 3$

22. $\dfrac{4x + 3}{15} - \dfrac{2x - 3}{9} = \dfrac{6x + 4}{6} - x$

23. $5x + \dfrac{6x - 8}{14} + \dfrac{10x + 6}{6} = 43$

24. $\dfrac{3x}{5} - \dfrac{9 - 3x}{10} = \dfrac{6}{10} - \dfrac{3x + 6}{10}$

25. $\dfrac{4x}{6} - \dfrac{x + 5}{2} = \dfrac{6x - 6}{8}$

26. $\dfrac{x}{3} + \dfrac{2x + 4}{4} = \dfrac{x - 1}{6} - \dfrac{3 - 2x}{2}$

27. $\dfrac{4}{x} = 6$

28. $\dfrac{2}{x} - 8 = -7$

29. $5 - \dfrac{1}{x} = 7$

30. $\dfrac{3}{x} - 6 = 8$

31. $\dfrac{5}{y} - 1 = 4$

32. $\dfrac{17}{x} = 8$

33. $\dfrac{3}{x} - 8 = 7$

34. $\dfrac{5}{2x} + 8 = 17$

35. $7 - \dfrac{6}{x} = 5$

36. $9 + \dfrac{3}{x} = 10\dfrac{1}{2}$

37. $\dfrac{6}{x} + 5 = 14$

38. $\dfrac{3}{x} - 3 = \dfrac{5}{2x} - 2$

39. $1 - \dfrac{2}{x} = \dfrac{14}{3x} - \dfrac{1}{3}$

40. $\dfrac{3}{x} + 2 = \dfrac{5}{x} - 4$

41. $\dfrac{7}{2x} + 14\dfrac{1}{2} = \dfrac{7}{x} - 10$

42. $\dfrac{8}{x} + \dfrac{1}{4} = \dfrac{5}{x} + \dfrac{1}{3}$

7.6 Formulas

A *formula* is a general rule written as an equation, usually expressed in letters, which shows the relationship between quantities.

Example 1. The formula $d = rt$ states that the distance traveled, d, equals the product of the rate, r, and the time, t.

Example 2. The formula $p = \dfrac{F}{A}$ states that the pressure, p, equals the quotient of the force, F, and the area, A.

To solve a formula for a given letter means to isolate the given letter on one side of the equation and express it in terms of all the remaining letters. This means that the given letter appears on one side of the equation by itself; all the other letters appear on the opposite side of the equation. We solve a formula using the same methods we use in solving any equation.

Example 3. Solve $d = rt$ for t.

$$d = rt$$

$$\frac{d}{r} = \frac{rt}{r} \qquad \text{Divide both sides by } r.$$

$$\frac{d}{r} = t$$

Example 4. Solve $p = \dfrac{F}{A}$ for F and then for A.

First, solve for F: $\quad p = \dfrac{F}{A}$

$$pA = \left(\frac{F}{A}\right)A \qquad \text{Multiply both sides by } A.$$

$$pA = F$$

Now, solve for A: $\quad p = \dfrac{F}{A}$

$$pA = F \qquad\qquad \text{Multiply both sides by } A.$$

$$\frac{pA}{p} = \frac{F}{p} \qquad\qquad \text{Divide both sides by } p.$$

$$A = \frac{F}{p}$$

<u>Example 5</u>. Solve $V = E - Ir$ for I.

One way:
$$V = E - Ir$$
$$V - E = E - Ir - E \qquad \text{Subtract } E \text{ from both sides.}$$
$$V - E = -Ir$$
$$\frac{V - E}{-r} = \frac{-Ir}{-r} \qquad \text{Divide both sides by } -r.$$
$$\frac{V - E}{-r} = I$$

Alternate way:
$$V = E - Ir$$
$$V + Ir = E - Ir + Ir \qquad \text{Add } Ir \text{ to both sides.}$$
$$V + Ir = E$$
$$V + Ir - V = E - V \qquad \text{Subtract } V \text{ from both sides.}$$
$$Ir = E - V$$
$$\frac{Ir}{r} = \frac{E - V}{r} \qquad \text{Divide both sides by } r.$$
$$I = \frac{E - V}{r}$$

Note that the two results are equivalent. Take the first result,

$$\frac{V - E}{-r}$$

and multiply numerator and denominator by -1.

$$\left(\frac{V - E}{-r}\right)\left(\frac{-1}{-1}\right) = \frac{-V + E}{r} = \frac{E - V}{r}$$

<u>Example 6</u>. Solve $S = \dfrac{n(a + \ell)}{2}$ for n and then for ℓ.

First, solve for n:

$$S = \frac{n(a + \ell)}{2}$$
$$2S = n(a + \ell) \qquad \text{Multiply both sides by 2.}$$
$$\frac{2S}{a + \ell} = n \qquad \text{Divide both sides by } (a + \ell).$$

Now, solve for ℓ:

$$S = \frac{n(a + \ell)}{2}$$

$2S = n(a + \ell)$ Multiply both sides by 2.

$2S = na + n\ell$ Remove parentheses.

$2S - na = n\ell$ Subtract na from both sides.

$\dfrac{2S - na}{n} = \ell$ Divide both sides by n.

Or, $\dfrac{2S}{n} - a = \ell$

EXERCISES 7.6   *Solve each formula for the given letter:*

1. $E = Ir$ for r 2. $A = bh$ for b

3. $F = ma$ for a 4. $w = mg$ for m

5. $C = \pi d$ for d 6. $V = IR$ for R

7. $V = \ell w h$ for w 8. $X_L = 2\pi f L$ for f

9. $A = 2\pi r h$ for h 10. $C = 2\pi r$ for r

11. $v^2 = 2gh$ for h 12. $V = \pi r^2 h$ for h

13. $I = \dfrac{Q}{t}$ for t 14. $I = \dfrac{Q}{t}$ for Q

15. $v = \dfrac{s}{t}$ for s 16. $I = \dfrac{E}{Z}$ for Z

17. $I = \dfrac{V}{R}$ for R 18. $P = \dfrac{w}{t}$ for w

19. $E = \dfrac{I}{4\pi r^2}$ for I 20. $R = \dfrac{\pi}{2P}$ for P

21. $X_C = \dfrac{1}{2\pi f C}$ for f 22. $R = \dfrac{\rho L}{A}$ for L

23. $A = \dfrac{1}{2}bh$ for b 24. $V = \dfrac{1}{3}\pi r^2 h$ for h

25. $Q = \dfrac{I^2 R t}{J}$ for R 26. $R = \dfrac{k\ell}{D^2}$ for ℓ

27. $F = \dfrac{9}{5}C + 32$ for C 28. $C = \dfrac{5}{9}(F - 32)$ for F

29. $C_T = C_1 + C_2 + C_3 + C_4$ for C_2

30. $R_T = R_1 + R_2 + R_3 + R_4$ for R_4

31. $Ax + By + C = 0$ for x

32. $A = P + Prt$ for r 33. $Q_1 = P(Q_2 - Q_1)$ for Q_2

34. $v_2 = v_1 + at$ for v_1

35. $A = \left(\dfrac{a+b}{2}\right)h$ for h 36. $A = \left(\dfrac{a+b}{2}\right)h$ for b

37. $\ell = a + (n-1)d$ for d

38. $A = ab + \dfrac{d}{2}(a+c)$ for d

39. $Ft = m(V_2 - V_1)$ for m

40. $\ell = a + (n-1)d$ for n

41. $Q = wc(T_1 - T_2)$ for c

42. $Ft = m(V_2 - V_1)$ for V_1

43. $V = \dfrac{2\pi(3960 + h)}{P}$ for h

44. $Q = wc(T_1 - T_2)$ for T_2

7.7 Substituting Data into Formulas

A basic tool needed in all technical fields is problem solving.

Problem Solving

A necessary part of problem solving includes:

1. Analyzing the given data.
2. Finding an equation or formula that relates the given quantities with the unknown quantity.
3. Solving the formula for the unknown quantity.
4. Substituting the given data into this solved formula.

Working with formulas in this manner is one of the most important tools that you will gain from this course.

Actually, you may solve the formula for the unknown quantity, and then substitute the data. Or, you may substitute the data into the formula first, and then solve for the unknown quantity. If you wish to use a calculator, the first method is more helpful.

Example 1. Given the formula $V = IR$, $V = 120$, and $R = 200$. Find I.

First, solve for I:

$$V = IR$$

$$\frac{V}{R} = \frac{IR}{R} \qquad \text{Divide both sides by } R.$$

$$\frac{V}{R} = I$$

Then, substitute the data:

$$I = \frac{V}{R} = \frac{120}{200} = 0.6$$

Example 2. Given $v = v_0 + at$, $v = 60$, $v_0 = 20$, and $t = 5$. Find a.

First, solve for a:

$$v = v_0 + at$$

$$v - v_0 = at \qquad \text{Subtract } v_0 \text{ from both sides.}$$

$$\frac{v - v_0}{t} = \frac{at}{t} \qquad \text{Divide both sides by } t.$$

$$\frac{v - v_0}{t} = a$$

Then, substitute the data:

$$a = \frac{v - v_0}{t} = \frac{60 - 20}{5} = \frac{40}{5} = 8$$

Example 3. Given the formula $S = \frac{MC}{\ell}$, $S = 47.5$, $M = 190$, and $C = 8$. Find ℓ.

First, solve for ℓ:

$$S = \frac{MC}{\ell}$$

$$S\ell = MC \qquad \text{Multiply both sides by } \ell.$$

$$\ell = \frac{MC}{S} \qquad \text{Divide both sides by } S.$$

Then, substitute the data:

$$\ell = \frac{MC}{S} = \frac{190(8)}{47.5} = 32$$

Example 4. Given the formula $Q = WC(T_1 - T_2)$, $Q = 15$, $W = 3$, $T_1 = 110$, and $T_2 = 60$. Find C.

First, solve for C:

$$Q = WC(T_1 - T_2)$$

$$\frac{Q}{W(T_1 - T_2)} = C \qquad \text{Divide both sides by } W(T_1 - T_2).$$

Then, substitute the data:

$$\frac{15}{3(110 - 60)} = C$$

$$C = \frac{15}{150} = 0.1$$

Example 5. Given the formula $V = \frac{1}{2}\ell w(D + d)$, $V = 156.8$, $D = 2.00$, $\ell = 8.37$, and $w = 7.19$. Find d.

First, solve for d:

$$V = \frac{1}{2}\ell w(D + d)$$

$$2V = \ell w(D + d) \qquad \text{Multiply both sides by 2.}$$

$$2V = \ell wD + \ell wd \qquad \text{Remove parentheses.}$$

$$2V - \ell wD = \ell wd \qquad \text{Subtract } \ell wD \text{ from both sides.}$$

$$\frac{2V - \ell wD}{\ell w} = d \qquad \text{Divide both sides by } \ell w.$$

$$\frac{2V}{\ell w} - D = d \qquad \text{Simplify the left side.}$$

Then, substitute the data:

$$d = \frac{2V}{\ell w} - D = \frac{2(156.8)}{(8.37)(7.19)} - 2.00 = 3.21$$

In problems like Example 5, a calculator is helpful.

EXERCISES 7.7   *Solve for the indicated letter. Then substitute the given values to find the value of the indicated letter:*

	Formula	Given	Find
1.	$A = \ell w$	$A = 414$, $w = 18$	ℓ
2.	$V = IR$	$I = 9.2$, $V = 5.52$	R
3.	$V = \dfrac{\pi r^2 h}{3}$	$V = 753.6$, $r = 6$, $\pi = 3.14$	h
4.	$I = \dfrac{V}{R}$	$R = 44$, $I = 2.5$	V
5.	$E = \dfrac{mv^2}{2}$	$E = 484{,}000$; $v = 22$	m
6.	$v = v_0 + at$	$v = 88$, $v_0 = 10$, $t = 12$	a
7.	$v = v_0 + at$	$v = 193.1$, $v_0 = 14.9$, $a = 18$	t
8.	$y = mx + b$	$x = 3$, $y = 2$, $b = 9$	m
9.	$v^2 = v_0{}^2 + 2gh$	$v = 192$, $v_0 = 0$, $g = 32$	h
10.	$A = P + Prt$	$r = 0.07$, $P = 1500$, $A = \$2025$	t
11.	$L = \pi(r_1 + r_2) + 2d$	$L = 37.68$, $d = 6.28$, $r_2 = 5$, $\pi = 3.14$	r_1
12.	$C = \dfrac{5}{9}(F - 32)$	$C = -5$	F
13.	$Fgr = Wv^2$	$F = 12{,}000$; $W = 24{,}000$; $v = 176$; $g = 32$	r
14.	$Q = WC(T_1 - T_2)$	$Q = 18.9$, $W = 3$, $C = 0.18$, $T_2 = 59$	T_1
15.	$A = \dfrac{1}{2}h(a + b)$	$A = 1159.4$, $h = 22$, $a = 56.5$	b
16.	$A = \dfrac{1}{2}h(a + b)$	$A = 5502$, $h = 28$, $b = 183$	a
17.	$V = \dfrac{1}{2}\ell w(D + d)$	$V = 226.8$, $\ell = 9$, $w = 6.3$, $D = 5$	d
18.	$S = \dfrac{n}{2}(a + \ell)$	$S = 575$, $n = 25$, $\ell = 15$	a
19.	$S = \dfrac{n}{2}(a + \ell)$	$S = 147.9$, $n = 14.5$, $\ell = 3.8$	a
20.	$S = \dfrac{n}{2}(a + \ell)$	$S = 96\frac{7}{8}$, $n = 15$, $a = 8\frac{2}{3}$	ℓ

21. A drill draws a current, I, of 4.50 A. Its resistance, R, is 16.0 Ω. Find its power in watts. $P = I^2R$.

22. A flashlight bulb is connected to a 1.50 V source. Its current, I, is 0.250 A. What is its resistance in ohms? $V = IR$.

23. The area of a rectangle is 84 ft^2. Its length is 12.5 ft. Find its width. $A = \ell w$.

24. The volume of a box is given by $V = \ell wh$. Find the width if the volume is 3780 ft^3, its length is 21.0 ft and its height is 15.0 ft.

25. The volume of a cylinder is given by the formula $V = \pi r^2 h$, where r is the radius and h is the height. Find the height in m if the volume is 8550 m^3 and the radius is 15.0 m.

26. The pressure at the bottom of a lake is found by the formula $P = hD$, where h is the depth of the water and D is the density of the water. Find the pressure in lb/in^2 at 175 ft below the surface. $D = 62.4$ lb/ft^3.

27. The equivalent resistance R of two resistances connected in parallel is given by $\dfrac{1}{R} = \dfrac{1}{R_1} + \dfrac{1}{R_2}$. Find the equivalent resistance for two resistances of 20.0 Ω and 60.0 Ω connected in parallel.

28. The R value of insulation is given by the formula $R = \dfrac{L}{K}$, where L is the thickness of the insulating material and K is the thermal conductivity of the material. Find the R value of 8.0 in. of mineral wool insulation. $K = 0.026$. Note: L must be in feet.

29. A steel railroad rail expands and contracts according to the formula $\Delta \ell = \alpha \ell \Delta T$, where $\Delta \ell$ is the change in length, α is a constant called the coefficient of linear expansion, ℓ is the original length, and ΔT is the change in temperature.

    If a 50.0 ft steel rail is installed at 0°F, how many inches will it expand when the temperature rises to 110°F? $\alpha = 6.5 \times 10^{-6}/\text{F}°$.

30. The inductive reactance, X_L, of a coil is given by $X_L = 2\pi f L$, where f is the frequency and L is the inductance. If the inductive reactance is 245 Ω and the frequency is 60.0 cycles/s, find the inductance L in henrys, H.

Chapter 7 Review

Solve each equation and check:

1. $2x + 4 = 7$

2. $11 - 3x = 23$

3. $\dfrac{x}{3} - 7 = 12$

4. $5 - \dfrac{x}{6} = 1$

5. $78 - 16y = 190$

6. $25 = -3x - 2$

7. Mary has 60 ft of fence to enclose a portion of her garden. She wants it in the shape of a rectangle with each end 8 ft wide. How long will her garden be?

8. Bill and Tony restore cars in their spare time. Bill is more experienced and is paid twice as much as Tony. For one job they received a total of $1440. How much did they each receive?

Solve each equation and check:

9. $2x + 9 = 5x - 15$ 10. $-6x + 5 = 2x - 19$

11. $3 - 2x = 9 - 3x$ 12. $4x + 1 = 4 - x$

13. $7 - (x - 5) = 11$ 14. $4x + 2(x + 3) = 42$

15. $3y - 5(2 - y) = 22$ 16. $6(x + 7) - 5(x + 8) = 0$

17. $3x - 4(x - 3) = 3(x - 4)$

18. $4(x + 3) - 9(x - 2) = x + 27$

19. $\dfrac{2x}{3} = \dfrac{16}{9}$ 20. $\dfrac{x}{3} - 2 = \dfrac{3x}{5}$

21. $\dfrac{3x}{4} - \dfrac{x - 1}{5} = \dfrac{3 + x}{2}$ 22. $\dfrac{7}{x} - 3 = \dfrac{1}{x}$

23. $5 - \dfrac{7}{x} = 3\dfrac{3}{5}$

Solve each formula for the given letter:

24. $F = Wg$ for g.   25. $P = \dfrac{W}{A}$ for A.

26. $L = A + B + \dfrac{1}{2}t$ for t.

27. $k = \dfrac{1}{2}mv^2$ for m.   28. $P_2 = \dfrac{P_1 T_2}{T_1}$ for T_1.

29. $v = \dfrac{v_t + v_0}{2}$ for v_0.

30. $K = \dfrac{5}{9}(F - 32) + 273$, find F if $K = 173$.

31. $P = 2(\ell + w)$, find w if $P = 112.5$ and $\ell = 36.9$.

32. $k = \dfrac{1}{2}mv^2$, find m if $k = 460$ and $v = 5$.

RATIO AND PROPORTION

8.1 Ratio*

The comparison of two numbers is a very important concept, and one of the most important of all comparisons is the ratio. The *ratio* of two numbers, a and b, is the first number divided by the second number. Ratios may be written in several different ways. For example, the ratio of 3 to 4 may be written as $\frac{3}{4}$, 3/4, $4\overline{)3}$, 3 : 4, or 3 ÷ 4. Each of these forms is read "the ratio of 3 to 4."

If the quantities to be compared include units, it is best if the units are the same. To find the ratio of 1 ft to 15 in., first express both quantities in inches and then find the ratio:

$$\frac{1 \text{ ft}}{15 \text{ in.}} = \frac{12 \text{ in.}}{15 \text{ in.}}, \quad \text{or} \quad \frac{12}{15}$$

When ratios are used, they are usually given in lowest terms. To reduce a ratio to lowest terms, express the ratio as a common fraction and reduce the common fraction to lowest terms. Thus, the ratio of 1 ft to 15 in. is $\frac{12}{15}$, or $\frac{4}{5}$.

Example 1. Express the ratio of 36 : 45 in lowest terms.

$$36 : 45 = \frac{36}{45} = \frac{\cancel{9} \cdot 4}{\cancel{9} \cdot 5} = \frac{4}{5}$$

*Note: Do not use rules for calculating with measurements in this chapter.

Example 2. Express the ratio of 3 ft to 18 in. in lowest terms.

$$\frac{3 \text{ ft}}{18 \text{ in.}} = \frac{36 \text{ in.}}{18 \text{ in.}} = \frac{18 \times 2 \text{ in.}}{18 \times 1 \text{ in.}} = \frac{2}{1} \text{ or } 2$$

Note. $\frac{2}{1}$ and 2 indicate the same ratio, "the ratio of 2 to 1."

Example 3. Express the ratio of 50 cm to 2 m in lowest terms.

First express the measurements in the same units.
1 m = 100 cm, so 2 m = 200 cm.

$$\frac{50 \text{ cm}}{2 \text{ m}} = \frac{50 \text{ cm}}{200 \text{ cm}} = \frac{1}{4}$$

When finding the ratio of two numbers which are fractions, keep in mind the technique for dividing fractions.

Example 4. Express the ratio of $\frac{2}{3} : \frac{8}{9}$ in lowest terms.

$$\frac{2}{3} : \frac{8}{9} = \frac{2}{3} \div \frac{8}{9} = \frac{2}{3} \times \frac{9}{8} = \frac{18}{24} = \frac{6 \cdot 3}{6 \cdot 4} = \frac{3}{4}$$

Example 5. Express the ratio of $2\frac{1}{2}$ to 10 in lowest terms.

$$2\frac{1}{2} : 10 = \frac{5}{2} : 10 = \frac{5}{2} \div 10 = \frac{5}{2} \times \frac{1}{10} = \frac{5}{20} = \frac{5 \cdot 1}{5 \cdot 4} = \frac{1}{4}$$

Example 6. Steel can be worked in a lathe at a cutting speed of 25 ft/min. Stainless steel can be worked in a lathe at a cutting speed of 15 ft/min. What is the ratio of the cutting speed of steel to the cutting speed of stainless steel?

$$\frac{\text{cutting speed of steel}}{\text{cutting speed of stainless steel}} = \frac{25 \text{ ft/min}}{15 \text{ ft/min}} = \frac{5}{3}$$

Example 7. A construction crew uses 4 buckets of cement and 12 buckets of sand to mix a supply of concrete. What is the ratio of cement to sand?

$$\frac{\text{amount of cement}}{\text{amount of sand}} = \frac{4 \text{ buckets}}{12 \text{ buckets}} = \frac{1}{3}$$

In a ratio, we compare like or related quantities, such as $\frac{18 \text{ ft}}{12 \text{ ft}} = \frac{3}{2}$ or $\frac{50 \text{ cm}}{2 \text{ m}} = \frac{1}{4}$. And, we saw that a ratio simplified into its lowest terms is a pair of unitless numbers.

Suppose that you drive 75 miles and use 3 gallons of gasoline. Your mileage would be found as follows:

$$\frac{75 \text{ mi}}{3 \text{ gal}} = \frac{25 \text{ mi}}{1 \text{ gal}}$$

We say that your mileage is 25 miles per gallon. Note that each of these two fractions compares unlike quantities: miles and gallons. A *rate* is the comparison of two unlike quantities whose units do not cancel.

<u>Example 8</u>. Express the rate of $\frac{250 \text{ gal}}{50 \text{ acres}}$ in lowest terms.

$$\frac{250 \text{ gal}}{50 \text{ acres}} = \frac{5 \text{ gal}}{1 \text{ acre}} \text{ or } 5 \text{ gal/acre}$$

The symbol "/" is read "per." The rate is read "5 gallons per acre."

EXERCISES 8.1   *Express each ratio in lowest terms:*

1. 3 to 15   2. 6 : 12   3. 7 to 21

4. $\frac{4}{22}$ 5. $\frac{80}{48}$ 6. 7 to 5

7. 3 in. to 15 in.   8. 3 ft to 15 in.

9. 3 cm to 15 mm   10. 1 in. to 8 ft

11. 9 in^2 : 2 ft^2 12. 4 m : 30 cm

13. $\frac{3}{4}$ to $\frac{7}{6}$ 14. $\frac{2}{3} : \frac{22}{9}$ 15. $2\frac{3}{4} : 4$

16. $\frac{18\frac{1}{2}}{2\frac{1}{4}}$ 17. $1\frac{3}{4}$ to 7   18. 6 to $4\frac{2}{9}$

19. 10 to $2\frac{1}{2}$ 20. $\frac{7}{8} : \frac{9}{16}$ 21. $3\frac{1}{2}$ to $2\frac{1}{2}$

22. $2\frac{2}{3} : 3\frac{3}{4}$

23. A bearing bronze mix includes 96 lb of copper and 15 lb of lead. Find the ratio of copper to lead.

 24. What is the alternator-to-engine-drive ratio if the alternator turns at 1125 rpm when the engine is idling at 500 rpm?

We see in parts (g) and (h) that the product of the means (12) equals the product of the extremes (12). Let us look at another proportion and see if this is true again.

Example 2. Given the proportion $\frac{5}{13} = \frac{10}{26}$, find the product of the means and the product of the extremes.

The extremes are 5 and 26, and the means are 13 and 10. The product of the extremes is 130 and the product of the means is 130. Here again, the product of the means equals the product of the extremes. As a matter of fact,

> In any proportion, the product of the means equals the product of the extremes.

To determine whether two ratios are equal, put the two ratios in the form of a proportion. If the product of the means equals the product of the extremes, the ratios are equal.

Example 3. Determine whether or not the ratios $\frac{13}{36}$ and $\frac{29}{84}$ are equal.

If $13 \times 84 = 29 \times 36$, then $\frac{13}{36} = \frac{29}{84}$. However, $13 \times 84 = 1092$ and $29 \times 36 = 1044$. Therefore, $\frac{13}{36} \neq \frac{29}{84}$.

("$\neq$" means "is not equal to.")

To solve a proportion means to find the missing term. To do this, form an equation by setting the product of the means equal to the product of the extremes. Then solve the resulting equation.

Example 4. Solve the proportion $\frac{x}{3} = \frac{8}{12}$.

$$\frac{x}{3} = \frac{8}{12}$$
$$12x = 24$$
$$x = 2$$

Example 5. Solve the proportion $\frac{5}{x} = \frac{10}{3}$.

$$\frac{5}{x} = \frac{10}{3}$$
$$10x = 15$$
$$x = \frac{3}{2} \text{ or } 1.5$$

A calculator will be helpful when the proportion involves decimal fractions.

<u>Example 6</u>. Solve $\dfrac{32.3}{x} = \dfrac{17.9}{25.1}$.

$$17.9x = (32.3)(25.1)$$

$$x = \frac{(32.3)(25.1)}{17.9} = 45.3,\text{ rounded to three}$$
$$\text{significant digits}$$

<u>Example 7</u>. If 125 bolts cost \$7.50, how much do 75 bolts cost?

First, let's find the rate of dollars/bolts in each case:

$$\frac{\$7.50}{125 \text{ bolts}} \quad\text{and}\quad \frac{x}{75 \text{ bolts}} \qquad \text{where } x = \text{the cost of 75 bolts}$$

Since these two rates are equal, we have the proportion:

$$\frac{7.5}{125} = \frac{x}{75}$$

$125x = (7.5)(75)$ The product of the means equals the product of the extremes.

$x = \dfrac{(7.5)(75)}{125}$ Divide both sides by 125.

$$x = 4.5$$

That is, the cost of 75 bolts is \$4.50.

<u>Example 8</u>. The pitch of a roof is the ratio of the rise to the run of a rafter. (See Fig. 8.1) The pitch of the roof below is 2:7. Find the rise if the run is 21 ft.

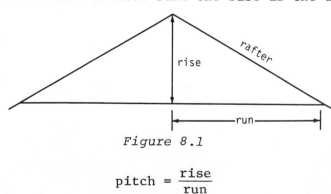

Figure 8.1

$$\text{pitch} = \frac{\text{rise}}{\text{run}}$$

$$\frac{2}{7} = \frac{x}{21 \text{ ft}}$$

$$7x = (2)(21 \text{ ft})$$

$$x = \frac{(2)(21 \text{ ft})}{7}$$

$$x = 6 \text{ ft, which is the rise.}$$

EXERCISES 8.2   *In each proportion find (a) the means, (b) the extremes,*
(c) the product of the means, and (d) the product of the
extremes:

1. $\dfrac{1}{2} = \dfrac{3}{6}$ 2. $\dfrac{3}{4} = \dfrac{6}{8}$ 3. $\dfrac{7}{9} = \dfrac{28}{36}$

4. $\dfrac{x}{3} = \dfrac{6}{9}$ 5. $\dfrac{x}{7} = \dfrac{w}{z}$ 6. $\dfrac{a}{b} = \dfrac{4}{5}$

Determine whether or not each pair of ratios are equal:

7. $\dfrac{2}{3}$, $\dfrac{10}{15}$ 8. $\dfrac{2}{3}$, $\dfrac{9}{6}$ 9. $\dfrac{3}{5}$, $\dfrac{18}{20}$

10. $\dfrac{3}{7}$, $\dfrac{9}{21}$ 11. $\dfrac{1}{3}$, $\dfrac{4}{12}$ 12. $\dfrac{125}{45}$, $\dfrac{25}{9}$

13. $\dfrac{9}{4}$, $\dfrac{82}{36}$ 14. $\dfrac{2}{3}$, $\dfrac{3}{2}$

Solve each proportion: (In Exercises 33—44, round each
result to three significant digits when necessary.)

15. $\dfrac{x}{4} = \dfrac{9}{12}$ 16. $\dfrac{1}{a} = \dfrac{4}{16}$ 17. $\dfrac{5}{7} = \dfrac{4}{y}$

18. $\dfrac{12}{5} = \dfrac{x}{10}$ 19. $\dfrac{2}{x} = \dfrac{4}{28}$ 20. $\dfrac{10}{15} = \dfrac{y}{75}$

21. $\dfrac{5}{7} = \dfrac{3x}{14}$ 22. $\dfrac{x}{18} = \dfrac{7}{9}$ 23. $\dfrac{5}{7} = \dfrac{25}{y}$

24. $\dfrac{1.1}{6} = \dfrac{x}{12}$ 25. $\dfrac{-5}{x} = \dfrac{2}{3}$ 26. $\dfrac{4x}{9} = \dfrac{12}{7}$

27. $\dfrac{1}{0.0004} = \dfrac{700}{x}$ 28. $\dfrac{x}{9} = \dfrac{2}{0.6}$ 29. $\dfrac{3x}{27} = \dfrac{0.5}{9}$

30. $\dfrac{0.25}{2x} = \dfrac{8}{48}$ 31. $\dfrac{17}{28} = \dfrac{153}{2x}$ 32. $\dfrac{3x}{10} = \dfrac{7}{50}$

33. $\dfrac{12}{y} = \dfrac{84}{144}$ 34. $\dfrac{13}{169} = \dfrac{27}{x}$ 35. $\dfrac{x}{48} = \dfrac{56}{72}$

36. $\dfrac{124}{67} = \dfrac{149}{x}$ 37. $\dfrac{472}{x} = \dfrac{793}{64.2}$ 38. $\dfrac{94.7}{6.72} = \dfrac{x}{19.3}$

39. $\dfrac{30.1}{442} = \dfrac{55.7}{x}$ 40. $\dfrac{9.4}{291} = \dfrac{44.1}{x}$ 41. $\dfrac{36.9}{104} = \dfrac{3210}{x}$

42. $\dfrac{0.0417}{0.355} = \dfrac{26.9}{x}$ 43. $\dfrac{x}{4.2} = \dfrac{19.6}{3.87}$ 44. $\dfrac{0.120}{3x} = \dfrac{0.575}{277}$

45. You need $2\dfrac{3}{4}$ ft^3 of sand to make 8 ft^3 of concrete. How
much sand would you need to make 128 ft^3 of concrete?

46. The pitch of a roof is $\dfrac{1}{3}$. If the run is 15 ft, find
the rise. See Example 8.

47. A builder sells a 1500-ft^2 home for $70,000. What would
be the price of a 2100-ft^2 home of similar quality?
Note. Assume the price per square foot remains constant.

48. Suppose 826 bricks are used in constructing a wall 14 ft long. How many bricks will be needed for a similar wall 35 ft long?

49. A buyer purchases 75 yd of material for $120. Then an additional 90 yd are ordered. What will the additional cost be?

50. A salesperson is paid a commission of $75 for selling $300 worth of goods. What should the commission be on $760 of sales at the same rate of commission?

51. A business realizes a profit of $780 on $6500 of sales. What would the volume of sales have to be to earn $1020 in profits? Note. Assume the same profit-to-sales ratio.

52. A building assessed at $32,000 is billed $638.40 for property taxes. A similar building next door is assessed at $56,000. How much should this building be billed for property taxes?

53. Suppose 20 gal of water and 3 lb of pesticide are applied per acre. How much pesticide do you put in a 350-gal spray tank?

54. A farmer uses 150 lb of Aatrex 80W on a 40-acre field. How many pounds will he need for a 220-acre field? Assume the same rate of application.

55. Suppose a yield of 100 bu of corn per acre removes 90 lb of nitrogen, potassium, and potash (N, P, and K). How many pounds of N, P, and K would be removed by a yield of 120 bu per acre?

56. A farmer has a total yield of 42,000 bu of corn from a 350-acre farm. What total yield should he expect from a similar 560-acre farm?

57. A copper wire 750 ft long has a resistance of 1.563 Ω. How long is a copper wire of the same area whose resistance is 2.605 Ω? Note. The resistance of these wires will be proportional to their length.

58. The voltage drop across a 28-Ω resistor is 52 V. What is the voltage drop across a 63-Ω resistor that is in series with the first one? Note. Resistors in series have voltage drops proportional to their resistances.

59. The ratio of secondary turns to primary turns in a transformer is 35 to 4. How many secondary turns are there if the primary coil has 68 turns?

60. A piece of cable 180 ft long costs $63. What will 450 ft cost at the same rate?

61. An engine with displacement of 380 in^3 develops 212 hp. How many horsepower would be developed by a 318 in^3 engine of the same design?

62. An 8-V automotive coil has 250 turns of wire in the primary circuit. The secondary voltage is 15,000 V. How many secondary turns are there in the coil? Note. The ratio of secondary voltage to primary voltage is equal to the ratio of secondary turns to primary turns.

63. The clutch linkage on a vehicle has an overall advantage of 24 : 1. The pressure plate applies a force of 504 lb. How much force must the driver apply to release the clutch?

64. A fuel pump delivers 35 mL of fuel in 420 strokes. How many strokes are needed to get 50 mL of fuel?

 65. A label reads: "4 cm^3 of solution contains gr X (10 gr) of potassium chloride." How many cubic centimetres would be needed to give gr XXXV (35 gr)?

 66. A multiple-dose vial has been mixed and labeled "200,000 units in 1 cm^3." How many cubic centimetres would be needed to give 900,000 units?

 67. You are to administer 150 mg of Aminophylline from a bottle marked 250 mg/10 cm^3. How many cubic centimetres should you draw?

 68. A label reads: "Gantrisin, 1.5 g in 20 cm^3." How much Gantrisin would be needed to give 10.5 g?

8.3 Percent and Proportion

 In Chapter 3 you studied percent, using the formula $P = BR$. Knowing this formula and knowing the fact that percent means per hundred, we can write the proportion

$$\frac{P}{B} = \frac{R}{100},$$

where R is the rate written as a percent. We can use this proportion to solve percent problems.

Example 1. A student answered 27 of 30 questions correctly. What percent of the answers were correct?

$$P \text{ (percentage)} = 27$$
$$B \text{ (base)} = 30$$

$$\frac{27}{30} = \frac{R}{100}$$

$$30R = 2700$$

$$R = 90$$

Therefore, the student answered 90% of the questions correctly.

Example 2. A factory produces bearings used in automobiles. After inspecting 4500 bearings, the inspectors find that 127 are defective. What percent are defective?

$$P \text{ (percentage)} = 127$$
$$B \text{ (base)} = 4500$$

$$\frac{127}{4500} = \frac{R}{100}$$

$$4500R = 12,700$$

$$R = 2.82$$

Therefore, 2.82% of the bearings are defective.

EXERCISES 8.3   *Round to the nearest tenth of a percent when necessary.*

1. 4 is what % of 20?
2. 70 is what % of 210?
3. 17 is what % of 32?
4. 63.4 is what % of 72.1?
5. 563 is what % of 1437?
6. 49 is what % of 78?
7. 0.014 is what % of 0.63?
8. 1471 is what % of 1948?
9. Bill bought a used car for $750. He paid $175 down. What percent of the price was his down payment?
10. A baseball team last year won 18 games and lost 12 games. What percent did they win? What percent did they lose?
11. A car is listed to sell at $10,200 and the salesman offers to sell it to you for $9600. What percent of the list price is the reduction?

12. A live hog weighs 254 lb and its carcass weighs 198 lb. What percent of the live hog is carcass? What percent is waste?

13. In a 100-g sample of beef there were 17 g of fat.
    (a) What is the percent of fat in the beef?
    (b) How many pounds of fat would there be in a 650-lb beef carcass? Assume the same percent of fat.

14. A store has a net profit of $8000, which is $4\frac{4}{9}$% of the total sales. How much were the total sales?

15. At the beginning of a trip a tire on a car has a pressure of 32 psi (lb/in^2). At the end of the trip the pressure is 38 psi. What is the percent increase in pressure?

16. A gasoline tank has 5.7 hectolitres (hL) in it when it is 30% full. What is the capacity of the tank?
17. Steve Davis had a pay raise from $920 to $1104 per month. Find the percent increase.
18. The population of a certain city is 756,000. Of these, 4539 are police officers. What percent of the population are police officers?

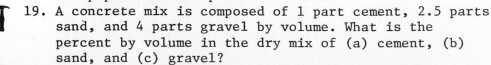

19. A concrete mix is composed of 1 part cement, 2.5 parts sand, and 4 parts gravel by volume. What is the percent by volume in the dry mix of (a) cement, (b) sand, and (c) gravel?

20. Put 4 qt of pure antifreeze in a tractor radiator and then fill the radiator with 5 gal of water. What percent of antifreeze will be in the radiator?

8.4 Direct Variation

When two quantities, x and y, change so that their **ratios** are constant, that is, $\dfrac{y_1}{x_1} = \dfrac{y_2}{x_2}$, they are said to *vary directly*. The relationship between the two quantities is called *direct variation*. If one quantity increases, the other increases by the same factor. Likewise, if one decreases, the other decreases by the same factor.

<u>Example 1</u>. Consider the following data.

y	6	24	15	18	9	30
x	2	8	5	6	3	10

Note that y varies directly with x because the ratio $\dfrac{y}{x}$ is always 3, a constant. This relationship may also be written $y = 3x$.

> **Direct Variation**
>
> $$\frac{y_1}{x_1} = \frac{y_2}{x_2}$$

Examples of direct variation are scale drawings such as maps and blueprints where

$$\frac{\text{scale measurement 1}}{\text{scale measurement 2}} = \frac{\text{actual measurement 1}}{\text{actual measurement 2}}$$

or

$$\frac{\text{scale measurement 1}}{\text{actual measurement 1}} = \frac{\text{scale measurement 2}}{\text{actual measurement 2}}$$

A *scale drawing* of an object has the same shape as the actual object, but the scale drawing may be smaller than, equal to, or larger than the actual object. The scale used in a drawing indicates what the ratio is between the size of the scale drawing and the size of the object drawn. Maps and blueprints are common examples of scale drawings.

A portion of a map of the state of Illinois is shown in Fig. 8.2. The scale is 1 in. = 32 mi.

Example 2. Find the approximate distance between Champaign and Kankakee on the map in Fig. 8.2.

The distance on the map measures $2\frac{3}{8}$ in. Set up a proportion which has as its first ratio the scale drawing ratio, and as its second ratio the length measured on the map to the actual distance.

$$\frac{1}{32} = \frac{2\frac{3}{8}}{d}$$

$$1d = 32\left(2\frac{3}{8}\right)$$

$$d = 32\left(\frac{19}{8}\right) = 76 \text{ miles}$$

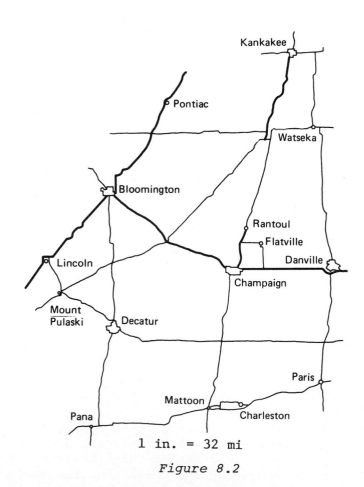

1 in. = 32 mi

Figure 8.2

Square-ruled paper may also be used to represent scale drawings. Each square represents a unit of length according to a scale.

Example 3. The scale drawing in Fig. 8.3 represents a metal plate a machinist is to make. (a) How long is the plate?

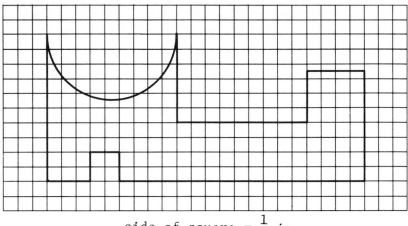

$$\text{side of square} = \frac{1}{4} \text{ in.}$$

Figure 8.3

Count the number of spaces, then set up the proportion

$$\frac{\text{scale measurement 1}}{\text{actual measurement 1}} = \frac{\text{scale measurement 2}}{\text{actual measurement 2}}$$

or

$$\frac{1 \text{ space}}{\frac{1}{4} \text{ in.}} = \frac{22 \text{ spaces}}{x \text{ in.}}$$

$$x = 22\left(\frac{1}{4}\right) = 5\frac{1}{2} \text{ in.}$$

(b) What is the width of the plate at its right end?

It is $7\frac{1}{2}$ spaces, so the proportion is

$$\frac{1 \text{ space}}{\frac{1}{4} \text{ in.}} = \frac{7\frac{1}{2} \text{ spaces}}{x \text{ in.}}$$

$$x = \left(7\frac{1}{2}\right)\left(\frac{1}{4}\right) = 1\frac{7}{8} \text{ in.}$$

(c) What is the diameter of the semicircle?

$$\frac{1 \text{ space}}{\frac{1}{4} \text{ in.}} = \frac{9 \text{ spaces}}{x \text{ in.}}$$

$$x = 9\left(\frac{1}{4}\right) = 2\frac{1}{4} \text{ in.}$$

Another example of direct variation is the hydraulic press, which allows one to exert a small force to move or raise a large object, such as a car. Other uses of hydraulics include compressing junk cars, stamping metal sheets to form car parts, and lifting truck beds.

When someone presses a force of 50 lb on the small piston of the hydraulic press in Fig. 8.4, a force of 5000 lb is exerted by the large piston. The *mechanical advantage* (MA) of the hydraulic press is the ratio of the force from the large piston (F_ℓ) to the force on the small piston (F_s). By formula,

$$MA = \frac{F_\ell}{F_s}$$

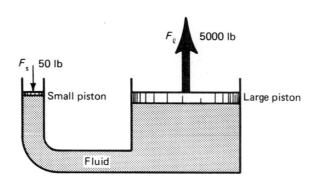

Figure 8.4

<u>Example 4</u>. Find the mechanical advantage of the press shown in Fig. 8.4.

$$MA = \frac{F_\ell}{F_s}$$

$$= \frac{5000 \text{ lb}}{50 \text{ lb}} = \frac{100}{1}$$

Thus, for every pound exerted on the small piston, 100 lb is exerted by the large piston.

The mechanical advantage can also be calculated when the radii of the pistons are known.

$$MA = \frac{r_\ell^2}{r_s^2}$$

Example 5. The radius of the large piston of a hydraulic press is 12 in. The radius of the small piston is 2 in. Find MA.

$$MA = \frac{r_\ell^2}{r_s^2}$$

$$= \frac{(12 \text{ in.})^2}{(2 \text{ in.})^2}$$

$$= \frac{144 \text{ in}^2}{4 \text{ in}^2}$$

$$= \frac{36}{1} \text{ ; that is, for every pound exerted on the small piston, 36 lb is exerted by the large piston.}$$

You now have two ways of finding mechanical advantage: when F_ℓ and F_s are known and when r_ℓ and r_s are known. From this knowledge you can find a relationship among F_ℓ, F_s, r_ℓ, and r_s:

Since $MA = \dfrac{F_\ell}{F_s}$, and $MA = \dfrac{r_\ell^2}{r_s^2}$, then $\dfrac{F_\ell}{F_s} = \dfrac{r_\ell^2}{r_s^2}$.

Example 6. Given F_s = 240 lb, r_ℓ = 16 in., and r_s = 2 in. Find F_ℓ .

$$\frac{F_\ell}{F_s} = \frac{r_\ell^2}{r_s^2}$$

$$\frac{F_\ell}{240 \text{ lb}} = \frac{(16 \text{ in.})^2}{(2 \text{ in.})^2}$$

$$F_\ell = \frac{(240 \text{ lb})(16 \text{ in.})^2}{(2 \text{ in.})^2}$$

$$= \frac{(240 \text{ lb})(256 \text{ in}^2)}{4 \text{ in}^2}$$

$$= 15,360 \text{ lb}$$

EXERCISES 8.4   *Use the map in Fig. 8.2 to find the approximate distance between each pair of cities. Find straight-line (air) distances only:*

1. Champaign and Bloomington
2. Bloomington and Decatur
3. Rantoul and Kankakee
4. Rantoul and Bloomington
5. Pana and Rantoul
6. Champaign and Mattoon
7. Charleston and Pontiac
8. Paris and Watseka
9. Lincoln and Danville
10. Flatville and Mt. Pulaski

Use the map in Fig. 8.5 to find the approximate air distance between each pair of cities:

11. St. Louis and Kansas City
12. Memphis and St. Louis
13. Memphis and Little Rock
14. Sedalia, Mo. and Tulsa, Okla.
15. Fort Smith, Ark. and Springfield, Mo.
16. Pine Bluff, Ark. and Jefferson City, Mo.

1 cm = 85 km

Figure 8.5

*Use the scale drawing of a metal plate cover in Fig. 8.6
in Exercises 17—26:*

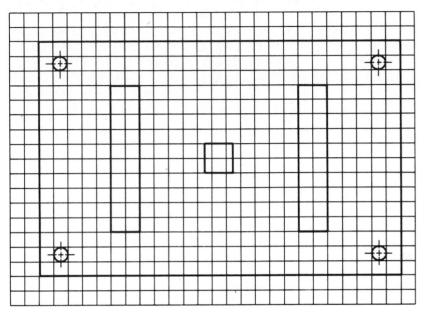

side of square = 0.5 cm

Figure 8.6

17. What is the length of the plate cover?
18. What is the width of the plate cover?
19. What is the area of the plate cover?
20. What is the diameter of the circular holes?
21. What are the dimensions of the square hole?
22. What are the dimensions of the rectangular holes?
23. What is the distance between the rectangular holes,
    center to center?
24. What is the distance between the centers of the upper
    pair of circular holes?
25. What is the distance between the centers of the right
    pair of circular holes?
26. Answer each of the questions 17—25 if the scale were

    changed so that the side of a square = $2\frac{1}{16}$ in.

*With pencil and ruler make line drawings on square-ruled
paper which fit each description:*

27. A rectangle 8 ft by 6 ft. Use the scale: side of a
    square = 1 ft.
28. A square 16 cm on a side. Use the scale: side of a
    square = 2 cm.
29. A circle 36 mm in diameter. Use the scale: side of
    a square = 3 mm.
30. A rectangle 12 in. by 8 in. with a circle in its center
    3 in. in diameter. Use the scale: side of a square
    = 2 in.

31. Can the actual circle in Exercise 29 be placed within the actual rectangle in Exercise 27? Can the scale drawing of the circle be placed within the scale drawing of the rectangle?

32. Can the actual circle in Exercise 29 be placed within the actual square in Exercise 28? Can the scale drawing of the circle be placed within the scale drawing of the square?

Use the formulas for the hydraulic press to find each value:

33. F_ℓ = 4000 lb, F_s = 200 lb. Find MA.

34. F_ℓ = 4800 lb, F_s = 160 lb. Find MA.

35. F_ℓ = 3600 lb, F_s = 400 lb. Find MA.

36. F_ℓ = 2400 lb, MA = $\frac{50}{1}$. Find F_s.

37. F_ℓ = 5100 lb, MA = $\frac{75}{1}$. Find F_s.

38. F_s = 2750 lb, MA = $\frac{36}{1}$. Find F_ℓ.

39. F_s = 2650 lb, MA = $\frac{90}{1}$. Find F_ℓ.

40. F_ℓ = 2450 lb, MA = $\frac{125}{1}$. Find F_s.

41. r_ℓ = 27 in., r_s = 3 in. Find MA.

42. r_ℓ = 36 in., r_s = 4 in. Find MA.

43. r_s = 3 in., r_ℓ = 33 in. Find MA.

44. r_s = 2 in., r_ℓ = 18 in. Find MA.

45. r_ℓ = 32.3 in., r_s = 2.44 in. Find MA.

46. r_ℓ = 137.6 in., r_s = 9.25 in. Find MA.

47. F_s = 25 lb, r_ℓ = 8 in., r_s = 2 in. Find F_ℓ.

48. F_s = 81 lb, r_ℓ = 9 in., r_s = 1 in. Find F_ℓ.

49. F_ℓ = 6400 lb, r_ℓ = 16 in., r_s = 4 in. Find F_s.

50. F_ℓ = 7500 lb, r_ℓ = 15 in., r_s = 3 in. Find F_s.

51. F_s = 40 lb, r_ℓ = 28 in., r_s = 7 in. Find F_ℓ.

52. F_ℓ = 8100 lb, r_ℓ = 30 in., r_s = 3 in. Find F_s.

8.5 Inverse Variation

If two quantities, y and x, change so that their product is constant, that is, if $y_1 x_1 = y_2 x_2$, they are said to *vary inversely*. This relationship between the two quantities is called *inverse variation*. This means that if one quantity increases, the other decreases so that their product is always the same. Compare this with direct variation, where the ratio of the two quantities is always the same.

Example 1. Consider the following data.

y	8	24	12	3	6	48
x	6	2	4	16	8	1

Note that y varies inversely with x because the product is always 48, a constant. This relationship may also be written $y = \dfrac{48}{x}$.

Inverse Variation

$$x_1 y_1 = x_2 y_2$$

An inverse variation relationship may also be written in the form

$$\frac{y_1}{y_2} = \frac{x_2}{x_1}$$

One example of inverse variation is the relationship between two rotating pulleys connected by a belt. This relationship is given by the formula:

(diameter of A)(rpm of A) = (diameter of B)(rpm of B)

See Fig. 8.7.

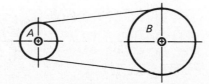

Figure 8.7

Example 2. A small pulley is 11 in. in diameter and a
larger one is 20 in. in diameter. How many rpm does the
smaller pulley make if the larger one revolves at 44 rpm?

(diameter of A)(rpm of A) = (diameter of B)(rpm of B)

$$11 \quad \cdot \quad x \quad = \quad 20 \quad \cdot \quad 44$$

$$x = \frac{(20)(44)}{11} = 80 \text{ rpm}$$

Another example of inverse variation is the relationship
between the number of teeth and the number of rpm of two
rotating gears, as shown in Fig. 8.8.

(no. of teeth in A)(rpm of A) = (no. of teeth in B)(rpm of B)

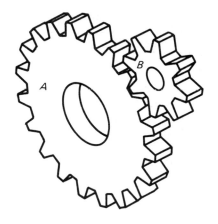

Figure 8.8

Example 3. A large gear with 14 teeth revolves at
40 rpm. It turns a small gear with 8 teeth. How fast
does the small gear rotate?

(no. of teeth in A)(rpm of A) = (no. of teeth in B)(rpm of B)

$$14 \quad \cdot \quad 40 \quad = \quad 8 \quad \cdot \quad x$$

$$\frac{14(40)}{8} = x$$

$$70 \text{ rpm} = x$$

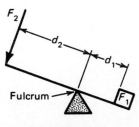

Figure 8.9

Figure 8.9 shows a lever, which is a rigid bar, pivoted
to turn on a point (or edge) called a *fulcrum*. The parts of
the lever on either side of the fulcrum are called *lever
arms*.

To lift the box, F_1, one pushes down on the other end
of the lever with a force, F_2. The distance from F_2 to
the fulcrum is d_2. The distance from F_1 to the fulcrum
is d_1. The principle of the lever is another example of
inverse variation and can be expressed by the formula

$$F_1 d_1 = F_2 d_2$$

When the equation is true, the lever is balanced.

Example 4. A man places one end of a lever under a large
rock, as in Fig. 8.10. He places a second rock under the
lever 2 ft from the first rock to act as a fulcrum. He
exerts a force of 180 pounds 6 ft from the fulcrum. What
is F_1, the maximum weight of the rock that could be
lifted?

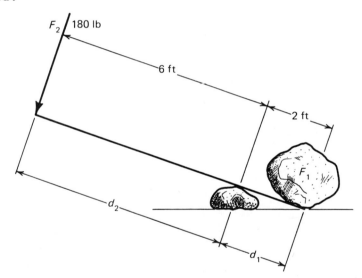

Figure 8.10

$$F_1 d_1 = F_2 d_2$$

$$F_1 \times 2 \text{ ft} = (180 \text{ lb})(6 \text{ ft})$$

$$F_1 = \frac{(180 \text{ lb})(\overset{3}{\cancel{6 \text{ ft}}})}{\cancel{2 \text{ ft}}}$$

$$F_1 = 540 \text{ lb}$$

EXERCISES 8.5   *Fill in the blanks:*

	Pulley A		Pulley B	
	Diameter	rpm	Diameter	rpm
1.	25 cm	72	50 cm	
2.	18 cm		12 cm	96
3.	10 cm	120	15 cm	
4.		84	8 in.	48
5.	34 cm	440		680
6.		225	15 in.	465
7.	25 cm	600	42 cm	
8.	98 cm	240		360

9. A small pulley is 13 in. in diameter and a larger one is 18 in. in diameter. How many rpm does the larger pulley make if the smaller one revolves at 720 rpm?

10. A 21-in. pulley, revolving at 65 rpm, turns a smaller pulley at 210 rpm. What is the diameter of the smaller pulley?

11. A large pulley turns at 15 rpm. A smaller pulley 8 in. in diameter turns at 300 rpm. What is the diameter of the larger pulley?

12. A large pulley 27 cm in diameter turns at 28 rpm. What is the diameter of a smaller pulley turning at 126 rpm?

13. A pulley 2.8 cm in diameter turns at 640 rpm. How many rpm will a pulley 2 cm in diameter turn?

14. A pulley 32 in. in diameter turns at 825 rpm. How many rpm will a pulley 25 in. in diameter turn?

15. One pulley is 7 cm larger in diameter than a second pulley. The larger pulley turns at 80 rpm and the smaller pulley turns at 136 rpm. What is the diameter of each pulley?

16. One pulley is twice as large in diameter as a second pulley. If the larger pulley turns at 256 rpm, what is the rpm of the smaller?

Fill in the blanks:

	Gear A		Gear B	
	Number of teeth	rpm	Number of teeth	rpm
17.	50	400		125
18.	220		45	440
19.	42	600	25	
20.	50	64		80
21.		$6\frac{1}{4}$	120	30
22.	80	$1\frac{1}{4}$		$3\frac{1}{3}$

23. A small gear with 25 teeth turns a large gear with 75 teeth at 32 rpm. How many rpm does the small gear make?
24. A large gear with 180 teeth running at 600 rpm turns a small gear 900 rpm. How many teeth does the small gear have?
25. A large gear with 60 teeth turning at 72 rpm turns a small gear with 30 teeth how many rpm?
26. A large gear with 80 teeth turning at 150 rpm turns a small gear with 12 teeth how many rpm?
27. A large gear with 120 teeth turning at 30 rpm turns a small gear at 90 rpm. How many teeth does the small gear have?
28. A large gear with 200 teeth turning at 17 rpm turns a small gear at 100 rpm. How many teeth does the small gear have?

Complete the table:

	F_1	d_1	F_2	d_2
29.	18	5	9	
30.	30		70	8
31.	40	9		3
32.		6.3	458.2	8.7

In Exercises 33—37, draw a sketch for each and solve:

33. An object is 6 ft from the fulcrum and balances a second object 8 ft from the fulcrum. The first object weighs 180 lb. How much does the second object weigh?

34. A block of steel weighing 1800 lb is to be raised by a lever extending under the block 9 in. from the fulcrum. How far from the fulcrum must a 150-lb man apply his weight to the bar to balance the steel?

35. A rocker arm raises oil from an oil well. On each stroke it lifts a weight of one ton on a weight arm 4 ft long. What force is needed to lift the oil on a force arm 8 ft long?

36. A carpenter needs to raise one side of a building with a lever 3.65 m in length. The lever, with one end under the building, is placed on a fulcrum 0.45 m from the building. A 90-kg force pulls down on the other end. What weight is being lifted when the building begins to rise?

37. A 1200-g weight is placed 72 cm from the fulcrum of a lever. How far from the fulcrum is a 1350-g weight which balances it?

38. A lever is in balance when a force of 2000 g is placed 28 cm from a fulcrum. An unknown force is placed 20 cm from the fulcrum on the other side. What is the unknown force?

39. A 210-lb object is placed on a lever. It balances a 190-lb weight which is 28 in. from the fulcrum. What distance from the fulcrum should the 210-lb weight be placed?

40. A piece of machinery weighs 3 tons. It is to be balanced by two men whose combined weight is 330 lb. The piece of machinery is placed 11 in. from the fulcrum. How far from the fulcrum must the two men exert their weight in order to balance it?

Chapter 8 Review

Write each ratio in lowest terms.

1. 7 to 28

2. 60 : 40

3. 1 g to 500 mg

4. $\dfrac{5 \text{ ft } 6 \text{ in.}}{9 \text{ ft}}$

Determine whether or not each pair of ratios is equal.

5. $\dfrac{7}{2}$, $\dfrac{35}{10}$

6. $\dfrac{5}{18}$, $\dfrac{30}{115}$

Solve each proportion.

7. $\dfrac{x}{4} = \dfrac{5}{20}$

8. $\dfrac{10}{25} = \dfrac{x}{75}$

9. $\dfrac{1}{x} = \dfrac{8}{64}$

10. $\dfrac{72}{96} = \dfrac{30}{x}$

 Use a calculator to solve each proportion. (Round each to three significant digits.)

11. $\dfrac{73.4}{x} = \dfrac{25.9}{37.4}$

12. $\dfrac{x}{19.7} = \dfrac{144}{68.7}$

13. $\dfrac{61.1}{81.3} = \dfrac{592}{x}$

14. $\dfrac{243}{58.3} = \dfrac{x}{127}$

15. A piece of cable 180 ft long costs $67.50. How much will 500 ft cost at the same unit price?

16. A copper wire 750 ft long has a resistance 1.89 Ω. How long is a copper wire of the same area whose resistance is 3.15 Ω?

17. A crew of electricians can wire 6 houses in 144 h. How many hours will it take them to wire 9 houses?

18. An automobile braking system has a 12 to 1 lever advantage on the master cylinder. A 25-lb force is applied to the pedal. What force will be put on the master cylinder?

19. Jones invests $6380 and Smith invests $4620 in a partnership business. What percent of the total investment does each have?

20. One gallon of a pesticide mixture weighs 7 lb 13 oz. It contains 11 oz of pesticide. What percent of the mixture is pesticide?

21. What kind of variation is shown by the equation $\dfrac{y_1}{x_1} = \dfrac{y_2}{x_2}$?

22. What kind of variation is indicated when one quantity increases while the other increases?

23. What kind of variation is shown by the equation $y_1 x_1 = y_2 x_2$?

24. Suppose $\dfrac{1}{4}$ in. on a map represents 25 mi. What distance is represented by $3\dfrac{5}{8}$ in.?

25. The scale on a map is 1 in. = 200 ft. Two places are known to be 2 mi apart. What distance will show between them on the map?

26. Two pulleys are connected by a belt. The numbers of rpm of the two pulleys vary inversely as their diameters. A pulley having a diameter of 25 cm is turning at 800 rpm. What is the number of rpm of the second pulley, which has a diameter of 40 cm?

27. A large gear with 42 teeth revolves at 25 rpm. It turns a small gear with 14 teeth. How fast does the small gear rotate?

28. In hydraulics the formula relating the forces and the radii of the pistons is $\dfrac{F_\ell}{F_s} = \dfrac{r_\ell^2}{r_s^2}$. Given $F_\ell = 6050$ kg, $r_\ell = 22$ cm, and $r_s = 2$ cm, find F_s.

29. An object 9 ft from the fulcrum of a lever balances a second object 12 ft from the fulcrum. The first object weighs 240 lb. How much does the second object weigh?

30. The current I varies directly as the voltage E. Suppose $I = 0.6$ A when $E = 30$ V. Find the value of I when $E = 100$ V.

SYSTEMS OF LINEAR EQUATIONS

9.1 Linear Equations in Two Variables

In Chapter 7 you studied solving equations with one variable. These equations are frequently used in solving technical problems. However, many problems require the use of more than one variable. In this chapter you will learn some methods for solving pairs of equations containing two variables. (It is possible, of course, to solve problems using many equations and many variables. Using a computer, this can be done very rapidly.)

A straight-line graph can be used to show at a glance all the possible solutions of an equation containing two variables. This involves the use of a number plane. You will study the number plane as the first step in learning to draw graphs and to solve pairs of equations in two variables.

To construct a number plane, draw a horizontal number line, which is called the *x axis*. Then draw a second number line intersecting the first line at right angles so that both number lines have the same zero point, called the *origin*. The vertical number line is called the *y axis*.

Each number line, or axis, has a scale. The numbers on the *x* axis are *positive at the right* of the origin and *negative at the left* of the origin. Similarly, the numbers on the *y* axis are *positive above* the origin and *negative below* the origin.

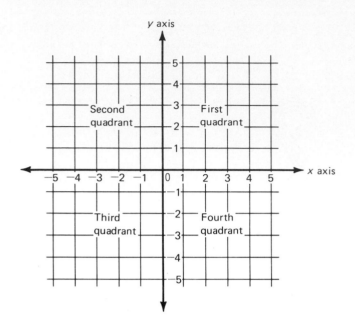

Figure 9.1

All the points in the plane determined by these two intersecting axes make up the *number plane*. The axes divide the number plane into four regions, called *quadrants*. The quadrants are numbered as shown in Fig. 9.1.

Points in the number plane are usually indicated by an *ordered pair* of numbers written in the form (*x*, *y*), where *x* is the first number in the ordered pair and *y* is the second number in the ordered pair. The numbers *x* and *y* are also called the *coordinates* of a point in the number plane.

Plotting Points

To locate the point in the number plane which corresponds to an ordered pair:

Step 1: Count right or left, from 0 (origin) along the x axis, the number of spaces corresponding to the first number of the ordered pair (right if positive, left if negative).

Step 2: Count up or down, from the point reached on the x axis in Step 1, the number of spaces corresponding to the second number of the ordered pair (up if positive, down if negative).

Step 3: Mark the last point reached with a dot or an "X."

<u>Example 1</u>. Plot the point corresponding to the ordered pair (3, −4) in Fig. 9.2.

First, count three spaces to the right along the **x** axis. Then, count down four spaces from that point. Mark the final point with an "X."

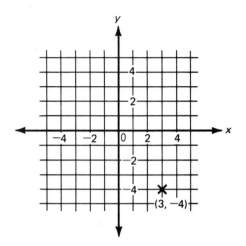

Figure 9.2

<u>Example 2</u>. Plot the points corresponding to the ordered pairs in the number plane in Fig. 9.3: (1, 2), (3, −2), (−4, 7), (5, 0), (−2, −3), (−5, −1), (−2, 4).

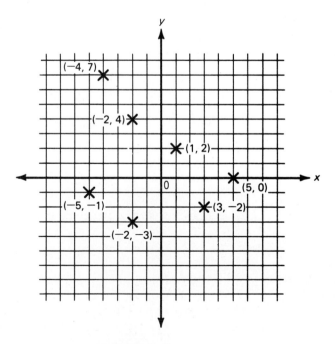

Figure 9.3

You have already studied linear equations in one variable, such as $2x + 4 = 10$ and $3x - 7 = 4$. A *linear equation in two variables*, such as $3x + 4y = 12$, is one that contains two variables, both having an exponent of one. The replacements for *both* variables that make the equation true are called *solutions*. Although most linear equations in *one variable* have only *one* root, most linear equations in *two variables* have *many possible solutions*. For example, the equation $2x + 4 = 10$ has only one root, 3. The equation $x + y = 7$ has many solutions. Some of these are 1 for x, 6 for y; 2 for x, 5 for y; -2 for x, 9 for y; $5\frac{1}{2}$ for x, $1\frac{1}{2}$ for y; and so on.

Since it is very time-consuming to write pairs of replacements in this manner, we use *ordered pairs* in the form (x, y) to write solutions of equations in two variables. Therefore, instead of writing the solutions of the equation $x + y = 7$ as above, we write them as $(1, 6)$, $(2, 5)$, $(-2, 9)$, $\left(5\frac{1}{2}, 1\frac{1}{2}\right)$.

To find solutions of a linear equation in two variables, replace one variable by a number you have chosen and then solve the resulting linear equation for the remaining variable.

<u>Example 3</u>. Complete the three ordered pair solutions of $2x + y = 5$.

(a) $(4, \quad)$

Replace x by 4. Any number could be used, but for this example we will use 4. The resulting equation is

$$2(4) + y = 5$$
$$8 + y = 5$$
$$8 + y - 8 = 5 - 8 \qquad \text{Subtract 8 from both sides.}$$
$$y = -3$$

<u>Check</u>: Replace x by 4 and y by -3.

$$2(4) + (-3) = 5 \qquad ?$$
$$8 - 3 = 5 \qquad \text{True}$$

Therefore, $(4, -3)$ is a solution.

(b) $(-2, \quad)$

Replace x by -2. The resulting equation is

$$2(-2) + y = 5$$
$$-4 + y = 5$$
$$-4 + y + 4 = 5 + 4 \qquad \text{Add 4 to both sides.}$$
$$y = 9$$

Check: Replace x by -2 and y by 9.

$$2(-2) + 9 = 5 \quad ?$$
$$-4 + 9 = 5 \quad \text{True}$$

Thus, $(-2, 9)$ is a solution.

(c) $(0, \quad)$

Replace x by 0. The resulting equation is

$$2(0) + y = 5$$
$$0 + y = 5$$
$$y = 5$$

Check: Replace x by 0 and y by 5.

$$2(0) + (5) = 5 \quad ?$$
$$0 + 5 = 5 \quad \text{True}$$

Therefore, $(0, 5)$ is a solution.

You may find it easier to first solve the equation for y and then make each replacement for x.

Example 4. Complete the three ordered pair solutions of $3x - y = 4$ by first solving the equation for y.

(a) $(5, \quad)$ (b) $(-2, \quad)$ (c) $(0, \quad)$

$$3x - y = 4$$

$$3x - y - 3x = 4 - 3x \qquad \text{Subtract } 3x \text{ from both sides.}$$
$$-y = 4 - 3x$$

$$y = -4 + 3x \qquad \text{Divide both sides by } -1.$$

You may make a table to keep your work in order.

	x	$3x - 4$	y
(a)	5	$3(5) - 4 = 15 - 4$	11
(b)	-2	$3(-2) - 4 = -6 - 4$	-10
(c)	0	$3(0) - 4 = 0 - 4$	-4

Therefore, the three solutions are $(5, 11)$, $(-2, -10)$, $(0, -4)$.

Example 5. Complete the three ordered pair solutions of $5x + 3y = 7$.

(a) $(2, \quad)$ (b) $(0, \quad)$ (c) $(-1, \quad)$

We will first solve for y.

$$5x + 3y = 7$$

$$5x + 3y - 5x = 7 - 5x \qquad \text{Subtract } 5x \text{ from both sides.}$$

$$3y = 7 - 5x$$

$$\frac{3y}{3} = \frac{7 - 5x}{3} \qquad \text{Divide both sides by 3.}$$

$$y = \frac{7 - 5x}{3}$$

	x	$\dfrac{7 - 5x}{3}$	y
(a)	2	$\dfrac{7 - 5(2)}{3} = \dfrac{-3}{3}$	-1
(b)	0	$\dfrac{7 - 5(0)}{3} = \dfrac{7}{3}$	$\dfrac{7}{3}$
(c)	-1	$\dfrac{7 - 5(-1)}{3} = \dfrac{12}{3}$	4

The three solutions are $(2, -1)$, $\left(0, \dfrac{7}{3}\right)$, $(-1, 4)$.

EXERCISES 9.1   *Write the ordered pair corresponding to each point below:*

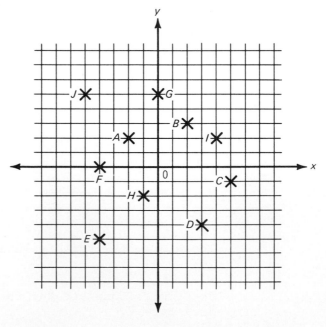

1. A	2. B	3. C	4. D	5. E
6. F	7. G	8. H	9. I	10. J

*Plot each point in the number plane. Label each point by
writing its ordered pair and letter:*

11. A (1, 3)   12. B (4, 0)   13. C (−6, −2)

14. D (−2, 4)   15. E (5, −4)   16. F (−4, −4)

17. G (0, 9)   18. H (3, 7)   19. I (5, −5)

20. J (−5, 5)   21. K (−6, −3)   22. L (3, −7)

23. M (−4, 5)   24. N (−2, −6)   25. O (1, −3)

26. P (5, 2)   27. Q $\left(3, \frac{1}{2}\right)$ 28. R $\left(4\frac{1}{2}, \; -3\frac{1}{2}\right)$

29. S $\left(-6, \frac{1}{2}\right)$ 30. T $\left(-4\frac{1}{2}, \; 6\frac{1}{2}\right)$

Complete the three ordered pair solutions of each equation:

Equation	Ordered Pairs
31. $x + y = 5$	(3,) (8,) (−2,)
32. $-2x + y = 8$	(2,) (7,) (−4,)
33. $5x + 3y = 7$	(2,) (0,) (−2,)
34. $6x - y = 0$	(3,) (5,) (−2,)
35. $3x - 4y = 8$	(0,) (2,) (−4,)
36. $5x - 3y = 8$	(1,) (0,) (−2,)
37. $-2x + 5y = 10$	(5,) (0,) (−3,)
38. $-4x - 7y = -3$	(−1,) (0,) (−8,)
39. $9x - 2y = 10$	(2,) (0,) (−4,)
40. $2x + 3y = 6$	(3,) (0,) (−6,)
41. $y = 3x + 4$	(2,) (0,) (−3,)
42. $y = 4x - 8$	(3,) (0,) (−4,)
43. $5x + y = 7$	(2,) (0,) (−4,)
44. $4x - y = 8$	(1,) (0,) (−3,)
45. $2x = y - 4$	(3,) (0,) (−1,)
46. $3y - x = 5$	(1,) (0,) (−4,)
47. $4x - 2y = -8$	(3,) (0,) (−2,)
48. $2x - 3y = 1$	(2,) (0,) (−4,)
49. $9x - 2y = 5$	(1,) (0,) (−3,)
50. $2x + 7y = -12$	(1,) (0,) (−8,)
51. $y = 3$	(2,) (0,) (−4,)
(Think: $0x + 1y = 3$)	

52. $y + 4 = 0$ (3,) (0,) (-7,)

53. $x = 5$ (, 4) (, 0) (, -2)
    (Think: $1x + 0y = 5$)

54. $x + 7 = 0$ (, 5) (, 0) (, -6)

Solve for y in terms of x:

55. $2x + 3y = 6$ 56. $4x + 5y = 10$ 57. $x + 2y = 7$

58. $2x + 2y = 5$ 59. $x - 2y = 6$ 60. $x - 3y = 9$

61. $2x - 3y = 9$ 62. $4x - 5y = 10$ 63. $-2x + 3y = 6$

64. $-3x + 5y = 25$ 65. $-2x - 3y = -15$ 66. $-3x - 4y = -8$

9.2 Graphing Linear Equations

In Sec. 9.1 you learned that a linear equation in two variables has many solutions. In Example 3, you found that three of the solutions of $2x + y = 5$ were (4, -3), (-2, 9), and (0, 5). Now, plot the points corresponding to these ordered pairs and connect the points, as shown in Fig. 9.4.

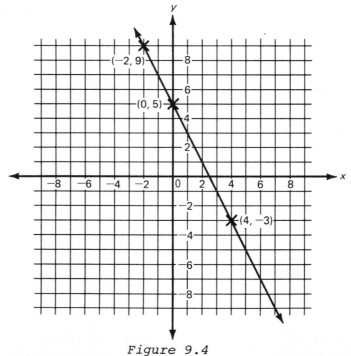

Figure 9.4

You can see from Fig. 9.4 that the three points lie on the same straight line. If you find another solution of $2x + y = 5$, say (1, 3), the point corresponding to this ordered pair also lies on the same straight line. The solutions of a linear equation in two variables always correspond to points lying on a straight line. Therefore, *the graph of the solutions of a linear equation in two variables is always a straight line.* Only part of this line can be shown on the graph, and the arrows at each end indicate that the line extends without limit in both directions.

> Graphing Linear Equations
>
> To draw the graph of a linear equation in two variables:
>
> Step 1: Find any three solutions of the equation. (Two solutions would be enough, since two points determine a straight line. However, a third solution gives a third point as a check. If the three points do not lie on the same straight line, you have made an error.)
>
> Step 2: Plot the points corresponding to the three ordered pairs that you have found in Step 1.
>
> Step 3: Draw a line through the three points. If it is not a straight line, check your solutions.

Example 1. Draw the graph of $3x + 4y = 12$.

Step 1: Find any three solutions of $3x + 4y = 12$. First, solve for y:

$$3x + 4y - 3x = 12 - 3x \qquad \text{Subtract } 3x \text{ from both sides.}$$

$$4y = 12 - 3x$$

$$\frac{4y}{4} = \frac{12 - 3x}{4} \qquad \text{Divide both sides by 4.}$$

$$y = \frac{12 - 3x}{4}$$

x	$\dfrac{12 - 3x}{4}$	y
4	$\dfrac{12 - 3(4)}{4} = \dfrac{0}{4}$	0
0	$\dfrac{12 - 3(0)}{4} = \dfrac{12}{4}$	3
−2	$\dfrac{12 - 3(-2)}{4} = \dfrac{18}{4}$	$\dfrac{9}{2}$

Three solutions are $(4, 0)$, $(0, 3)$, and $\left(-2, \frac{9}{2}\right)$.

Step 2: Plot the points corresponding to (4, 0), (0, 3), and $\left(-2, \dfrac{9}{2}\right)$.

Step 3: Draw a straight line through these three points. (See Fig. 9.5.)

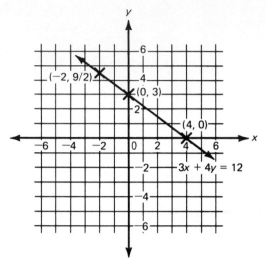

Figure 9.5

An alternative method is shown in Example 2.

Example 2. Draw the graph of $2x - 3y = 6$.

Step 1: Set up a table as shown and write the values you choose for x, say 3, 0, and −3.

x	3	0	−3
y			

Step 2: Substitute the chosen values of x in the given equation and solve for y.

$2x - 3y = 6$	$2x - 3y = 6$	$2x - 3y = 6$
$2(3) - 3y = 6$	$2(0) - 3y = 6$	$2(-3) - 3y = 6$
$6 - 3y = 6$	$0 - 3y = 6$	$-6 - 3y = 6$
$-3y = 0$	$-3y = 6$	$-3y = 12$
$y = 0$	$y = -2$	$y = -4$

Step 3: Write the values for *y* which correspond to the chosen values for *x* in the table, thus:

x	3	0	-3
y	0	-2	-4

That is, three solutions of $2x - 3y = 6$ are the ordered pairs (3, 0), (0, -2), and (-3, -4).

Step 4: Plot the points from Step 3 and draw a straight line through them as in Fig. 9.6.

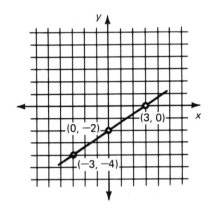

Figure 9.6

EXERCISES 9.2   *Draw the graph of each equation:*

1. $x + y = 7$ 2. $x + 3y = 9$ 3. $y = 2x + 3$

4. $y = 4x - 5$ 5. $4y = x$ 6. $2x + y = 6$

7. $6x - 2y = 10$ 8. $2x + 3y = 9$ 9. $3x - 4y = 12$

10. $3x - 5y = 15$ 11. $5x + 4y = 20$ 12. $2x - 3y = 18$

13. $2x + 7y = 14$ 14. $2x - 5y = 20$ 15. $y = 2x$

16. $y = -3x$ 17. $3x + 5y = 11$ 18. $4x - 7y = 15$

19. $y = -\frac{1}{2}x + 4$ 20. $y = \frac{2}{3}x - 6$ 21. $y = 3$

22. $y + 2 = 0$ 23. $x = -4$ 24. $x - 5 = 0$

9.3 Solving Pairs of Linear Equations by the Graphic Method

You will find that many problems can be solved by using two equations in two variables and solving them simultaneously (at the same time). *To solve a pair of linear equations in two variables simultaneously, you must find an ordered pair that will make both equations true at the same time.*

As you know, the graph of a linear equation in two variables is a straight line. As shown in Fig. 9.7, two straight lines (the graphs of *two* linear equations in two variables) in the same plane may be arranged as follows:

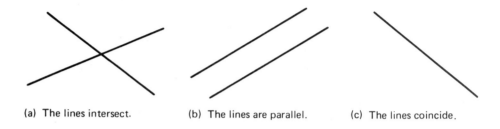

(a) The lines intersect.   (b) The lines are parallel.   (c) The lines coincide.

Figure 9.7

(a) The lines may intersect. If so, they have one point in common. The equations have one common solution. The coordinates of the point of intersection is the common solution.

(b) The lines may be parallel. If so, they have no point in common. The equations have no common solution.

(c) The lines may coincide. That is, one line lies on top of the other. If so, any solution on one line or of one equation is also a solution of the other. Hence, there are infinitely many points of intersection and infinitely common solutions. Every solution of one equation is a solution of the other.

<u>Example 1</u>. Draw the graphs of $x - y = 2$ and $x + 3y = 6$ in the same number plane. Find the common solution of the equations.

Step 1: Draw the graph of $x - y = 2$. First, solve for y:

$$x - y = 2$$

$$x - y - x = 2 - x \qquad \text{Subtract } x \text{ from both sides.}$$

$$-y = 2 - x$$

$$\frac{-y}{-1} = \frac{2 - x}{-1} \qquad \text{Divide both sides by } -1.$$

$$y = -2 + x$$

Then, find three solutions:

x	$-2 + x$	y
2	$-2 + 2$	0
0	$-2 + 0$	-2
-2	$-2 + (-2)$	-4

Three solutions are $(2, 0)$, $(0, -2)$, and $(-2, -4)$. Plot the three points which correspond to the three ordered pairs above. Then draw a straight line through these three points, as in Fig. 9.8.

Step 2: Draw the graph of $x + 3y = 6$. First, solve for y:

$$x + 3y = 6$$

$$x + 3y - x = 6 - x \qquad \text{Subtract } x \text{ from both sides.}$$

$$3y = 6 - x$$

$$\frac{3y}{3} = \frac{6 - x}{3} \qquad \text{Divide both sides by 3.}$$

$$y = \frac{6 - x}{3}$$

x	$\dfrac{6 - x}{3}$	y
6	$\dfrac{6 - 6}{3} = \dfrac{0}{3}$	0
0	$\dfrac{6 - 0}{3} = \dfrac{6}{3}$	2
-3	$\dfrac{6 - (-3)}{3} = \dfrac{9}{3}$	3

Three solutions are $(6, 0)$, $(0, 2)$, and $(-3, 3)$. Plot the three points which correspond to the three ordered pairs. Then draw a straight line through these three points. The point that corresponds to $(3, 1)$ is the point of intersection. See Fig. 9.8. Therefore, $(3, 1)$ is the common solution. That is, $x = 3$ and $y = 1$ are the only values that satisfy both equations.

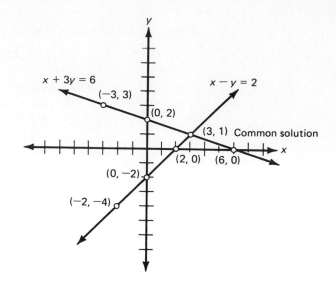

Figure 9.8

<u>Example 2</u>. Draw the graphs of $2x + 3y = 6$ and $4x + 6y = 30$ in the same number plane. Find the common solution of the equations.

<u>Step 1</u>: Draw the graph of $2x + 3y = 6$. First, solve for y:

$$2x + 3y = 6$$

$$2x + 3y - 2x = 6 - 2x \qquad \text{Subtract } 2x \text{ from both sides.}$$

$$3y = 6 - 2x$$

$$\frac{3y}{3} = \frac{6 - 2x}{3} \qquad \text{Divide both sides by 3.}$$

$$y = \frac{6 - 2x}{3}$$

Then, find three solutions:

x	$\dfrac{6 - 2x}{3}$	y
3	$\dfrac{6 - 2(3)}{3} = \dfrac{0}{3}$	0
0	$\dfrac{6 - 2(0)}{3} = \dfrac{6}{3}$	2
-3	$\dfrac{6 - 2(-3)}{3} = \dfrac{12}{3}$	4

Three solutions are (3, 0), (0, 2), and (-3, 4). Plot the three points which correspond to the three ordered pairs. Then draw a straight line through these three points.

Step 2: Draw the graph of $4x + 6y = 30$. First, solve for y:

$$4x + 6y = 30$$

$$4x + 6y - 4x = 30 - 4x \qquad \text{Subtract } 4x \text{ from both sides.}$$

$$6y = 30 - 4x$$

$$\frac{6y}{6} = \frac{30 - 4x}{6} \qquad \text{Divide both sides by 6.}$$

$$y = \frac{30 - 4x}{6}$$

Then, find three solutions:

x	$\dfrac{30 - 4x}{6}$	y
3	$\dfrac{30 - 4(3)}{6} = \dfrac{18}{6}$	3
0	$\dfrac{30 - 4(0)}{6} = \dfrac{30}{6}$	5
-6	$\dfrac{30 - 4(-6)}{6} = \dfrac{54}{6}$	9

Three solutions are (3, 3), (0, 5), and (-6, 9). Plot the three points which correspond to these ordered pairs in the same number plane as the points found in Step 1. Then draw a straight line through these three points. As you see in Fig. 9.9, the lines are parallel. The lines have no points in common, and, therefore, there is no common solution.

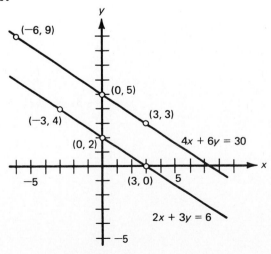

Figure 9.9

<u>Example 3</u>. Draw the graphs of $-2x + 8y = 24$ and $-3x + 12y = 36$ in the same number plane. Find the common solution of the equations.

<u>Step 1</u>: Draw the graph of $-2x + 8y = 24$. First, solve for y:

$$-2x + 8y = 24$$

$$-2x + 8y + 2x = 24 + 2x \qquad \text{Add } 2x \text{ to both sides.}$$

$$8y = 24 + 2x$$

$$\frac{8y}{8} = \frac{24 + 2x}{8} \qquad \text{Divide both sides by 8.}$$

$$y = \frac{24 + 2x}{8}$$

Then, find three solutions:

x	$\dfrac{24 + 2x}{8}$	y
-4	$\dfrac{24 + 2(-4)}{8} = \dfrac{16}{8}$	2
0	$\dfrac{24 + 2(0)}{8} = \dfrac{24}{8}$	3
8	$\dfrac{24 + 2(8)}{8} = \dfrac{40}{8}$	5

Three solutions are $(-4, 2)$, $(0, 3)$, and $(8, 5)$. Plot the three points which correspond to the three ordered pairs. Then draw a straight line through these three points.

<u>Step 2</u>: Draw the graph of $-3x + 12y = 36$. First, solve for y:

$$-3x + 12y = 36$$

$$-3x + 12y + 3x = 36 + 3x \qquad \text{Add } 3x \text{ to both sides.}$$

$$12y = 36 + 3x$$

$$\frac{12y}{12} = \frac{36 + 3x}{12} \qquad \text{Divide both sides by by 12.}$$

$$y = \frac{36 + 3x}{12}$$

Then, find three solutions:

x	$\dfrac{36 + 3x}{12}$	y
4	$\dfrac{36 + 3(4)}{12} = \dfrac{48}{12}$	4
-8	$\dfrac{36 + 3(-8)}{12} = \dfrac{12}{12}$	1
-5	$\dfrac{36 + 3(-5)}{12} = \dfrac{21}{12}$	$\dfrac{7}{4}$ or $1\dfrac{3}{4}$

Three solutions are $(4, 4)$, $(-8, 1)$, $\left(-5,\ 1\dfrac{3}{4}\right)$. Plot the three points which correspond to these ordered pairs in the same number plane as the points found in Step 1. Then draw a straight line through these points.

Note that the lines coincide. Any solution of one equation is also a solution of the other. Hence, there are infinitely many points of intersection and infinitely many common solutions (see Fig. 9.10).

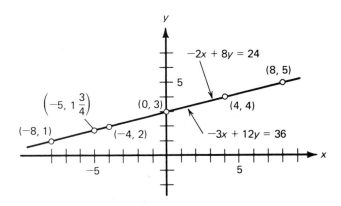

Figure 9.10

EXERCISES 9.3   *Draw the graphs of each pair of linear equations. Find the point of intersection. If the lines do not intersect, tell whether the lines are parallel or coincide:*

1. $y = 3x$
   $y = x + 4$

2. $x - y = 2$
   $x + 3y = 6$

3. $y = -x$
   $y - x = 2$

4. $x + y = 3$
   $2x + 2y = 6$

5. $x - 3y = 6$
   $2x - 6y = 18$

6. $x + y = 4$
   $2x + y = 5$

7. $2x - 4y = 8$
   $3x - 6y = 12$

8. $4x - 3y = 7$
   $6x + 5y = 8$

9. $3x - 6y = 12$
   $4x - 8y = 12$

10. $2x + y = 6$
    $2x - y = 6$

11. $3x + 2y = 10$
    $2x - 3y = 11$

12. $5x - y = 10$
    $x - 3y = -12$

13. $5x + 8y = -58$
    $2x + 2y = -18$

14. $6x + 2y = 24$
    $3x - 4y = 12$

15. $3x + 2y = 17$
    $x = 3$

16. $5x - 4y = 28$
    $y = -2$

17. $y = 2x$
    $y = -x + 2$

18. $y = -5$
    $y = x + 3$

19. $2x + y = 6$
    $y = -2x + 1$

20. $3x + y = -5$
    $2x + 5y = 1$

21. $4x + 3y = 2$
    $5x - y = 12$

22. $4x - 6y = 10$
    $2x - 3y = 5$

23. $2x - y = 9$
    $-2x + 3y = -11$

24. $x - y = 5$
    $2x - 3y = 5$

25. $8x - 3y = 0$
    $4x + 3y = 3$

26. $2x + 8y = 9$
    $4x + 4y = 3$

9.4 Solving Pairs of Linear Equations by Addition

Exercise 13 in Sec. 9.3 clearly illustrates that solving linear equations by graphing gives only approximate solutions. To obtain an exact solution, use the *addition method* outlined below.

Solving a Pair of Linear Equations by the Addition Method

<u>Step 1</u>: If necessary, multiply both sides of one or both equations by a number (or numbers) so that the numerical coefficients of one of the variables are negatives of each other.

<u>Step 2</u>: Add the two equations from Step 1 to obtain an equation containing one variable.

<u>Step 3</u>: Solve the equation from Step 2 for the one remaining variable.

<u>Step 4</u>: Solve for the second variable by substituting the solution from Step 3 in either of the original equations.

<u>Step 5</u>: Check your solution by substituting the ordered pair in the original equation not chosen in Step 4.

<u>Example 1</u>. Solve the following pair of linear equations by addition. Check your solution.

$$2x - y = 6$$
$$x + y = 9$$

Step 1 of the preceding rules is unnecessary since you can eliminate the y variable by adding the two equations as they are.

$$
\begin{array}{rcl}
2x - y &=& 6 \\
\underline{x + y} &=& \underline{9} \\
3x + 0 &=& 15 \\
3x &=& 15 \\
x &=& 5
\end{array}
$$

Now, substitute 5 for x in either of the original equations to solve for y. (Choose the simpler equation to make the arithmetic easier.)

$$x + y = 9$$
$$5 + y = 9$$
$$y = 4$$

The solution should be (5, 4).

Check: Use the remaining original equation, $2x - y = 6$, and substitute 5 for x and 4 for y.

$$2x - y = 6$$
$$2(5) - 4 = 6 \quad ?$$
$$10 - 4 = 6 \quad \text{True}$$

The solution checks. Thus, the solution is (5, 4).

Example 2. Solve the following pair of linear equations by addition. Check your solution.

$$2x + y = 5$$
$$x + y = 4$$

First, multiply both sides of the second equation by -1 to eliminate y by addition.

$$\begin{array}{l} 2x + y = 5 \\ \underline{(-1)(x + y) = (-1)(4)} \end{array} \quad \text{or} \quad \begin{array}{l} 2x + y = 5 \\ \underline{-x - y = -4} \\ x = 1 \end{array}$$

Now, substitute 1 for x in the equation $x + y = 4$ to solve for y.

$$x + y = 4$$
$$1 + y = 4$$
$$y = 3$$

The solution is (1, 3). The check is left to the student.

Example 3. Solve the following pair of linear equations by addition. Check your solution.

$$4x + 2y = 2$$
$$3x - 4y = 18$$

Multiply both sides of the first equation by 2 to eliminate y by addition.

$$\begin{array}{l} 2(4x + 2y) = 2(2) \\ \underline{3x - 4y = 18} \end{array} \quad \text{or} \quad \begin{array}{l} 8x + 4y = 4 \\ \underline{3x - 4y = 18} \\ \begin{array}{ll} 11x & = 22 \\ x & = 2 \end{array} \end{array}$$

Now, substitute 2 for x in the equation $4x + 2y = 2$ to solve for y.

$$4x + 2y = 2$$
$$4(2) + 2y = 2$$
$$8 + 2y = 2$$
$$2y = -6$$
$$y = -3$$

The solution should be $(2, -3)$.

Check: Use the equation $3x - 4y = 18$.

$$3x - 4y = 18$$
$$3(2) - 4(-3) = 18 \quad ?$$
$$6 + 12 = 18 \quad \textbf{True}$$

The solution checks. Thus, the solution is $(2, -3)$.

Example 4. Solve the following pair of **linear equations** by addition. Check your solution.

$$3x - 4y = 11$$
$$4x - 5y = 14$$

Multiply both sides of the first equation by 4. Then multiply both sides of the second equation by -3 to eliminate x by addition.

$$4(3x - 4y) = 4(11)$$
$$-3(4x - 5y) = -3(14)$$

or

$$12x - 16y = 44$$
$$-12x + 15y = -42$$
$$0 - 1y = 2$$
$$y = -2$$

Now, substitute -2 for y in the equation $3x - 4y = 11$ to solve for x.

$$3x - 4y = 11$$
$$3x - 4(-2) = 11$$
$$3x + 8 = 11$$
$$3x = 3$$
$$x = 1$$

The solution is $(1, -2)$.

In the preceding examples, we considered only pairs of linear equations with one common solution. Thus, the **graphs** of these equations intersect at a point, and the ordered pair that names this point is the common solution for the pair of equations. Sometimes, when solving a pair of linear equations by addition, the final statement is that two unequal numbers are equal. For example, the statement may be $0 = -2$. If so, the pair of equations does not have a common solution and the graphs of these equations are **parallel lines**.

Example 5. Solve the following pair of linear equations by addition.

$$2x + 3y = 7$$
$$4x + 6y = 12$$

Multiply both sides of the first equation by -2 to eliminate x by addition.

$$\begin{array}{r} -2(2x + 3y) = -2(7) \\ 4x + 6y = 12 \end{array} \quad \text{or} \quad \begin{array}{r} -4x - 6y = -14 \\ 4x + 6y = 12 \\ \hline 0 + 0 = -2 \\ 0 = -2 \end{array}$$

Since $0 \ne -2$, there is no common solution, and the graphs of these two equations are parallel lines.

If addition is used to solve a pair of linear equations and the resulting statement is $0 = 0$, then there are many common solutions. In fact, any solution of one equation is also a solution of the other equation. In this case, the graphs of the two equations coincide.

Example 6. Solve the following pair of linear equations by addition.

$$2x + 5y = 7$$
$$4x + 10y = 14$$

Multiply both sides of the first equation by -2 to eliminate x by addition.

$$\begin{array}{r} -2(2x + 5y) = -2(7) \\ 4x + 10y = 14 \end{array} \quad \text{or} \quad \begin{array}{r} -4x - 10y = -14 \\ 4x + 10y = 14 \\ \hline 0 + 0 = 0 \\ 0 = 0 \end{array}$$

Since $0 = 0$, there are many common solutions, and the graphs of the two equations coincide.

EXERCISES 9.4   *Solve each pair of linear equations by addition. If there is one common solution, give the ordered pair that names the point of intersection. If there is no one common solution, tell whether the lines are parallel or whether they coincide:*

1. $3x + y = 7$
   $x - y = 1$

2. $x + y = 8$
   $x - y = 4$

3. $2x + 5y = 18$
   $4x - 5y = 6$

4. $3x - y = 9$
   $2x + y = 6$

5. $-2x + 5y = 39$
   $2x - 3y = -25$

6. $-4x + 2y = 12$
   $-3x - 2y = 9$

7. $x + 3y = 6$
$x - y = 2$

8. $3x - 2y = 10$
$3x + 4y = 20$

9. $2x + 5y = 15$
$7x + 5y = -10$

10. $4x + 5y = -17$
$4x - y = 13$

11. $5x + 6y = 31$
$2x + 6y = 16$

12. $6x + 7y = 0$
$2x - 3y = 32$

13. $4x - 5y = 14$
$2x + 3y = -4$

14. $6x - 4y = 10$
$2x + y = 4$

15. $3x - 2y = -11$
$7x - 10y = -47$

16. $3x + 2y = 10$
$x + 5y = -27$

17. $x + 2y = -3$
$2x + y = 9$

18. $5x - 2y = 6$
$3x - 4y = 12$

19. $3x + 5y = 7$
$2x - 7y = 15$

20. $12x + 5y = 21$
$13x + 6y = 21$

21. $8x - 7y = -51$
$12x + 13y = 41$

22. $5x - 7y = -20$
$3x - 19y = -12$

23. $5x - 12y = -5$
$9x - 16y = -2$

24. $2x + 3y = 2$
$3x - 2y = 3$

25. $2x + 3y = 8$
$x + y = 2$

26. $4x + 7y = 9$
$12x + 21y = 12$

27. $3x - 5y = 7$
$9x - 15y = 21$

28. $2x - 3y = 8$
$4x - 3y = 0$

29. $2x + 5y = -1$
$3x - 2y = 8$

30. $3x - 7y = -9$
$2x + 14y = -6$

31. $16x - 36y = 70$
$4x - 9y = 17$

32. $8x + 12y = 36$
$16x + 15y = 45$

33. $4x + 3y = 17$
$2x - y = -4$

34. $12x + 15y = 36$
$7x - 12y = 187$

35. $2x - 5y = 8$
$4x - 10y = 16$

36. $3x - 2y = 5$
$7x + 3y = 4$

37. $5x - 8y = 10$
$-10x + 16y = 8$

38. $-3x + 2y = 5$
$-30x + y = 12$

39. $16x + 5y = 6$
$7x + \frac{5}{8}y = 2$

40. $3x - 10y = -21$
$5x + 4y = 27$

41. $8x - 5y = 426$
$7x - 2y = 444$

42. $\frac{1}{4}x - \frac{2}{5}y = 1$
$5x - 8y = 20$

43. $7x + 8y = 47$
$5x - 3y = 51$

44. $2x - 5y = 13$
$5x + 7y = 13$

9.5 Applications Involving Pairs of Linear Equations*

Often a technical application can be expressed mathematically as a system of linear equations. The procedure is similar to that outlined in Sec. 7.2, except that here you need to write two equations which express the information given in the problem and which involve both unknowns.

To solve applications involving equations with two variables:

1. Choose a different variable for each of the two unknowns you are asked to find. Write what each variable represents.

2. Write the problem as two equations using both variables. To obtain these two equations, look for two different relationships that express the two unknown quantities in equation form.

3. Solve this resulting system of equations using the methods given in this chapter.

4. Answer the question or questions asked in the problem.

5. Check your answers using the original problem.

<u>Example 1</u>. The sum of two voltages is 120 V. The difference between them is 24 V. Find each voltage.

$$\text{Let } x = \text{large voltage}$$
$$y = \text{small voltage}$$

The sum of two voltages is 120 V; that is,

$$x + y = 120$$

The difference between them is 24 V; that is,

$$x - y = 24$$

The system of equations is

$$
\begin{array}{rl}
x + y = & 120 \\
x - y = & 24 \\
\hline
2x = & 144 \\
x = & 72
\end{array}
\qquad \text{Add the equations.}
$$

*Do not use in this section the rules for calculating with measurements.

Substitute $x = 72$ in the equation $x + y = 120$ and solve for y.

$$x + y = 120$$
$$72 + y = 120$$
$$y = 48$$

Thus, the voltages are 72 V and 48 V.

Check: The sum of the voltages, 72 V + 48 V, is 120 V. The difference between them, 72 V - 48 V, is 24 V.

Example 2. How many pounds of feed mix A that is 75% corn and how many pounds of feed mix B that is 50% corn will need to be mixed to make a 400-lb mixture that is 65% corn?

Let x = number of pounds of mix A (75% corn), and
y = number of pounds of mix B (50% corn)

The sum of the two mixtures is 400 lb; that is,

$$x + y = 400$$

Then, 75% of x is corn plus 50% of y is corn results in a 400-lb final mixture that is 65% corn; that is,

$$0.75x + 0.50y = (0.65)(400)$$

or $\qquad 0.75x + 0.50y = 260$

The system of equations is

$$x + y = 400$$

$$0.75x + 0.50y = 260$$

First, let's multiply both sides of the second equation by 100 to eliminate the decimals.

$$x + y = 400$$

$$75x + 50y = 26,000$$

Then, multiply both sides of the first equation by -50 to eliminate y by addition.

$$
\begin{array}{r}
-50x - 50y = -20,000 \\
75x + 50y = 26,000 \\
\hline
25x = 6,000
\end{array}
$$

$$\frac{25x}{25} = \frac{6000}{25} \qquad \text{Divide both sides by 25.}$$

$$x = 240$$

Now, substitute 240 for x in the equation $x + y = 400$ to solve for y.

$$x + y = 400$$
$$240 + y = 400$$
$$240 + y - 240 = 400 - 240$$
$$y = 160$$

Therefore, we need 240 lb of mix A and 160 lb of mix B.

Check: Use the remaining original equation, $0.75x + 0.50y = 260$, and substitute 240 for x and 160 for y.

$$0.75x + 0.50y = 260$$
$$0.75(240) + 0.50(160) = 260 \quad ?$$
$$180 + 80 = 260 \quad \text{True}$$

Example 3. A company sells two grades of sand. One grade sells for 15¢/lb and the other sells for 25¢/lb. How much of each grade needs to be mixed to obtain a 1000 lb mixture worth 18¢/lb?

Let x = amount of sand selling at 15¢/lb
y = amount of sand selling at 25¢/lb

The total amount of sand is 1000 lb; that is,

$$x + y = 1000$$

One grade sells at 15¢/lb and the other sells for 25¢/lb to obtain a 1000 lb mixture worth 18¢/lb. Here we need to write an equation that relates the cost of the sand; that is, the cost of the sand separately equals the cost of the sand mixed.

$$15x + 25y = 18(1000)$$

That is, the cost of x pounds of sand at 15¢/lb is $15x$ cents.

And, the cost of y pounds of sand at 25¢/lb is $25y$ cents.

And, the cost of 1000 pounds of sand at 18¢/lb is 18(1000).

Therefore, the system of equations is

$$x + y = 1000$$
$$15x + 25y = 18000$$

Multiply the first equation by −15 to eliminate x by addition.

$$-15x - 15y = -15,000$$
$$\underline{15x + 25y = 18,000}$$
$$10y = 3,000$$

$$y = 300$$

Substitute $y = 300$ in the equation $x + y = 1000$ and solve for x.

$$x + y = 1000$$
$$x + 300 = 1000$$
$$x = 700$$

That is, 700 lb of sand selling at 15¢/lb and 300 lb of sand selling at 25¢/lb are needed to be mixed to obtain a 1000-lb mixture worth 18¢/lb.

Check: Left to the student.

EXERCISES 9.5

1. A board 96 cm long is cut into two pieces so that one piece is 12 cm longer than the other. Find the length of each piece.

2. Find the capacity of two trucks if 6 trips of the smaller and 4 trips of the larger make a total haul of 36 tons. And, 8 trips of the larger and 4 trips of the smaller make a total haul of 48 tons.

3. A plumbing contractor decides to field-test two new pumps. One is rated at 180 gal/h and the other at 250 gal/h. He tests one, then the other. Over a period of 6 h he pumps a total of 1325 gal. Assume both pumps operated as rated. How long was each in operation?

4. A bricklayer works at a job, laying an average of 150 bricks per hour. During the job he is called away and replaced by a less experienced man who averages only 120 bricks an hour. The two men laid 930 bricks in 7 h. How long did each one work?

5. In a certain concrete mix there was four times as much gravel as cement. The total volume of gravel and cement in the mix was 22.5 ft^3. How much of each was in the mix?

6. A contractor going over receipts finds a bill for $225, for 720 ceiling tiles. He knows that there were actually two types of tiles used; one selling at 25¢ a tile and the other at 40¢ a tile. How many of each type were used?

7. A farmer has two types of feed. One has 5% digestible protein and the other 15% digestible protein. How much of each type will he need to mix 100 lb of 12%-digestible-protein feed?

8. A dairyman wants to make 125 lb of 12%-butterfat cream. How many pounds of 40%-butterfat cream and how many pounds of 2%-butterfat milk would he have to mix?

9. A farmer sells part of his corn and soybean crop to an elevator. He gets $3.20/bu for his corn and $5.80/bu for his beans. The entire 3150 bu brought him $11,250. How much of each crop did he sell?

10. A farmer has a 1.4% solution and a 2.9% solution of a certain pesticide. How much of each would he mix to get 2000 gal of 2% solution for his sprayer?

11. A farmer has a 6% solution and a 12% solution of pesticide. How much of each must he mix to have 300 gal of an 8% solution for his sprayer?

12. The sum of two capacitors is 85 microfarads (μF). The difference between them is 25 μF. Find the size of each capacitor.

13. Nine batteries are hooked in series to provide a 33-V power source. Some of the batteries are 3 V and some are 4.5 V. How many of each type are used?

14. In a parallel circuit the total current is 1.25 A through the two branches. One branch has a resistance of 50 Ω and the other a resistance of 200 Ω. What current is flowing through each of the branches? Note. In a parallel circuit, the products of the current in amperes and the resistance in ohms are equal in all branches.

15. How much of an 8% solution and a 12% solution would you use to make 140 mL of a 9%-electrolyte solution?

16. In a parallel circuit with seven branches the total current is 1.95 A. Some of the branches have currents of 0.25 A and others 0.35 A. How many of each type of branch are there in the circuit? Note. The total current in a parallel circuit is equal to the sum of the currents in each branch.

17. A small, single-cylinder engine is operated on a test stand for 14 min. It was run first at 850 rpm, then run at 1250 rpm, during the test. A total of 15,500 revolutions was counted during the test. How long was the engine operated at each speed?

18. In testing a hybrid engine, various mixtures of gasoline and methanol are being tried. How much of a 95%-gasoline mixture and how much of an 80%-gasoline mixture would be needed to make 240 gal of a 90%-gasoline mixture?

19. An engine on a test stand is operated at two fixed settings, each with an appropriate load. At the first setting, fuel consumption was 1 gal every 12 min. At the second setting it was 1 gal every 15 min. The test took 5 h, and 22 gal of fuel were used. How long did the engine run at each setting?

20. A mechanic stores a parts cleaner as a 65% solution, which is to be diluted to a 25% solution for use. Someone accidentally prepares a 15% solution. How much of the 65% solution and the 15% solution should be mixed to make 100 gal of the 25% solution?

21. A salesman turns in a ticket on two carpets for $1210. He sold a total of 75 yd^2 of carpet. One type was worth $14.50/yd^2 and the second was worth $18/yd^2. He neglects to note, however, how much of each type he sold. How much did he sell of each type?

22. A certain apartment owner rents one-bedroom apartments for $325 and two-bedroom apartments for $410. He has a total of 13 apartments rented for $4905 a month. How many of each type does he have?

23. A sporting goods store carries two types of snorkels. One sells for $14.95 and another sells for $21.75. Records for July show that 23 snorkels were sold, for $357.45. How many of each type were sold?

24. A butcher has some 65%-lean hamburger and some 85%-lean hamburger. How much of each should he use to make 160 lb of 80%-lean hamburger?

25. A hospital has 35%-saline solution on hand. How much water and how much of this solution would you use to prepare 140 mL of a 20%-saline solution?

26. A nurse is to administer 1000 cm^3 of an intravenous (I.V.) solution over a period of 8 h. At first it is to be given at a rate of 140 cm^3/h, then given at a reduced rate of 100 cm^3/h. How long should it be administered at each rate?

27. A hospital has a 4%-saline solution and an 8%-saline solution on hand. How much of each should be used to prepare 1000 cm^3 of 5%-saline solution?

28. A certain medication is available in 2-cm^3 vials and in 5-cm^3 vials. In a certain month 42 vials were used, totaling 117 cm^3 of medication. How many of each type of vial were used?

9.6 Solving Pairs of Linear Equations by Substitution

For many problems, there is an easier method than addition for finding exact solutions. This method is called *substitution*. Use substitution when one or both equations have one variable alone as one member.

Solving a Pair of Linear Equations by the Substitution Method

1. From either of the two given equations, solve for one variable in terms of the other.
2. Substitute this result from Step 1 into the remaining equation. Note that this step eliminates one variable.
3. Solve the equation from Step 2 for the remaining variable.
4. Solve for the second variable by substituting the solution from Step 3 into the equation resulting from Step 1.
5. Check your solution by substituting the ordered pair in the original equation not used in Step 1.

<u>Example 1</u>. Solve the following pair of linear equations by substitution. Check your solution.

$$x + 3y = 15$$
$$x = 2y$$

First, substitute $2y$ for x in the first equation.

$$x + 3y = 15$$
$$2y + 3y = 15$$
$$5y = 15$$
$$y = 3$$

Now, substitute 3 for y in the equation $x = 2y$ to solve for x.

$$x = 2y$$
$$x = 2(3)$$
$$x = 6$$

The solution should be (6, 3).

<u>Check</u>: Use the equation $x + 3y = 15$. Substitute 6 for x and 3 for y.

$$x + 3y = 15$$
$$6 + 3(3) = 15 \qquad ?$$
$$6 + 9 = 15 \qquad ?$$
$$15 = 15 \qquad \text{True}$$

Thus, the solution is (6, 3).

The addition method is preferred if the pair of linear equations has no numerical coefficients equal to 1, or if one variable alone is not one member of one of the equations. For example, the addition method should be used to solve the following pair of linear equations.

$$3x + 4y = 7$$
$$5x + 7y = 12$$

<u>Example 2</u>. A rectangular yard (Fig. 9.10) is to be fenced in so that the length is twice the width. The length of the 80-ft house is used to enclose part of one side of the yard. If 580 ft of fencing are used, what are the dimensions of the yard?

$$\text{Let } x = \text{length of the yard, and}$$
$$y = \text{width of the yard}$$

The amount of fencing used (two lengths plus two widths minus the length of the house) is 580 ft; that is,

$$2x + 2y - 80 = 580$$

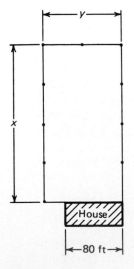

Figure 9.10

or

$$2x + 2y = 660$$

The length of the yard is twice the width; that is,

$$x = 2y$$

The system of equations is

$$2x + 2y = 660$$
$$x = 2y$$

Substitute $2y$ for x in the equation $2x + 2y = 660$ and solve for y.

$$2(2y) + 2y = 660$$
$$4y + 2y = 660$$
$$6y = 660$$
$$y = 110$$

Now, substitute 110 for y in the equation $x = 2y$ and solve for x.

$$x = 2y$$
$$x = 2(110)$$
$$x = 220$$

Therefore, the length is 220 ft and the width is 110 ft. The check is left to the student.

EXERCISES 9.6   *Solve, using the substitution method, and check:*

1. $2x + y = 12$
   $y = 3x$

2. $3x + 4y = -8$
   $x = 2y$

3. $5x - 2y = 46$
   $x = 5y$

4. $2x - y = 4$
   $y = -x$

5. $3x + 2y = 30$
   $x = y$

6. $3x - 2y = 49$
   $y = -2x$

7. $5x - y = 18$
   $y = \frac{1}{2}x$

8. $15x + 3y = 9$
   $y = -2x$

9. $x - 6y = 3$
   $3y = x$

10. $4x + 5y = 10$
    $4x = -10y$

11. $3x + y = 7$
    $4x - y = 0$

12. $5x + 2y = 1$
    $y = -3x$

13. $4x + 3y = -2$
    $x + y = 0$

14. $7x + 8y + 93 = 0$
    $y = 3x$

15. $6x - 8y = 115$
    $x = -\frac{y}{5}$

16. $2x + 8y = 12$
    $x = -4y$

17. $3x + 8y = 27$
    $y = 2x + 1$

18. $4x - 5y = -40$
    $x = 3 - 2y$

19. $8y - 2x = -34$
    $x = 1 - 4y$

20. $2y + 7x = 48$
    $y = 3x - 2$

21. The sum of two resistors is 550 Ω. One is 4.5 times the other. Find the size of each resistor.

22. In one concrete mix there is four times as much gravel as cement. The total volume is 15 yd³. How much of each ingredient is in the mix?

23. A wire 120 cm long is to be cut into two pieces so that one piece is three times as long as the other. Find the length of each piece.

24. The sum of two resistances is 1500 Ω. The larger is four times the smaller. Find the size of each resistance.

25. A rectangle is twice as long as it is wide. Its perimeter is 240 cm. Find the length and the width of the rectangle.

26. The sum of two voltages is 84 V. One voltage is 12 V larger than the other. Find each voltage.

27. The sum of three currents is 210 mA. Two currents are the same. The third is five times either of the other two. Find the third current.

28. Together Bryan and George make $480 to complete a drafting project. George receives $\frac{2}{3}$ of what Bryan does. How much does each receive?

Chapter 9 Review

Write the ordered pair corresponding to each point below:

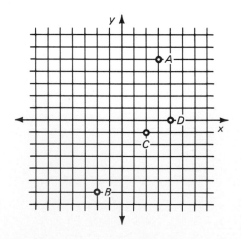

1. A   2. B   3. C   4. D

Plot each point in the number plane. Label each point by writing its ordered pair:

  5. (3, −2)   6. (−7, −4)   7. (−1, 5)   8. (0, −5)

Complete the ordered pair solutions of each equation:

  9. $x + 2y = 8$ (3,) (0,) (−4,)

 10. $2x − 3y = 12$ (3,) (0,) (−3,)

 11. Solve for y: $6x + y = 15$.
 12. Solve for y: $3x − 5y = −10$.
 13. Graph the equation $2x − 3y = 6$.
 14. Draw the graphs of $x + y = 6$ and $2x − y = 3$ on the same set of coordinate axes. Find the point of intersection.

Solve each system of equations:

 15. $x + y = 7$ 16. $3x + 2y = 11$
      $2x − y = 2$ $x + 2y = 5$

 17. $3x − 5y = −3$ 18. $2x − 3y = 1$
     $2x − 3y = −1$ $4x − 6y = 5$

 19. $x + 2y = 3$ 20. $3x + 5y = 52$
      $3x + 6y = 9$ $y = 2x$

 21. $5y − 4x = −6$ 22. $3x − 7y = −69$
     $x = \dfrac{1}{2}y$ $y = 4x + 5$

 23. You can buy 20 resistors and 8 capacitors for $3.60, or 60 resistors and 40 capacitors for $14. Find the price of each.
 24. The sum of the length and width of a rectangular lot is 190 ft. The lot is 75 ft deeper than it is wide. Find the length and width of the lot.

10

FACTORING ALGEBRAIC EXPRESSIONS

10.1 Finding Monomial Factors

Factoring an algebraic expression, like finding the prime factors of a number, means *writing the expression as a product*. The prime factorization of 12 is 2 • 2 • 3. Other factorizations of 12 are 2 • 6 and 4 • 3.

To factor the expression $2x + 2y$, notice that 2 is a factor common of both terms of the expression. In other words, 2 is a factor of $2x + 2y$. To find the other factor, divide by 2.

$$\frac{2x + 2y}{2} = \frac{2x}{2} + \frac{2y}{2} = x + y$$

Therefore, a factorization of $2x + 2y$ is $2(x + y)$.

When a factor such as 2 divides *each* term of an algebraic expression, it is called a *monomial factor*. (Recall that a monomial is an expression that is a single term.) When factoring any algebraic expression, *always look first for monomial factors* that are common to all terms.

Example 1. Factor $3a + 6b$.

First, look for a common monomial factor. Since 3 divides both $3a$ and $6b$, 3 is a common monomial factor of $3a + 6b$. Divide $3a + 6b$ by 3.

$$\frac{3a + 6b}{3} = \frac{3a}{3} + \frac{6b}{3} = a + 2b$$

Thus, a factorization of $3a + 6b$ is $3(a + 2b)$. Check this result by multiplication: $3(a + 2b) = 3a + 6b$.

<u>Example 2</u>. Factor $4x^2 + 8x + 12$.

Since 4 divides each term of the expression, divide $4x^2 + 8x + 12$ by 4 to obtain the other factor.

$$\frac{4x^2 + 8x + 12}{4} = \frac{4x^2}{4} + \frac{8x}{4} + \frac{12}{4}$$

$$= x^2 + 2x + 3$$

Thus, a factorization of $4x^2 + 8x + 12$ is $4(x^2 + 2x + 3)$.

To check your work, multiply 4 and $(x^2 + 2x + 3)$. You should get the original expression as this product.

In Example 2, note that 2 is also a common factor of each term of the expression. However, 4 is the greatest common factor. *When factoring, always choose the monomial factor that is the greatest common factor.*

<u>Example 3</u>. Factor $9ax - 6ay$.

Note that 3 divides both $9ax$ and $6ay$, so 3 is a common factor of $9ax - 6ay$. However, a also divides $9ax$ and $6ay$ so a is also a common factor of $9ax - 6ay$. We are looking for the greatest common factor (GCF), which in this case is $3a$, so we divide $9ax - 6ay$ by $3a$ to obtain the other factor.

$$\frac{9ax - 6ay}{3a} = \frac{9ax}{3a} - \frac{6ay}{3a}$$

$$= 3x - 2y$$

Thus, a factorization of $9ax - 6ay$ is $3a(3x - 2y)$. Note that $3(3ax - 2ay)$ or $a(9x - 6y)$ are also factored forms of $9ax - 6ay$. However, we must always use the monomial factor that is the greatest common factor.

EXERCISES 10.1   *Factor:*

1. $4a + 4$ 2. $3x - 6$ 3. $bx + by$
4. $9 - 18y$ 5. $15b - 20$ 6. $12ab + 12ac$
7. $x^2 - x$ 8. $3x^2 - 6x$ 9. $a^2 - 4a$
10. $7xy - 21y$ 11. $4n^2 - 8n$ 12. $x^2 + 5x$
13. $10x^2 + 25x$ 14. $y^2 - 8y$ 15. $3r^2 - 6r$
16. $x^3 + 13x^2 + 25x$
17. $4x^4 + 8x^3 + 12x^2 - 24x$
18. $3a^2b^2c^2 + 27a^3b^3c^2 - 81abc$
19. $9a^2 - 9ax^2$ 20. $a - a^3$

21. $10x + 10y - 10z$ 22. $2x^2 - 2x$

23. $3y - 6$ 24. $y - 3y^2$

25. $14xy - 7x^2y^2$ 26. $25a^2 - 25b^2$

27. $12x^2m - 7m$ 28. $90r^2 - 10R^2$

29. $60ax - 12a$ 30. $2x^2 - 100x^3$

31. $52m^2n^2 - 13mn$ 32. $40x - 8x^3 + 4x^4$

33. $52m^2 - 14m + 2$ 34. $27x^3 - 54x$

35. $36y^2 - 18y^3 + 54y^4$ 36. $20y^3 - 10y^2 + 5y$

37. $6m^4 - 12m^2 + 3m$ 38. $-16x^3 - 32x^2 - 16x$

39. $-4x^2y^3 - 6x^2y^4 - 10x^2y^5$ 40. $18x^3y - 30x^4y + 48xy$

10.2   Finding the Product of Two Binomials Mentally

In Sec. 6.5, you learned how to multiply two binomials such as $(2x + 3)(4x - 5)$ by the following method:

<u>Long Method</u>

$$
\begin{array}{r}
2x + 3 \\
4x - 5 \\
\hline
-10x - 15 \\
8x^2 + 12x \phantom{{}-15} \\
\hline
8x^2 + 2x - 15
\end{array}
$$

This process of multiplying two binomials can be shortened by following the three steps below:

Finding the Product of Two Binomials Mentally

1. The <u>first term</u> of the product is the product of the first terms of the binomials.
2. The <u>middle term</u> of the product is the sum of the outer product and the inner product of the binomials.
3. The <u>last term</u> of the product is the product of the last terms of the binomials.

Let's look again at the product $(2x + 3)(4x - 5)$ by using the above steps:

inner
product

$$(2x + 3)(4x - 5)$$

outer
product

Step 1: Product of the first terms: $(2x)(4x) = 8x^2$

Step 2: Outer product: $(2x)(-5) = -10x$
        Inner product: $(3)(4x) = \underline{12x}$
        Sum: $= \quad 2x$

Step 3: Product of the last terms: $(3)(-5) = \underline{\quad\quad -15}$

    Therefore, $(2x + 3)(4x - 5)$ $= 8x^2 + 2x - 15$

Note that in each method we found the exact same terms. The second method is much quicker, especially when you become more familiar and successful with it. The reason for the second method is that it is commonly used to factor polynomials. Factoring polynomials is the content of the rest of this chapter and is also a necessary part of the next chapter. Therefore, it is very important that you can successfully and quickly find the product of two binomials, mentally, before proceeding with the next section.

    <u>Example 1.</u> Find the product $(2x - 7)(3x - 4)$ mentally.

inner
product

$$(2x - 7)(3x - 4)$$

outer
product

Step 1: Product of first terms: $(2x)(3x) = 6x^2$

Step 2: Outer product: $(2x)(-4) = -8x$
        Inner product: $(-7)(3x) = \underline{-21x}$
        Sum: $= \quad -29x$

Step 3: Product of last terms: $(-7)(-4) = \underline{\quad\quad 28}$

    Therefore, $(2x - 7)(3x - 4)$ $= 6x^2 - 29x + 28$

Example 2. Find the product $(x + 4)(3x + 5)$ mentally.

inner
product

$$(x + 4)(3x + 5)$$

outer
product

Step 1: $(x)(3x)$ $\qquad = 3x^2$

Step 2: $(x)(5) = 5x$
$\qquad (4)(3x) = 12x$
$\qquad$ Sum $\qquad = 17x$

Step 3: $(4)(5)$ $\qquad = 20$

Therefore, $(x + 4)(3x + 5) = 3x^2 + 17x + 20$

By now, you should be writing only the final result of each product. If you need some help, refer to the three steps on page 354 and the outline shown in Examples 1 and 2.

Example 3. Find the product $(x + 8)(x + 5)$ mentally.

$$(x + 8)(x + 5) = x^2 + (5x + 8x) + 40$$
$$= x^2 + 13x + 40$$

Example 4. Find the product $(x - 6)(x - 9)$ mentally.

$$(x - 6)(x - 9) = x^2 + (-9x - 6x) + 54$$
$$= x^2 - 15x + 54$$

Example 5. Find the product $(x + 2)(x - 5)$ mentally.

$$(x + 2)(x - 5) = x^2 + (-5x + 2x) - 10$$
$$= x^2 - 3x - 10$$

Example 6. Find the product $(4x + 1)(5x + 8)$ mentally.

$$(4x + 1)(5x + 8) = 20x^2 + (32x + 5x) + 8$$
$$= 20x^2 + 37x + 8$$

Example 7. Find the product $(6x + 5)(2x - 3)$ mentally.

$$(6x + 5)(2x - 3) = 12x^2 + (-18x + 10x) - 15$$
$$= 12x^2 - 8x - 15$$

$\underline{\text{Example 8.}}$ Find the product $(4x - 5)(4x - 5)$ mentally.

$$(4x - 5)(4x - 5) = 16x^2 + (-20x - 20x) + 25$$
$$= 16x^2 - 40x + 25$$

EXERCISES 10.2   *Find each product mentally:*

1. $(x + 5)(x + 2)$ 2. $(x + 3)(2x + 7)$

3. $(2x + 3)(3x + 4)$ 4. $(x + 3)(x + 18)$

5. $(x - 5)(x - 6)$ 6. $(x - 9)(x - 8)$

7. $(x - 12)(x - 2)$ 8. $(x - 9)(x - 4)$

9. $(x + 8)(2x + 3)$ 10. $(3x - 7)(2x - 5)$

11. $(x + 6)(x - 2)$ 12. $(x - 7)(x - 3)$

13. $(x - 9)(x - 10)$ 14. $(x - 9)(x + 10)$

15. $(x - 12)(x + 6)$ 16. $(2x + 7)(4x - 5)$

17. $(2x - 7)(4x + 5)$ 18. $(2x - 5)(4x + 7)$

19. $(2x + 5)(4x - 7)$ 20. $(6x + 5)(5x - 1)$

21. $(7x + 3)(2x + 5)$ 22. $(5x - 7)(2x + 1)$

23. $(x - 9)(3x + 8)$ 24. $(x - 8)(2x + 9)$

25. $(6x + 5)(x + 7)$ 26. $(16x + 3)(x - 1)$

27. $(13x - 4)(13x - 4)$ 28. $(12x + 1)(12x + 5)$

29. $(10x + 7)(12x - 3)$ 30. $(10x - 7)(12x + 3)$

31. $(10x - 7)(10x - 3)$ 32. $(10x + 7)(10x + 3)$

33. $(2x - 3)(2x - 5)$ 34. $(2x + 3)(2x + 5)$

35. $(2x - 3)(2x + 5)$ 36. $(2x + 3)(2x - 5)$

37. $(3x - 8)(2x + 7)$ 38. $(3x + 8)(2x - 7)$

39. $(3x + 8)(2x + 7)$ 40. $(3x - 8)(2x - 7)$

41. $(8x - 5)(2x + 3)$ 42. $(x - 7)(x + 5)$

43. $(y - 7)(2y + 3)$ 44. $(m - 9)(m + 2)$

45. $(3n - 6y)(2n + 5y)$ 46. $(6a - b)(2a + 3b)$

47. $(4x - y)(2x + 7y)$ 48. $(8x - 12)(2x + 3)$

49. $(\frac{1}{2}x - 8)(\frac{1}{4}x - 6)$ 50. $(\frac{2}{3}x - 6)(\frac{1}{3}x + 9)$

10.3 Finding Binomial Factors

The factors of a trinomial are often *binomial factors*. To find these binomial factors you must "undo" the process of multiplication as presented in Sec. 10.2. The following steps will enable you to undo the multiplication in a trinomial such as $x^2 + 7x + 10$.

Step 1: Factor any common monomial; $x^2 + 7x + 10$ has no common factor.

Step 2: If $x^2 + 7x + 10$ can be factored, the two factors will probably be binomials. Write parentheses for the binomials.

$$x^2 + 7x + 10 = (\qquad)(\qquad)$$

Step 3: The product of the two first terms of the binomials is the first term of the trinomial. So, the first term in each binomial must be x.

$$x^2 + 7x + 10 = (x\qquad)(x\qquad)$$

Step 4: Here all the signs of the trinomial are positive. So, the signs in the binomials are also positive.

$$x^2 + 7x + 10 = (x + \quad)(x + \quad)$$

Step 5: Find the last terms of the binomials by finding two numbers that have a product of +10 and a sum of +7. The only possible factorizations of 10 are 10 · 1 and 5 · 2. The sums of the pairs of factors are 10 + 1 = 11 and 5 + 2 = 7. Thus, the numbers you want are 5 and 2.

$$x^2 + 7x + 10 = (x + 5)(x + 2)$$

Step 6: Multiply the two binomials as a check to see if their product is the same as the original trinomial.

Example 1. Factor the trinomial $x^2 + 15x + 56$.

Step 1: $x^2 + 15x + 56$ has no common monomial factor.

Step 2: $x^2 + 15x + 56 = (\qquad)(\qquad)$

Step 3: $x^2 + 15x + 56 = (x\qquad)(x\qquad)$

Step 4: $x^2 + 15x + 56 = (x + \quad)(x + \quad)$ All the signs of the trinomial are positive.

To determine which factors of 56 to use, list all possible pairs.

$$
\begin{array}{ll}
1 \cdot 56 = 56 & 1 + 56 = 57 \\
2 \cdot 28 = 56 & 2 + 28 = 30 \\
4 \cdot 14 = 56 & 4 + 14 = 18 \\
7 \cdot 8 = 56 & 7 + 8 = 15
\end{array}
$$

Since the coefficient of x in the trinomial is 15, you choose 7 and 8 for the second terms of the binomial factors. There are no other pairs of positive whole numbers with a product of 56 and a sum of 15.

Step 5: $x^2 + 15x + 56 = (x + 7)(x + 8)$

In actual work, all five steps above are completed in one or two lines, depending on whether or not there is a common monomial.

Step 6: Check:

$$(x + 7)(x + 8) = x^2 + 15x + 56$$

Example 2. Factor the trinomial $x^2 - 13x + 36$.

Note that the only difference between this trinomial and the ones we have considered previously is the sign of the second term. Here, the sign of the second term is negative instead of positive. Thus, the steps for factoring will be the same except for Step 4.

Step 1: $x^2 - 13x + 36$ has no common monomial factor.

Step 2:
$$
\begin{aligned}
x^2 - 13x + 36 &= x^2 + (-13)x + 36 \\
&= (\quad)(\quad)
\end{aligned}
$$

Step 3: $x^2 + (-13)x + 36 = (x\quad)(x\quad)$

Note that the sign of the third term (+36) is positive and the coefficient of the second term (-13x) is negative. Since 36 is positive, the two factors of +36 must have like signs; and since the coefficient of -13x is negative, the signs in the two factors must be negative.

Step 4: $x^2 + (-13)x + 36 = (x - \quad)(x - \quad)$

Since $(-9)(-4) = 36$ and $(-9) + (-4) = -13$, these are the factors of 36 to be used.

Step 5: $x^2 - 13x + 36 = (x - 9)(x - 4)$

Step 6: Check: $(x - 9)(x - 4) = x^2 - 13x + 36$

Example 3. Factor the trinomial $3x^2 + 12x - 36$.

Step 1: $\qquad\qquad 3x^2 + 12x - 36 = 3(x^2 + 4x - 12)$

Step 2: $\qquad\qquad x^2 + 4x + (-12) = (\quad)(\quad)$

Step 3: $\qquad\qquad x^2 + 4x + (-12) = (x\quad)(x\quad)$

Note that the last term of the trinomial (−12) is negative. This means that the two factors of −12 must have unlike signs since a positive number times a negative number gives a negative number.

Step 4: $\qquad\qquad x^2 + 4x - 12 = (x +\quad)(x -\quad)$

You need to find two integers with a product of −12 and a sum of +4. All possible pairs of factors are shown below.

$$
\begin{array}{ll}
(-12)(+1) = -12 & (-12) + (+1) = -11 \\
(+12)(-1) = -12 & (+12) + (-1) = 11 \\
(+6)(-2) = -12 & (+6) + (-2) = 4 \\
(-6)(+2) = -12 & (-6) + (+2) = -4 \\
(-4)(+3) = -12 & (-4) + (+3) = -1 \\
(+4)(-3) = -12 & (+4) + (-3) = 1 \\
\end{array}
$$

From these possibilities, you see that the two integers with a product of −12 and a sum of +4 are +6 and −2. Write these numbers as the last terms of the binomials.

Step 5: $\qquad\qquad x^2 + 4x - 12 = (x + 6)(x - 2)$

Step 6: Check: $(x + 6)(x - 2) = x^2 + 4x - 12$

Thus, $3(x^2 + 4x - 12)$ equals the original trinomial, $3x^2 + 12x - 36$.

And, $\qquad\qquad 3x^2 + 12x - 36 = 3(x + 6)(x - 2)$

Example 4. Factor the trinomial $x^2 - 11x - 12$.

The signs of the factors of −12 must be different. From the list in Example 3, choose the two factors with a sum of −11.

$$x^2 - 11x - 12 = (x - 12)(x + 1)$$

Check: $\qquad\qquad (x - 12)(x + 1) = x^2 - 11x - 12$

To factor a trinomial $x^2 + bx + c$, use the following steps. Assume that b and c are both positive numbers.

Step 1: First, look for any common monomial factors.

Step 2: For the trinomial $x^2 + bx + c$, use the following form.

$$x^2 + bx + c = (x + \quad)(x + \quad)$$

Step 3: For the trinomial $x^2 - bx + c$, use the following form.

$$x^2 - bx + c = (x - \quad)(x - \quad)$$

Step 4: For the trinomials $x^2 - bx - c$ and $x^2 + bx - c$, use the forms:

$$x^2 - bx - c = (x + \quad)(x - \quad)$$
$$x^2 + bx - c = (x + \quad)(x - \quad)$$

EXERCISES 10.3   *Factor each trinomial completely:*

1. $x^2 + 6x + 8$

2. $x^2 + 8x + 15$

3. $y^2 + 9y + 20$

4. $2w^2 + 20w + 32$

5. $3r^2 + 30r + 75$

6. $a^2 + 14a + 24$

7. $b^2 + 11b + 30$

8. $c^2 + 21c + 54$

9. $x^2 + 17x + 72$

10. $y^2 + 18y + 81$

11. $5a^2 + 35a + 60$

12. $r^2 + 12r + 27$

13. $x^2 - 7x + 12$

14. $y^2 - 6y + 9$

15. $2a^2 - 18a + 28$

16. $c^2 - 9c + 18$

17. $3x^2 - 30x + 63$

18. $r^2 - 12r + 35$

19. $w^2 - 13w + 42$

20. $x^2 - 14x + 49$

21. $x^2 - 19x + 90$

22. $4x^2 - 84x + 80$

23. $t^2 - 12t + 20$

24. $b^2 - 15b + 54$

25. $x^2 + 2x - 8$

26. $x^2 - 2x - 15$

27. $y^2 + y - 20$

28. $2w^2 - 12w - 32$

29. $a^2 + 5a - 24$

30. $b^2 + b - 30$

31. $c^2 - 15c - 54$ 32. $b^2 - 6b - 72$

33. $3x^2 - 3x - 36$ 34. $a^2 + 5a - 14$

35. $c^2 + 3c - 18$ 36. $x^2 - 4x - 21$

37. $y^2 + 17y + 42$ 38. $m^2 - 18m + 72$

39. $r^2 - 2r - 35$ 40. $x^2 + 11x - 42$

41. $m^2 - 22m + 40$ 42. $y^2 + 17y + 70$

43. $x^2 - 9x - 90$ 44. $x^2 - 8x + 15$

45. $a^2 + 27a + 92$ 46. $x^2 + 17x - 110$

47. $2a^2 - 12a - 110$ 48. $y^2 - 14y + 40$

49. $a^2 + 29a + 100$ 50. $y^2 + 14y - 120$

51. $y^2 - 14y - 95$ 52. $b^2 + 20b + 36$

53. $y^2 - 18y + 32$ 54. $x^2 - 8x - 128$

55. $7x^2 + 7x - 14$ 56. $2x^2 - 6x - 36$

57. $6x^2 + 12x - 6$ 58. $4x^2 + 16x + 16$

59. $y^2 - 12y + 35$ 60. $a^2 + 16a + 63$

61. $a^2 + 2a - 63$ 62. $y^2 - y - 42$

63. $x^2 + 18x + 56$ 64. $x^2 + 11x - 26$

65. $2y^2 - 36y + 90$ 66. $ax^2 + 2ax + a$

67. $3xy^2 - 18xy + 27x$ 68. $x^2 - x - 156$

69. $x^2 + 30x + 225$ 70. $x^2 - 2x - 360$

71. $x^2 - 26x + 153$ 72. $x^2 + 8x - 384$

73. $x^2 + 28x + 192$ 74. $x^2 + 3x - 154$

75. $x^2 + 14x - 176$ 76. $x^2 - 59x + 798$

77. $2a^2b + 4ab - 48b$ 78. $x^2 - 15x + 44$

79. $y^2 - y - 72$ 80. $x^2 + 19x + 60$

10.4   Special Products

The square of a number is the product of that number times itself. The square of 3 is $3 \cdot 3$ or 3^2 or 9. The square of a is $a \cdot a$ or a^2 (read "a squared"). The square of the binomial $x + y$ is $(x + y)(x + y)$ or $(x + y)^2$, which is read "the quantity $x + y$ squared." When the multiplication is performed, the product is

$$(x + y)^2 = (x + y)(x + y) = x^2 + 2xy + y^2$$

A trinomial in this form is called a *perfect square trinomial*.

The Square of a Binomial

The square of the <u>sum</u> of two terms equals the square of the first term <u>plus</u> twice the product of the terms plus the square of the second term.

$$(a + b)(a + b) = (a + b)^2$$
$$= a^2 + 2ab + b^2$$

Similarly, the square of the <u>difference</u> of two terms equals the square of the <u>first term</u> <u>minus</u> twice the product of the terms plus the square of the second term. That is,

$$(a - b)(a - b) = (a - b)^2$$
$$= a^2 - 2ab + b^2$$

<u>Example 1.</u> Find $(x + 12)^2$.

The square of the first term is x^2. Twice the product of the terms is $2(12 \cdot x)$, or $24x$. The square of the second term is 144. Thus,

$$(x + 12)^2 = x^2 + 24x + 144$$

<u>Example 2.</u> Find $(xy - 3)^2$.

The square of the first term is x^2y^2. Twice the product of the term is $2(3)(xy)$, or $6xy$. The square of the second term is 9. Thus,

$$(xy - 3)^2 = x^2y^2 - 6xy + 9$$

Finding the *product of the sum and difference of two terms*, $(a + b)(a - b)$, is another special case in which the product appears to be a binomial. (Actually, it is a trinomial with zero for the coefficient of the middle term.)

The product of the sum and difference of two terms is the square of the first term minus the square of the second term. This product is the difference of two squares.

$$(a + b)(a - b) = a^2 - b^2$$

Example 3. Find the product $(x + 3)(x - 3)$.

The square of the first term is x^2. The square of the second term is 9. Thus,

$$(x + 3)(x - 3) = x^2 - 9$$

Example 4. Find the product $(y + 7)(y - 7)$.

The square of the first term is y^2. The square of the second term is 49. Thus,

$$(y + 7)(y - 7) = y^2 - 49$$

Example 5. Find the product $(3x - 8y)(3x + 8y)$.

The square of the first term is $9x^2$. The square of the second term is $64y^2$. Thus,

$$(3x - 8y)(3x + 8y) = 9x^2 - 64y^2$$

EXERCISES 10.4   *Find each product:*

1. $(x + 3)(x - 3)$ 2. $(x + 3)^2$

3. $(a + 5)(a - 5)$ 4. $(y^2 + 9)(y^2 - 9)$

5. $(2b + 11)(2b - 11)$ 6. $(x - 6)^2$

7. $(100 + 3)(100 - 3)$ 8. $(90 + 2)(90 - 2)$

9. $(3y^2 + 14)(3y^2 - 14)$ 10. $(y + 8)^2$

11. $(r - 12)^2$ 12. $(t + 10)^2$

13. $(4y + 5)(4y - 5)$ 14. $(200 + 5)(200 - 5)$

15. $(xy - 4)^2$ 16. $(x^2 + y)(x^2 - y)$

17. $(ab + d)^2$ 18. $(ab + c)(ab - c)$

19. $(z - 11)^2$ 20. $(x^3 + 8)(x^3 - 8)$

21. $(st - 7)^2$ 22. $(w + 14)(w - 14)$

23. $(x + y^2)(x - y^2)$ 24. $(1 - x)^2$

25. $(x + 5)^2$ 26. $(x - 6)^2$

27. $(x + 7)(x - 7)$ 28. $(y - 12)(y + 12)$

29. $(x - 3)^2$ 30. $(x + 4)^2$

31. $(ab + 2)(ab - 2)$ 32. $(m - 3)(m + 3)$

33. $(x^2 + 2)(x^2 - 2)$ 34. $(m + 15)(m - 15)$

35. $(r - 15)^2$ 36. $(t + 7a)^2$

37. $(y^3 - 5)^2$ 38. $(4 - x^2)^2$

39. $(10 - x)(10 + x)$ 40. $(ay^2 - 3)(ay^2 + 3)$

10.5   Finding Factors of Special Products

Before proceeding, you need to understand square root. The square root of 25 is 5 and is written as $\sqrt{25}$. The symbol $\sqrt{}$ is called a radical. The *square root* of a number is that positive number which, when multiplied by itself, gives the original number.

Example 1. Find each square root.

(a) $\sqrt{16} = 4$ because $4 \times 4 = 16$

(b) $\sqrt{64} = 8$ because $8 \times 8 = 64$

(c) $\sqrt{100} = 10$ because $10 \times 10 = 100$

(d) $\sqrt{144} = 12$ because $12 \times 12 = 144$

Numbers whose square roots are whole numbers are called *perfect squares*. That is, 1, 4, 9, 16, 25, 36, 49, 64, . . . are called perfect squares.

To find the square root of a variable raised to a power, divide the exponent by 2 and use this result as the exponent of the root of the given variable. Study the following example.

Example 2. Find each square root. (Assume x and y are positive.)

(a) $\sqrt{x^2} = x$ (b) $\sqrt{x^4} = x^2$

(c) $\sqrt{x^6} = x^3$ (d) $\sqrt{x^2 y^2} = xy$

To factor a trinomial, first look for a common monomial factor. Then inspect the remaining trinomial to see if it is one of the special products. If it is not a perfect square trinomial and if it can be factored, use the methods shown in Sec. 10.3. If it is a perfect square trinomial, it may be factored using the reverse of the rule in Sec. 10.4.

Factoring Perfect Square Trinomials

Each of the two factors of a perfect square trinomial in which the middle term is <u>positive</u> is the square root of the first term <u>plus</u> the square root of the third term. That is,

$$a^2 + 2ab + b^2 = (a + b)(a + b)$$

Similarly, each of the two factors of a perfect square trinomial in which the middle term is <u>negative</u> is the square root of the first term <u>minus</u> the square root of the third term. That is,

$$a^2 - 2ab + b^2 = (a - b)(a - b)$$

__Example 3.__ Factor $4x^2 + 8xy + 4y^2$.

First, find the common monomial factor, 4.

$$4x^2 + 8xy + 4y^2 = 4(x^2 + 2xy + y^2)$$

Since $x^2 + 2xy + y^2$ is a perfect square trinomial and the middle term is positive, each of its two factors is the square root of the first term plus the square root of the third term. The square root of the first term is x; the square root of the third term is y. The sum is $x + y$. Therefore,

$$4x^2 + 8xy + 4y^2 = 4(x + y)(x + y)$$

__Example 4.__ Factor $x^2 - 12x + 36$.

This trinomial is a perfect square trinomial that has no common monomial factor. Since the middle term is negative, each of its two factors is the square root of the first term minus the square root of the third term. The square root of the first term is x; the square root of the third term is 6. The difference is $x - 6$. Therefore,

$$x^2 - 12x + 36 = (x - 6)(x - 6)$$

__Note__: If you do not recognize $x^2 - 12x + 36$ as a perfect square trinomial and factor it by trial and error as you would any trinomial as in Sec. 10.3, your result should be the same.

To factor a binomial that is the difference of two squares, use the reverse of the rule in Sec. 10.4; that is,

---

Factoring the Difference of Two Squares

$$a^2 - b^2 = (a + b)(a - b)$$

Note that a is the square root of a^2 and b is the square root of b^2.

__Example 5.__ Factor $x^2 - 4$.

First, find the square root of each term of the expression. The square root of x^2 is x and the square root of 4 is 2. Thus, $x + 2$ is the sum of the square roots and $x - 2$ is the difference of the square roots.

$$x^2 - 4 = (x + 2)(x - 2)$$

Check: $(x + 2)(x - 2) = x^2 - 4$

Example 6. Factor $1 - 36y^4$.

The square root of 1 is 1 and the **square root of $36y^4$ is** $6y^2$. Thus, the sum, $1 + 6y^2$, and the **difference**, $1 - 6y^2$, of the square roots are the factors.

$$1 - 36y^4 = (1 + 6y^2)(1 - 6y^2)$$

Check: $(1 + 6y^2)(1 - 6y^2) = 1 - 36y^4$

Example 7. Factor $81y^4 - 1$.

The square root of $81y^4$ is $9y^2$ and the **square root of** 1 is 1. The sum, $9y^2 + 1$, and the **difference**, $9y^2 - 1$, of the square roots are the factors.

$$81y^4 - 1 = (9y^2 + 1)(9y^2 - 1)$$

However, $9y^2 - 1$ is also the difference of **two squares** and its factors are: $9y^2 - 1 = (3y + 1)(3y - 1)$. Therefore,

$$81y^4 - 1 = (9y^2 + 1)(3y + 1)(3y - 1)$$

Example 8. Factor $2x^2 - 18$.

First, find the common monomial factor, 2.

$$2x^2 - 18 = 2(x^2 - 9)$$

Then $x^2 - 9$ is the difference of **two squares whose** factors are $x + 3$ and $x - 3$. Therefore,

$$2x^2 - 18 = 2(x + 3)(x - 3)$$

EXERCISES 10.5   *Factor completely. Check by multiplying the factors:*

1. $a^2 + 8a + 16$
2. $b^2 - 2b + 1$
3. $b^2 - c^2$
4. $m^2 - 1$
5. $x^2 - 4x + 4$
6. $2c^2 - 4c + 2$
7. $4 - x^2$
8. $4x^2 - 1$
9. $y^2 - 36$
10. $a^2 - 64$
11. $5a^2 + 10a + 5$
12. $9x^2 - 25$
13. $1 - 81y^2$
14. $16x^2 - 100$
15. $49 - a^4$
16. $m^2 - 2mn + n^2$
17. $49x^2 - 64y^2$
18. $x^2y^2 - 1$
19. $1 - x^2y^2$
20. $c^2d^2 - 16$
21. $4x^2 - 12x + 9$
22. $16x^2 - 1$
23. $R^2 - r^2$
24. $36x^2 - 12x + 1$

25. $49x^2 - 25$

26. $1 - 100y^2$

27. $y^2 - 10y + 25$

28. $x^2 + 6x + 9$

29. $b^2 - 9$

30. $16 - c^2d^2$

31. $m^2 + 22m + 121$

32. $n^2 - 30n + 225$

33. $4m^2 - 9$

34. $16b^2 - 81$

35. $4x^2 + 24x + 36$

36. $-2y^2 + 12y - 18$

37. $27x^2 - 3$

38. $-9x^2 + 225x^4$

39. $am^2 - 14am + 49a$

40. $-bx^2 - 12bx - 36b$

10.6 Factoring General Trinomials

In previous examples such as $x^2 + 7x + 10 = (x + 2)(x + 5)$, there was only one possible choice for the first terms of the binomials x and x. When the coefficient of x^2 is greater than 1, however, there may be more than one possible choice. The idea is still the same: find two binomial factors whose product equals the trinomial. Use the relationships as outlined on page 361 for finding the signs.

Example 1. Factor $6x^2 - x - 2$.

The first terms of the binomial factors are either $6x$ and x or $3x$ and $2x$. The (−) sign of the constant term of the trinomial tells you that one sign will be (+) and the other will be (−) in the last terms of the binomials. The last terms are either +2 and −1 or −2 and +1.

The eight possibilities are:

1. $(6x + 2)(x - 1) = 6x^2 - 4x - 2$
2. $(6x - 1)(x + 2) = 6x^2 + 11x - 2$
3. $(6x - 2)(x + 1) = 6x^2 + 4x - 2$
4. $(6x + 1)(x - 2) = 6x^2 - 11x - 2$
5. $(3x + 2)(2x - 1) = 6x^2 + x - 2$
6. $(3x - 1)(2x + 2) = 6x^2 + 4x - 2$
7. $(3x - 2)(2x + 1) = 6x^2 - x - 2$
8. $(3x + 1)(2x - 2) = 6x^2 - 4x \ - 2$

Only (7) gives the desired middle term, $-x$. Therefore,

$$6x^2 - x - 2 = (3x - 2)(2x + 1)$$

When factoring trinomials of this type, sometimes you may have to make several guesses or look at several combinations until you find the correct one. It is a trial-and-error process. The rules for the signs, as outlined on page 361, simply reduce the possibilities to try.

Another way to reduce the possibilities is to eliminate any combination in which either binomial contains a common factor. In the above list, numbers 1, 3, 6, and 8 can be eliminated, since $6x + 2$, $6x - 2$, $2x + 2$, and $2x - 2$ all have the common factor 2 and cannot be correct. We find it important to look for common monomial factors as the first step in factoring any trinomial.

Example 2. Factor $12x^2 - 23x + 10$.

First terms of the binomial: $12x$ and x, $6x$ and $2x$, or $4x$ and $3x$.

Signs of the last term of the binomial: both $(-)$.

Last terms of the binomial: -1 and -10 or -2 and -5.

You cannot use $12x$, $6x$, $4x$, or $2x$ with -10 or -2 since they contain the common factor 2, so the list of possible combinations is narrowed to:

1. $(12x - 1)(x - 10) = 12x^2 - 121x + 10$
2. $(12x - 5)(x - 2) = 12x^2 - 29x + 10$
3. $(4x - 5)(3x - 2) = 12x^2 - 23x + 10$
4. $(4x - 1)(3x - 10) = 12x^2 - 43x + 10$

As you can see, (3) is the correct one since it gives the desired middle term, $-23x$. Therefore,

$$12x^2 - 23x + 10 = (4x - 5)(3x - 2)$$

Example 3. Factor $12x^2 - 2x - 4$.

First look for a common factor, which in this case is 2; so we write

$$12x^2 - 2x - 4 = 2(6x^2 - x - 2)$$

Next we try to factor the trinomial $6x^2 - x - 2$ into two binomial factors as we did in Examples 1 and 2. Since the third term is negative (-2), we know that the signs of the second terms of the binomials are different. For the first terms of the binomials we can try $6x$ and x or $3x$ and $2x$. The second terms can be $+2$ and -1 or -2 and $+1$. After eliminating all combinations with common factors we have the following possibilities:

1. $(6x + 1)(x - 2) = 6x^2 - 11x - 2$
2. $(6x - 1)(x + 2) = 6x^2 + 11x - 2$
3. $(3x + 2)(2x - 1) = 6x^2 + x - 2$
4. $(3x - 2)(2x + 1) = 6x^2 - x - 2$

The correct one, of course, is (4), since the middle term, $-x$, is the one we want. Therefore,

$$12x^2 - 2x - 4 = 2(3x - 2)(2x + 1)$$

EXERCISES 10.6   *Factor completely:*

1. $5x^2 - 28x - 12$
2. $4x^2 - 4x - 3$
3. $10x^2 - 29x + 21$
4. $4x^2 + 4x + 1$
5. $12x^2 - 28x + 15$
6. $9x^2 - 36x + 32$
7. $8x^2 + 26x - 45$
8. $4x^2 + 15x - 4$
9. $16x^2 - 11x - 5$
10. $6x^2 + 3x - 3$
11. $12x^2 - 4x - 8$
12. $10x^2 - 35x + 15$
13. $15y^2 - y - 6$
14. $6y^2 + y - 2$
15. $8m^2 - 10m - 3$
16. $2m^2 - 7m - 30$
17. $35a^2 - 2a - 1$
18. $12a^2 - 28a + 15$
19. $16y^2 - 8y + 1$
20. $25y^2 + 20y + 4$
21. $3x^2 + 20x - 63$
22. $4x^2 + 7x - 15$
23. $12b^2 + 5b - 2$
24. $10b^2 - 7b - 12$
25. $15y^2 - 14y - 8$
26. $5y^2 + 11y + 2$
27. $90 + 17c - 3c^2$
28. $10x^2 - x - 2$
29. $6x^2 - 13x + 5$
30. $56x^2 - 29x + 3$
31. $2y^4 + 9y^2 - 35$
32. $2y^2 + 7y - 99$
33. $4b^2 + 52b + 169$
34. $6x^2 - 19x + 15$
35. $14x^2 - 51x + 40$
36. $42x^4 - 13x^2 - 40$
37. $28x^3 + 140x^2 + 175x$
38. $-24x^3 - 54x^2 - 21x$
39. $10ab^2 - 15ab - 175a$
40. $40bx^2 - 72bx - 70b$

Chapter 10 Review

Find each product mentally:

1. $(c + d)(c - d)$
2. $(x - 6)(x + 6)$
3. $(y + 7)(y - 4)$
4. $(2x + 5)(2x - 9)$
5. $(x + 8)(x - 3)$
6. $(x - 4)(x - 9)$
7. $(x - 3)^2$
8. $(2x - 6)^2$
9. $(1 - 5x^2)^2$

Factor each expression completely:

10. $6a + 6$
11. $5x - 15$
12. $xy + 2xz$
13. $y^4 + 17y^3 - 18y^2$
14. $y^2 - 6y - 7$
15. $z^2 + 18z + 81$

16. $x^2 + 10x + 16$

17. $4a^2 + 4x^2$

18. $x^2 - 17x + 72$

19. $x^2 - 18x + 81$

20. $x^2 + 19x + 60$

21. $y^2 - 2y + 1$

22. $x^2 - 3x - 28$

23. $x^2 - 4x - 96$

24. $x^2 + x - 110$

25. $x^2 - 49$

26. $16y^2 - 9x^2$

27. $x^2 - 144$

28. $25x^2 - 81y^2$

29. $4x^2 - 24x - 364$

30. $5x^2 - 5x - 780$

31. $2x^2 + 11x + 14$

32. $12x^2 - 19x + 4$

33. $30x^2 + 7x - 15$

34. $12x^2 + 143x - 12$

35. $4x^2 - 6x + 2$

36. $36x^2 - 49y^2$

37. $28x^2 + 82x + 30$

38. $30x^2 - 27x - 21$

39. $4x^3 - 4x$

40. $25y^2 - 100$

11

QUADRATIC EQUATIONS

11.1 Solving Quadratic Equations by Factoring

A *quadratic equation* in one variable is an equation in which at least one term is second-degree and no other term is of higher degree. A quadratic equation is usually in the form $ax^2 + bx + c = 0$, where $a \neq 0$.

Recall that linear equations, like $2x + 3 = 0$, have at most *one* solution. Quadratic equations have at most *two* solutions. One way to solve quadratic equations is by factoring and using the rule:

If $ab = 0$, then either $a = 0$ or $b = 0$.

That is, if you multiply two factors and the product is 0, then one or both factors are 0.

<u>Example 1</u>. Solve $4(x - 2) = 0$.

If $4(x - 2) = 0$, then $4 = 0$ or $x - 2 = 0$. However, the first statement, $4 = 0$, is false; thus the solution is $x - 2 = 0$, or $x = 2$.

Example 2. Solve $(x - 2)(x + 3) = 0$.

If $(x - 2)(x + 3) = 0$, then either

$$x - 2 = 0 \quad \text{or} \quad x + 3 = 0,$$

so $\qquad\qquad x = 2 \quad \text{or} \quad x = -3.$

To Solve Quadratic Equations by Factoring:

1. If necessary, write an equivalent equation in descending order which has zero as one member.
2. Factor the other member of the equation.
3. Write an equation by setting each factor containing a variable equal to zero.
4. Solve the two resulting first-degree equations.
5. Check.

Example 3. Solve $x^2 + 6x + 5 = 0$ for x.

Step 1: Not needed.

Step 2: $\quad (x + 5)(x + 1) = 0$

Step 3: $\qquad\qquad x + 5 = 0 \quad \text{or} \quad x + 1 = 0$

Step 4: $\qquad\qquad\qquad x = -5 \quad \text{or} \quad x = -1$

Step 5: Check:

Replace x by -5. $\qquad\qquad$ Replace x by -1.

$$x^2 + 6x + 5 = 0 \qquad\qquad x^2 + 6x + 5 = 0$$
$$(-5)^2 + 6(-5) + 5 = 0 \ ? \qquad (-1)^2 + 6(-1) + 5 = 0 \ ?$$
$$25 - 30 + 5 = 0 \qquad\qquad 1 - 6 + 5 = 0$$
$$0 = 0 \qquad\qquad\qquad\qquad 0 = 0$$

The roots are -5 and -1.

Example 4. Solve $x^2 + 5x = 36$ for x.

Step 1: $\qquad x^2 + 5x - 36 = 0$

Step 2: $\qquad (x + 9)(x - 4) = 0$

Step 3: $\qquad\qquad\qquad x + 9 = 0 \quad \text{or} \quad x - 4 = 0$

Step 4: $\qquad\qquad\qquad\qquad x = -9 \quad \text{or} \quad x = 4$

Step 5: *Check*:

Replace x by -9.   Replace x by 4.

$$x^2 + 5x = 36 \qquad\qquad x^2 + 5x = 36$$
$$(-9)^2 + 5(-9) = 36 \ ? \qquad 4^2 + 5(4) = 36 \ ?$$
$$81 - 45 = 36 \qquad\qquad 16 + 20 = 36$$
$$36 = 36 \qquad\qquad\qquad 36 = 36$$

The roots are -9 and 4.

Example 5. Solve $3x^2 + 9x = 0$ for x.

Step 1: Not needed.

Step 2: $3x(x + 3) = 0$

Step 3: $3x = 0 \quad$ or $\quad x + 3 = 0$

Step 4: $x = 0 \quad$ or $\quad x = -3$

Step 5: *Check*: Left to the student.

Example 6. Solve $x^2 = 4$ for x.

Step 1: $x^2 - 4 = 0$

Step 2: $(x + 2)(x - 2) = 0$

Step 3: $x + 2 = 0 \quad$ or $\quad x - 2 = 0$

Step 4: $x = -2 \quad$ or $\quad x = 2$

Step 5: *Check*: Left to the student.

Example 7. Solve $6x^2 = 7x + 20$.

Step 1: $6x^2 - 7x - 20 = 0$

Step 2: $(3x + 4)(2x - 5) = 0$

Step 3: $3x + 4 = 0 \quad$ or $\quad 2x - 5 = 0$

Step 4: $3x = -4 \qquad\qquad 2x = 5$

$$x = -\frac{4}{3} \qquad\qquad x = \frac{5}{2}$$

So, the possible roots are $-\frac{4}{3}$ and $\frac{5}{2}$.

Step 5: _Check_: Replace x by $-\frac{4}{3}$ and by $\frac{5}{2}$ in the original
equation.

$$6x^2 = 7x + 20 \qquad\qquad 6x^2 = 7x + 20$$

$$6\left(-\frac{4}{3}\right)^2 = 7\left(-\frac{4}{3}\right) + 20 \qquad 6\left(\frac{5}{2}\right)^2 = 7\left(\frac{5}{2}\right) + 20$$

$$6\left(\frac{16}{9}\right) = -\frac{28}{3} + 20 \qquad 6\left(\frac{25}{4}\right) = \frac{35}{2} + 20$$

$$\frac{32}{3} = -\frac{28}{3} + \frac{60}{3} \qquad\qquad \frac{75}{2} = \frac{35}{2} + \frac{40}{2}$$

$$\frac{32}{3} = \frac{32}{3} \qquad\qquad\qquad \frac{75}{2} = \frac{75}{2}$$

So, the roots are $-\frac{4}{3}$ and $\frac{5}{2}$.

EXERCISES 11.1   _Solve each equation_:

1. $x^2 + x = 12$

2. $x^2 - 3x + 2 = 0$

3. $x^2 + x - 20 = 0$

4. $d^2 + 2d - 15 = 0$

5. $x^2 - 2 = x$

6. $x^2 - 15x = -54$

7. $x^2 - 1 = 0$

8. $4x^2 = 25$

9. $x^2 - 4 = 0$

10. $4n^2 = 64$

11. $w^2 + 5w + 6 = 0$

12. $x^2 - 6x = 0$

13. $y^2 - 4y = 21$

14. $c^2 + 2 = 3c$

15. $n^2 - 6n - 40 = 0$

16. $x^2 - 17x + 16 = 0$

17. $9m = m^2$

18. $6n^2 - 15n = 0$

19. $x^2 = 108 + 3x$

20. $x^2 - x = 42$

21. $c^2 + 6c = 16$

22. $4x^2 + 4x - 3 = 0$

23. $10x^2 + 29x + 10 = 0$

24. $2x^2 = 17x - 8$

25. $4x^2 = 25$

26. $25x = x^2$

27. $9x^2 + 16 = 24x$

28. $24x^2 + 10 = 31x$

29. $3x^2 + 9x = 0$

30. A rectangle is 5 ft longer than it is wide. The area of the rectangle is 84 ft^2. Use a quadratic equation to find the dimensions of the rectangle.

31. The area of a triangle is 66 m^2. The base is 1 m more than the height. Find the base and height of the triangle. (Use a quadratic equation.)

32. A rectangle is 9 ft longer than it is wide. Its area is 360 ft^2. Use a quadratic equation to find the length and width.

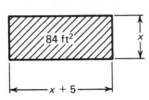 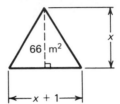 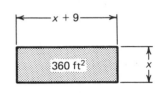

Exercise 30   Exercise 31   Exercise 32

11.2 The Quadratic Formula

Many quadratic equations cannot be solved by factoring. In this section you will study a method by which any quadratic equation can be solved. You may find it helpful to review the square root discussion in Sec. 10.5.

The quadratic equation is written in the general form

$$ax^2 + bx + c = 0 \qquad (\text{where } a \neq 0)$$

The roots of this equation are found by using the following formula.

> Quadratic Formula
>
> $$x = \frac{-b \pm \sqrt{b^2 - 4ac}}{2a}$$

where a is the coefficient of the x^2 term,
b is the coefficient of the x term, and
c is the constant term.

The symbol ($\pm$) is used to combine two expressions into one. For example, $a \pm 4$ means $a + 4$ or $a - 4$. Similarly,

$$x = \frac{-b \pm \sqrt{b^2 - 4ac}}{2a} \qquad \text{means}$$

$$x = \frac{-b + \sqrt{b^2 - 4ac}}{2a} \qquad \text{or} \qquad x = \frac{-b - \sqrt{b^2 - 4ac}}{2a}$$

<u>Example 1</u>. In the quadratic equation $3x^2 - x - 7 = 0$, find the values of a, b, and c.

$$a = 3, \ b = -1, \ \text{and} \ c = -7$$

<u>Example 2</u>. Solve $x^2 + 5x - 14 = 0$ using the quadratic formula.

$$x = \frac{-b \pm \sqrt{b^2 - 4ac}}{2a}, \quad a = 1, \ b = 5, \ c = -14$$

So,

$$x = \frac{-5 \pm \sqrt{5^2 - 4(1)(-14)}}{2(1)}$$

$$= \frac{-5 \pm \sqrt{25 + 56}}{2}$$

$$= \frac{-5 \pm \sqrt{81}}{2}$$

$$= \frac{-5 \pm 9}{2}$$

$$= \frac{-5 + 9}{2} \quad \text{or} \quad \frac{-5 - 9}{2}$$

$$= 2 \quad \text{or} \quad -7$$

Check:

Replace x by 2.   Replace x by -7.

$x^2 + 5x - 14 = 0$	$x^2 + 5x - 14 = 0$
$2^2 + 5(2) - 14 = 0$?	$(-7)^2 + 5(-7) - 14 = 0$?
$4 + 10 - 14 = 0$	$49 - 35 - 14 = 0$
$0 = 0$	$0 = 0$

The roots are 2 and -7.

Before using the quadratic formula, make certain the equation is written in the form $ax^2 + bx + c = 0$, such that one member is zero.

<u>Example 3</u>. Solve $2x^2 = x + 21$ by using the quadratic formula.

First, write the equation in the form $2x^2 - x - 21 = 0$.

$$x = \frac{-b \pm \sqrt{b^2 - 4ac}}{2a}, \quad a = 2, \ b = -1, \ c = -21$$

So,
$$x = \frac{-(-1) \pm \sqrt{(-1)^2 - 4(2)(-21)}}{2(2)}$$

$$= \frac{1 \pm \sqrt{1 + 168}}{4}$$

$$= \frac{1 \pm \sqrt{169}}{4}$$

$$= \frac{1 \pm 13}{4}$$

$$= \frac{1 + 13}{4} \quad \text{or} \quad \frac{1 - 13}{4}$$

$$= \frac{7}{2} \quad \text{or} \quad -3$$

Check: Left to the student.

If the number under the radical sign is not a perfect square, find the square root of the number by using a calculator and proceed as before. Round each final result to three significant digits.

Example 4. Solve $3x^2 + x - 5 = 0$ using the quadratic formula.

$$x = \frac{-b \pm \sqrt{b^2 - 4ac}}{2a}, \quad a = 3, \ b = 1, \ c = -5$$

So,
$$x = \frac{-1 \pm \sqrt{1^2 - 4(3)(-5)}}{2(3)}$$

$$= \frac{-1 \pm \sqrt{1 + 60}}{6}$$

$$= \frac{-1 \pm \sqrt{61}}{6}$$

$$= \frac{-1 \pm 7.81}{6}$$

$$= \frac{-1 + 7.81}{6} \quad \text{or} \quad \frac{-1 - 7.81}{6}$$

$$= \frac{6.81}{6} \quad \text{or} \quad \frac{-8.81}{6}$$

$$= 1.14 \quad \text{or} \quad -1.47$$

The roots are 1.14 and -1.47.

The check will not work out *exactly* when the **number under** the radical is not a perfect square.

EXERCISES 11.2 *Find the values of a, b, and c in each equation:*

1. $x^2 - 7x + 4 = 0$

2. $2x^2 + x - 3 = 0$

3. $3x^2 + 4x + 9 = 0$

4. $2x^2 - 14x + 37 = 0$

5. $-3x^2 + 4x + 7 = 0$

6. $17x^2 - x + 34 = 0$

7. $3x^2 - 14 = 0$

8. $2x^2 + 7x = 0$

Solve each equation using the quadratic formula. Check your solutions:

9. $x^2 + x - 6 = 0$

10. $x^2 - 4x - 21 = 0$

11. $x^2 + 8x - 9 = 0$

12. $2x^2 + 5x - 12 = 0$

13. $5x^2 + 2x = 0$

14. $3x^2 - 75 = 0$

15. $48x^2 - 32x - 35 = 0$

16. $13x^2 + 178x - 56 = 0$

Solve each equation using the quadratic formula. (When necessary, round results to three significant digits:)

17. $2x^2 + x - 5 = 0$

18. $-3x^2 + 2x + 5 = 0$

19. $3x^2 - 5x = 0$

20. $7x^2 + 9x + 2 = 0$

21. $-2x^2 + x + 3 = 0$

22. $5x^2 - 7x + 2 = 0$

23. $6x^2 + 9x + 1 = 0$

24. $16x^2 - 25 = 0$

25. $-4x^2 - 5x - 1 = 0$

26. $9x^2 - 21x + 10 = 0$

27. $3x^2 - 17 = 0$

28. $8x^2 - 11x + 3 = 0$

29. $x^2 - 15x - 7 = 0$

30. $x^2 + x - 1 = 0$

31. $3x^2 - 31 = 5x$

32. $-3x^2 - 5 = -7x^2$

33. $52.3x = -23.8x^2 + 11.8$

34. $18.9x^2 - 44.2x = 21.5$

 35. A variable voltage in a given electrical circuit is given by the formula $V = t^2 - 12t + 40$, where t is in seconds. At what two values of t is the voltage V (a) equal to 8 volts? (b) Equal to 25 volts? (c) Equal to 104 volts?

 36. A variable electric circuit is given by the formula $i = t^2 - 7t + 12$. Suppose t is given in seconds. At what times is the current i (a) equal to 2 A? (b) Equal to 0 A? (c) Equal to 4 A?

37. A rectangular piece of sheet metal is 4 ft longer than it is wide. The area of the piece of sheet metal is 21 ft^2. Find its length and width.

38. A square is to be cut out of each corner of a rectangular sheet of aluminum which is 40 cm by 60 cm. The sides are to be folded up to form a rectangular container with no top. The area of the bottom of the container is to be 1500 cm^2. What are the dimensions of each cut-out square? Also, find the volume of the container. ($V = \ell wh$)

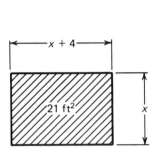

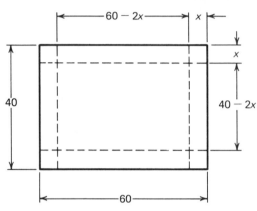

Exercise 37   Exercise 38

39. A square, 4 in. on a side, is cut out of each corner of a square sheet of aluminum. The sides are folded up to form a rectangular container with no top. The volume of the resulting container is 400 in^3. What was the size of the original sheet of aluminum?

40. The area of a rectangular lot 80 m by 100 m is to be increased by 4000 m^2. The length and the width will be increased by the same amount. What are the dimensions of the larger lot?

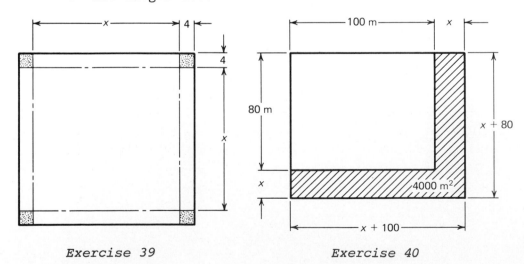

Exercise 39   Exercise 40

41. A border of uniform width is to be built around a rectangular garden which measures 16 ft by 20 ft. The area of the border is to be 160 ft^2. Find the width of the border.

42. A border of uniform width is to be allowed on a printed page which is 11 in. by 14 in. The area of the border is 66 in^2. Find the width of the border.

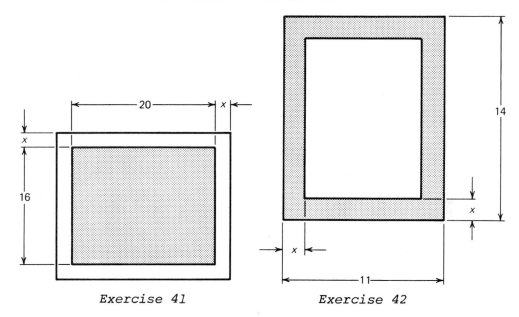

Exercise 41   *Exercise 42*

11.3 Graphs of Quadratic Equations

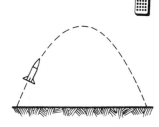

Figure 11.1

Many physical phenomena follow a path called a *parabola*. The trajectory of a rocket (Fig. 11.1) is one such phenomenon. The parabolic curve is really the graph of a quadratic equation in two variables. This relationship is very important since it allows us to use computers in analyzing rocket launches and to test the launches against the planned flight path.

The quadratic equation that represents a parabola is written in the form $y = ax^2 + bx + c$, where a, b, and c are real numbers and $a \neq 0$. The quadratic equation in the form $x = ay^2 + by + c$ also represents a parabola, but we will not be concerned with this equation in this book.

To draw the graph of a parabola, find points whose ordered pairs satisfy the equation (make the equation a true statement). Since this graph is not a straight line, you will need to find many points to get an accurate graph of the curve. A table is helpful for listing these ordered pairs.

<u>Example 1</u>. Graph the equation $y = x^2$. First, set up a table as follows:

x	-4	-3	-2	-1	0	1	2	3	4
$y = x^2$	16	9	4	1	0	1	4	9	16

Then, using a rectangular coordinate system, plot these points.

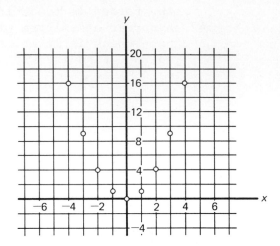

Figure 11.2

Notice that with only the points shown in Fig. 11.2, there isn't a definite outline of the curve. So, let's look closer at values of *x* between 0 and 1.

x	$\frac{1}{6}$	$\frac{1}{5}$	$\frac{1}{4}$	$\frac{1}{3}$	$\frac{2}{5}$	$\frac{1}{2}$	$\frac{3}{5}$	$\frac{3}{4}$	$\frac{2}{3}$	$\frac{4}{5}$	$\frac{5}{6}$
$y = x^2$	$\frac{1}{36}$	$\frac{1}{25}$	$\frac{1}{16}$	$\frac{1}{9}$	$\frac{4}{25}$	$\frac{1}{4}$	$\frac{9}{25}$	$\frac{9}{16}$	$\frac{4}{9}$	$\frac{16}{25}$	$\frac{25}{36}$

Plotting these additional ordered pairs, we get a better graph (Fig. 11.3).

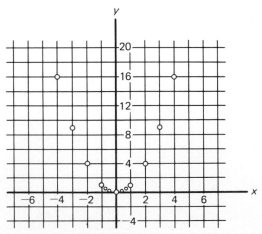

Figure 11.3

Now the graph looks almost solid between 0 and 1. If we were to continue to choose more and more values, the graph would appear as a solid line. Since it is impossible to find *all* the ordered pairs that satisfy the equation, we will assume that all the points between any two of the ordered pairs already located could be found, and they do lie on the graph. Thus, assume the graph of $y = x^2$ looks like the graph in Fig. 11.4.

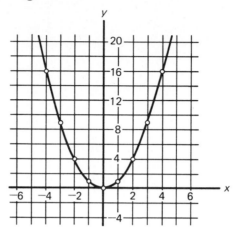

Figure 11.4

In summary, to draw the graph of a quadratic equation, form a table to find many ordered pairs that satisfy the equation. Then, plot these ordered pairs and connect them with a smooth curved line.

Example 2. Plot the graph of $y = 2x^2 - 4x + 5$.

Let $x = -7$; then $y = 2(-7)^2 - 4(-7) + 5 = 131$.
Let $x = -6$; then $y = 2(-6)^2 - 4(-6) + 5 = 101$.
Let $x = -5$; then $y = 2(-5)^2 - 4(-5) + 5 = 75$.
Let $x = -4$; then $y = 2(-4)^2 - 4(-4) + 5 = 53$.
Let $x = -3$; then $y = 2(-3)^2 - 4(-3) + 5 = 35$.
Let $x = -2$; then $y = 2(-2)^2 - 4(-2) + 5 = 21$.
Let $x = -1$; then $y = 2(-1)^2 - 4(-1) + 5 = 11$.
Let $x = 0$; then $y = 2(0)^2 - 4(0) + 5 = 5$.
Let $x = 1$; then $y = 2(1)^2 - 4(1) + 5 = 3$.
Let $x = 2$; then $y = 2(2)^2 - 4(2) + 5 = 5$.
Let $x = 3$; then $y = 2(3)^2 - 4(3) + 5 = 11$.
Let $x = 4$; then $y = 2(4)^2 - 4(4) + 5 = 21$.
Let $x = 5$; then $y = 2(5)^2 - 4(5) + 5 = 35$.
Let $x = 6$; then $y = 2(6)^2 - 4(6) + 5 = 53$.
Let $x = 7$; then $y = 2(7)^2 - 4(7) + 5 = 75$.
Let $x = 8$; then $y = 2(8)^2 - 4(8) + 5 = 101$.
Let $x = 9$; then $y = 2(9)^2 - 4(9) + 5 = 131$.

x	-7	-6	-5	-4	-3	-2	-1	0	1	2	3	4	5	6	7	8	9
y	131	101	75	53	35	21	11	5	3	5	11	21	35	53	75	101	131

Then plot the points from the table. Disregard those coordinates which cannot be plotted. Then, connect the points with a smooth curved line as shown in Fig. 11.5.

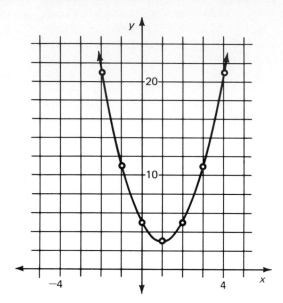

Figure 11.5

All parabolas have a property called *symmetry*, which means that a line can be drawn through a parabola dividing it into two parts that are mirror images of each other. (See Fig. 11.6.)

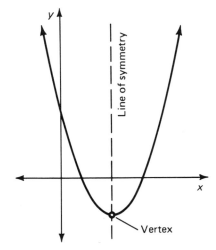

Figure 11.6

The intersection of this line of symmetry and the parabola is called the *vertex*. In this section, the vertex is either the highest or lowest point of the parabola. Locating the vertex is most helpful in drawing the graph of a parabola.

The vertex of a parabola whose equation is in the form $y = ax^2 + bx + c$ may be found as follows:

> 1. The x-coordinate is the value of $-\dfrac{b}{2a}$.
>
> 2. The y-coordinate is then found by substituting the x-coordinate from Step 1 into the parabola equation and solving for y.

<u>Example 3.</u> Find the vertex of the parabola $y = 2x^2 - 4x + 5$ in Example 2.

Note that $a = 2$ and $b = -4$.

1. The x-coordinate is $-\dfrac{b}{2a} = -\dfrac{-4}{2(2)} = 1$.

2. Substitute $x = 1$ into $y = 2x^2 - 4x + 5$ and solve for y.

$$y = 2(1)^2 - 4(1) + 5$$
$$= 2 \quad\quad - \quad 4 \;\; + 5 = 3$$

Thus, the vertex is (1, 3).

EXERCISES 11.3   *Draw the graph of each equation and label each vertex:*

1. $y = 2x^2$

2. $y = -2x^2$

3. $y = \dfrac{1}{2}x^2$

4. $y = -\dfrac{1}{2}x^2$

5. $y = x^2 + 3$

6. $y = x^2 - 4$

7. $y = 2(x - 3)^2$

8. $y = -(x + 2)^2$

9. $y = (x - 1)^2$

10. $y = 2(x + 1)^2 - 3$

11. $y = 2x^2 - 5$

12. $y = -3x^2 - 2x$

13. $y = x^2 - 2x - 5$

14. $y = -3x^2 + 6x + 15$

15. $y = x^2 - 2x - 15$

16. $y = 2x^2 - x - 15$

17. $y = -4x^2 - 5x + 9$

18. $y = 4x^2 - 12x + 9$

19. $y = \dfrac{1}{5}x^2 - \dfrac{2}{5}x + 4$

20. $y = -0.4x^2 + 2.4x + 0.7$

11.4   Imaginary Numbers

 What is the meaning of $\sqrt{-4}$? What number squared is -4? Try to find its value on your calculator.

As you can see, we have a different kind of number here. Up to now, we have considered only real numbers. The number $\sqrt{-4}$ is not a real number. The square root of a negative number is called an *imaginary number*. The imaginary unit is defined as $\sqrt{-1}$ and in many mathematics texts is given by the symbol i. However, in technical work, i is commonly used for current. To avoid confusion, many technical books use j for $\sqrt{-1}$, which is what we will use here. That is,

$$\boxed{\begin{array}{c} \text{Imaginary Unit} \\[4pt] \sqrt{-1} = j \end{array}}$$

Then, $\sqrt{-4} = \sqrt{(-1)(4)} = (\sqrt{-1})(\sqrt{4}) = (j)(2)$ or $2j$.

Example 1. Express each number in terms of j.

(a) $\sqrt{-25} =$
  $\sqrt{(-1)(25)} =$
  $(\sqrt{-1})(\sqrt{25}) =$
  $(j)(5)$ or $5j$

(b) $\sqrt{-45} =$
  $\sqrt{(-1)(45)} =$
  $(\sqrt{-1})(\sqrt{45}) =$
  $(j)(6.71)$ or $6.71j$

(c) $\sqrt{-183} =$
  $\sqrt{(-1)(183)} =$
  $(\sqrt{-1})(\sqrt{183}) =$
  $(j)(13.5)$ or $13.5j$

Now, let's consider powers of j or $\sqrt{-1}$. Using the rules of exponents and the definition of j, carefully study the following powers of j:

$$j = j$$
$$j^2 = (\sqrt{-1})^2 = -1$$
$$j^3 = (j^2)(j) = (-1)(j) = -j$$
$$j^4 = (j^2)(j^2) = (-1)(-1) = 1$$
$$j^5 = (j^4)(j) = (1)(j) = j$$
$$j^6 = (j^4)(j^2) = (1)(-1) = -1$$
$$j^7 = (j^4)(j^3) = (1)(-j) = -j$$
$$j^8 = (j^4)^2 = 1^2 = 1$$
$$j^9 = (j^8)(j) = (1)(j) = j$$
$$j^{10} = (j^8)(j^2) = (1)(-1) = -1$$

As you can see, the powers of j are repeating in the order of j, -1, $-j$, 1, j, -1, $-j$, 1, . . . Also note that any power of j divisible by four is equal to one.

Example 2. Simplify.

(a) $j^{15} =$

$(j^{12})(j^3) =$

$(1)(-j) = -j$

(b) $j^{21} =$

$(j^{20})(j) =$

$(1)(j) = j$

(c) $j^{72} = 1$

When solving the quadratic equation $ax^2 + bx + c = 0$, its solutions are given by the quadratic formula

$$x = \frac{-b \pm \sqrt{b^2 - 4ac}}{2a}$$

The part under the radical sign, $b^2 - 4ac$, is called the *discriminant*. That is, the value of $b^2 - 4ac$ determines what kind and the number of solutions the quadratic equation has. In general,

If $b^2 - 4ac$ is	Roots
positive and a perfect square,	both roots are rational.
positive and not a perfect square,	both roots are irrational.
zero,	there is only one rational root.
negative,	both roots are imaginary.

Example 3. Determine the nature of the roots of $3x^2 + 5x - 2 = 0$ without solving the equation.

$a = 3$ The value of the discriminant is $b^2 - 4ac =$
$b = 5$ $(5)^2 - 4(3)(-2) = 25 + 24 = 49.$
$c = -2$ Since 49 is a perfect square, both roots are rational.

Example 4. Determine the nature of the roots of $4x^2 - 12x + 9 = 0$ without solving the equation.

$a = 4$ The value of the discriminant is $b^2 - 4ac =$
$b = -12$ $(-12)^2 - 4(4)(9) = 144 - 144 = 0.$
$c = 9$ Therefore, there is only one rational root.

Example 5. Determine the nature of the roots of $x^2 - 3x + 8 = 0$ without solving the equation.

$a = 1$ The value of the discriminant is $b^2 - 4ac =$
$b = -3$ $(-3)^2 - 4(1)(8) = 9 - 32 = -23.$
$c = 8$ Since -23 is negative, both roots are imaginary.

Example 6. Solve $4x^2 - 6x + 5 = 0$ using the quadratic formula.

$$x = \frac{-b \pm \sqrt{b^2 - 4ac}}{2a}, \quad a = 4, \ b = -6, \ c = 5$$

So, $$x = \frac{-(-6) \pm \sqrt{(-6)^2 - 4(4)(5)}}{2(4)} = \frac{6 \pm \sqrt{36 - 80}}{8}$$

$$= \frac{6 \pm \sqrt{-44}}{8} = \frac{6 \pm 6.63j}{8} \quad [\sqrt{-44} = (\sqrt{-1})(\sqrt{44})$$

$$= \frac{6 + 6.63j}{8} \quad \text{or} \quad \frac{6 - 6.63j}{8} \qquad = 6.63j]$$

$$= 0.75 + 0.829j \quad \text{or} \quad 0.75 - 0.829j$$

The roots are $0.75 + 0.829j$ and $0.75 - 0.829j$.

EXERCISES 11.4

Express each number in terms of j: (When necessary, round results to three significant digits.)

1. $\sqrt{-49}$ 2. $\sqrt{-64}$ 3. $\sqrt{-14}$ 4. $\sqrt{-5}$

5. $\sqrt{-2}$ 6. $\sqrt{-3}$ 7. $\sqrt{-56}$ 8. $\sqrt{-121}$

9. $\sqrt{-169}$ 10. $\sqrt{-60}$ 11. $\sqrt{-27}$ 12. $\sqrt{-40}$

Simplify:

13. j^3 14. j^6 15. j^{13} 16. j^{16}

17. j^{19} 18. j^{31} 19. j^{24} 20. j^{26}

21. j^{38} 22. j^{81} 23. $\dfrac{1}{j}$ 24. $\dfrac{1}{j^6}$

Determine the nature of the roots of each quadratic equation without solving it:

25. $x^2 + 3x - 10 = 0$ 26. $2x^2 - 7x + 3 = 0$

27. $5x^2 + 4x + 1 = 0$ 28. $9x^2 + 12x + 4 = 0$

29. $3x + 1 = 2x^2$ 30. $3x^2 = 4x - 8$

31. $2x^2 + 6 = x$ 32. $2x^2 + 7x = 4$

33. $x^2 + 25 = 0$ 34. $x^2 - 4 = 0$

Solve each quadratic equation using the quadratic formula. When necessary, round results to three significant digits:

35. $x^2 - 6x + 10 = 0$ 36. $x^2 - x + 2 = 0$

37. $x^2 - 14x + 53 = 0$ 38. $x^2 + 10x + 34 = 0$

39. $x^2 + 8x + 41 = 0$ 40. $x^2 - 6x + 13 = 0$

41. $6x^2 + 5x + 8 = 0$ 42. $4x^2 + 3x - 1 = 0$

43. $3x^2 = 6x - 7$ 44. $5x^2 + 2x = -3$

45. $5x^2 + 8x + 4 = 0$ 46. $2x^2 + x + 3 = 0$

47. $5x^2 + 14x - 3 = 0$ 48. $2x^2 - x + 1 = 0$

49. $x^2 + x + 1 = 0$ 50. $12x^2 + 23x + 10 = 0$

Chapter 11 Review

1. If $ab = 0$, what is known about either a or b?
2. Solve for x: $3x(x - 2) = 0$.

Solve each equation by factoring:

3. $x^2 - 4 = 0$ 4. $x^2 - x = 6$ 5. $5x^2 - 6x = 0$

6. $x^2 - 3x - 28 = 0$ 7. $x^2 - 14x = -45$

8. $x^2 - 18 - 3x = 0$ 9. $3x^2 + 20x + 32 = 0$

Solve each equation using the quadratic formula. (When necessary, round results to three significant digits.)

10. $3x^2 - 16x - 12 = 0$ 11. $x^2 + x - 5 = 0$

12. $2x^2 + x = 15$ 13. $x^2 - 4x = 2$ 14. $3x^2 - x = 5$

15. The area of a piece of plywood is 36 ft^2. Its length is 5 ft more than its width. Find its length and width.
16. A variable electric current is given by the formula $i = t^2 - 12t + 36$, where t is in μs. At what times is the current i (a) equal to 4 A? (b) Equal to 0 A? (c) Equal to 10 A?

Draw the graph of each equation and label each vertex:

17. $y = x^2 - x - 6$ 18. $y = -3x^2 + 2$

Express each number in terms of j:

19. $\sqrt{-36}$ 20. $\sqrt{-73}$

Simplify:

21. j^{12} 22. j^{27}

*Determine the nature of the roots of each quadratic
equation without solving it:*

23. $9x^2 + 30x + 25 = 0$ 24. $3x^2 - 2x + 4 = 0$

*Solve each equation using the quadratic formula. When
necessary, round results to three significant digits:*

25. $x^2 - 4x + 5 = 0$ 26. $5x^2 - 6x + 4 = 0$

12

GEOMETRY

12.1 Angles and Polygons

Some fundamentals of geometry must be known in order to
solve many technical applications. Geometry is also needed
to understand some of the mathematical developments in
technical mathematics courses, in technical support courses,
and in on-the-job training programs. We will show the most
basic and most often used geometric terms and relationships.

An *angle* is determined by the parts of two lines that
have a common endpoint. The common endpoint is called the
vertex of the angle. The parts of the lines are called the
sides of the angle. An angle is designated either by a
single letter or by three letters. For example, the angle
in Fig. 12.1 is referred to as ∠A or ∠BAC. The middle letter of the three letters is always the one at the vertex.

The measure of an angle is the amount of rotation needed
to make one side coincide with the other side. The measure
can be expressed in any one of many different units. The
standard metric unit of plane angle is the radian (rad).
While the radian is the metric unit of angle measurement,
many ordinary measurements will continue to be made in degrees (°). Although some trades will subdivide the degree
into the traditional minutes and seconds, most others will
use tenths and hundredths of degrees because most calculators accommodate these decimal units with ease. The
radian measure is developed in Sec. 12.6.

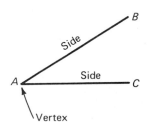

Figure 12.1

One degree is $\frac{1}{360}$ of one complete revolution; that is, 360° = one revolution. The protractor in Fig. 12.2 is an instrument, marked in degrees, which is used to measure angles.

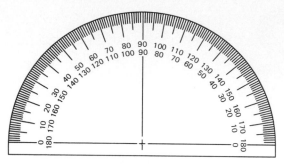

Figure 12.2

To find the degree measure of an angle using a protractor:

<u>Step 1</u>: Place the protractor so that the center mark on its base coincides with the vertex of the angle, and so that the 0° mark is on one side of the angle.

<u>Step 2</u>: Read the mark on the protractor that is on the other side of the angle (extended, if necessary).
(a) If the side of the angle under the 0° mark extends to the <u>right</u> from the vertex, read the inner scale to find the degree measure.
(b) If the side of the angle under the 0° mark extends to the <u>left</u> from the vertex, read the outer scale to find the degree measure.

<u>Example 1</u>. Find the degree measure of the angle in Fig. 12.3.

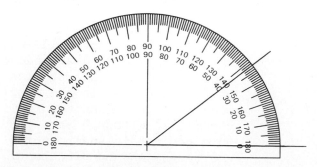

Figure 12.3

The measure of the angle is 38°.

Example 2. Find the degree measure of the angle in Fig. 12.4.

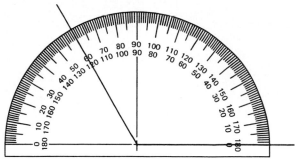

Figure 12.4

The measure of the angle is 120°.

Angles are often classified by degree measure.

A *right* angle is an angle that has a measure of 90°. In a sketch or diagram a 90° angle is denoted by placing ⌐ or Γ in the angle, as shown in Fig. 12.5.

An *acute* angle (Fig. 12.6) is an angle that has a measure less than 90°.

An *obtuse* angle (Fig. 12.7) is an angle that has a measure greater than 90° but less than 180°.

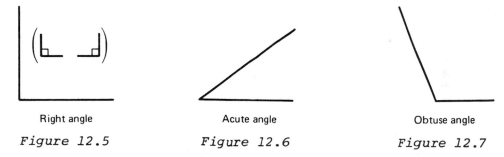

Right angle	Acute angle	Obtuse angle
Figure 12.5	*Figure 12.6*	*Figure 12.7*

We will first study some geometric relationships of angles and lines. All the lines here are in the same plane.

Two lines *intersect* if they have only one point in common. (See Fig. 12.8.)

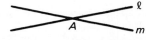

Figure 12.8 Lines ℓ and m intersect at point A.

Two lines are *parallel* (‖) if they do not intersect even if they are extended. (See Fig. 12.9.)

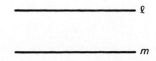

Figure 12.9 Lines ℓ and m are parallel.

Two angles are said to be *adjacent* if they have a common vertex and a common side between them. (See Fig. 12.10.)

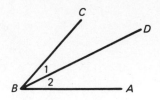

Figure 12.10

Here ∠1 and ∠2 are adjacent angles because both have the common vertex *B* and the common side $\overline{BD}$ between them.

Two lines are *perpendicular* (⊥) if they intersect and form equal adjacent angles. Each of these equal adjacent angles is called a right angle. (See Fig. 12.11.)

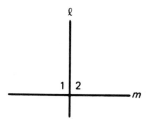

Figure 12.11

Here ℓ ⊥ *m* because ∠1 = ∠2. Angles 1 and 2 are called *right* angles.

Two angles are *complementary* if the sum of their measures is 90°. (See Figs. 12.12, 12.13.)

Figure 12.12

Angles *A* and *B* in Fig. 12.12 are complementary: 52° + 38° = 90°. Angles *LMN* and *NMP* in Fig. 12.13 are complementary: 25° + 65° = 90°.

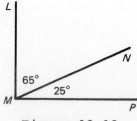

Figure 12.13

Two angles are *supplementary* if the sum of their measures is 180°. (See Fig. 12.14.)

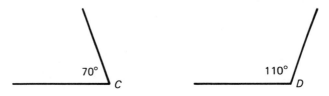

Figure 12.14

Angles *C* and *D* are supplementary: 70° + 110° = 180°.

Two adjacent angles with their exterior sides in a straight line are *supplementary*. (See Fig. 12.15.)

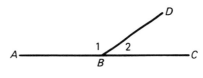

Figure 12.15

Angles 1 and 2 have their exterior sides in a straight line. So, they are supplementary. $\angle 1 + \angle 2 = 180°$.

When two lines intersect, the angles opposite each other are called *vertical angles*. (See. Fig. 12.16.)

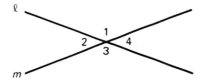

Figure 12.16

Angles 1 and 3 are vertical angles.
Angles 2 and 4 are vertical angles.

If two straight lines intersect, the vertical angles formed are equal. (See Fig. 12.16.)

Angles 1 and 3 are vertical angles, so $\angle 1 = \angle 3$.
Angles 2 and 4 are vertical angles, so $\angle 2 = \angle 4$.

A *transversal* is a line which intersects two or more lines in different points. (See Fig. 12.17.)

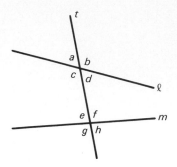

Figure 12.17

Here *t* is a transversal of lines *ℓ* and *m*.
Above, angles *a*, *b*, *g*, *h* are *exterior* angles.
Angles *c*, *d*, *e*, *f* are *interior* angles.

Exterior-interior angles on the same side of the transversal are *corresponding* angles. For example, angles *c* and *g* are corresponding angles. Angles *b* and *f* are also corresponding angles.

Angles on opposite sides of the transversal with different vertices are *alternate* angles. Angles *a* and *f* are alternate angles. Angles *a* and *h* are also alternate angles.

If two *parallel* lines are cut by a transversal, then

(a) *corresponding* angles are *equal*,
(b) *alternate-interior* angles are *equal*,
(c) *alternate-exterior* angles are *equal*, and
(d) *interior* angles on the same side of the transversal are *supplementary*.

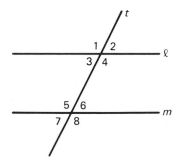

Figure 12.18

Lines *ℓ* and *m* are parallel and *t* is a transversal in Fig. 12.18. So, corresponding angles are equal. That is,

$\angle 1 = \angle 5$, $\angle 2 = \angle 6$, $\angle 3 = \angle 7$, and $\angle 4 = \angle 8$.

And, alternate-interior angles are equal. That is,

$\angle 3 = \angle 6$ and $\angle 4 = \angle 5$.

And, alternate-exterior angles are equal. That is,

$\angle 1 = \angle 8$ and $\angle 2 = \angle 7$.

And, interior angles on the same side of the transversal are supplementary. That is,

$\angle 3 + \angle 5 = 180°$ and

$\angle 4 + \angle 6 = 180°$.

Note: We will let $\overline{AB}$ be the line segment with endpoints at A and B •———•; $\overleftrightarrow{AB}$ be the line containing A and B ←•——•→; and AB be the length of $\overline{AB}$.

A *polygon* is a closed figure whose sides are straight lines. A polygon is shown in Fig. 12.19.

Polygons are named according to the number of sides they have. A *triangle* (Fig. 12.20) is a polygon with three sides. A *quadrilateral* (Fig. 12.21) is a polygon with four sides. A *pentagon* (Fig. 12.22) is a polygon with five sides.

A *regular* polygon has *all* its sides and angles equal (Fig. 12.22).

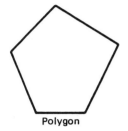

Polygon

Figure 12.19

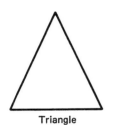

Triangle

Figure 12.20

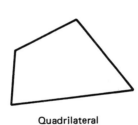

Quadrilateral

Figure 12.21

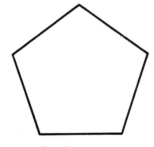

Regular pentagon

Figure 12.22

Some polygons with more than five sides are named as follows:

Number of sides	Name of polygon
6	Hexagon
7	Heptagon
8	Octagon
9	Nonagon

EXERCISES 12.1   *Use a protractor to measure each angle. Classify each angle as right, acute, or obtuse:*

1.

2.

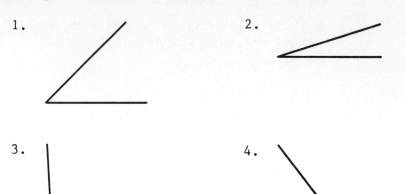

3.

4.

5.

6.

7.

8.

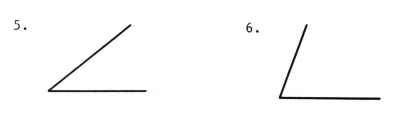

9. Line ℓ intersects line *m* and forms a right angle. Then ∠1 is a __?__ angle. Lines ℓ and *m* are __?__ .

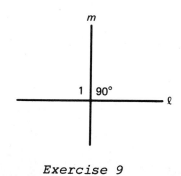

m

1 │ 90°

ℓ

Exercise 9

10. Suppose ℓ ∥ *m* and *t* ⊥ ℓ. Is *t* ⊥ *m*? Why or why not?

11. In the figure at the right:

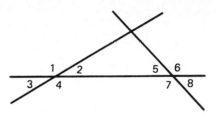

(a) Name the pairs of adjacent angles.
(b) Name the pairs of vertical angles.

Exercises 11 and 12

12. Suppose ∠3 = 40° and ∠7 =
97°. Find the measures of
the other angles.

13. Suppose ℓ ‖ m and ∠1 = 57°.
What are the measures of the
other angles?

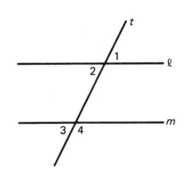

Exercise 13

14. Suppose b ‖ a, c ⊥ a, and
∠1 = 37°. What are the
measures of angles 2 and 3?

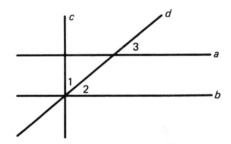

Exercise 14

15. Suppose $\overleftrightarrow{AOB}$ is a straight
line and ∠AOC = 119°. What
is the measure of ∠COB?

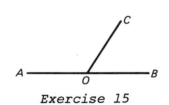

Exercise 15

16. Suppose angles 1 and 2 are supplementary and ∠1 = 63°.
Then ∠2 = ?
17. Suppose angles 3 and 4 are complementary and ∠3 = 38°.
Then ∠4 = ?
18. Suppose ℓ ‖ m, $\overleftrightarrow{AOB}$ is a straight line, and ∠3 = ∠6 = 68°.
Find the measure of each of the other angles.

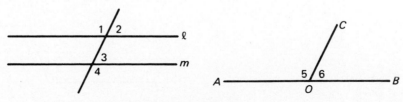

Exercise 18

19. Suppose $a \parallel b$, $t \parallel x$,
    $\angle 3 = 38°$, and $\angle 1 = 52°$.

    (a) Is $x \perp a$?
    (b) Is $x \perp b$?

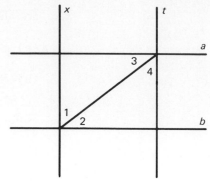

Exercise 19

20. Suppose angles 1 and 2 are complementary and $\angle 1 = \angle 2$.
    Find the measure of each angle.

21. Suppose $\angle 1$ and $\angle 3$ are supplementary. Find the measure
    of each angle.

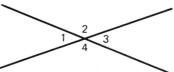

Exercise 21

22. Suppose $\ell \parallel m$, $\angle 1 = 3x - 50$,
    and $\angle 2 = x + 60$. Find the
    value of x.

23. Suppose $\ell \parallel m$, $\angle 1 = 10x - 35$,
    and $\angle 3 = 4x + 55$. Find the
    value of x.

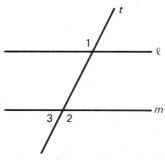

Exercises 22 and 23

24. A plumber wishes to add a pipe parallel to an existing
    pipe as shown. Find angle x.

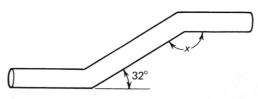

Exercise 24

25. A machinist is to weld a piece of iron parallel to an
    existing piece of iron as shown. What is angle y?

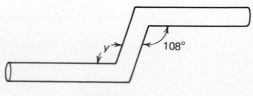

Exercise 25

26. What is angle z in the figure if m ∥ n?

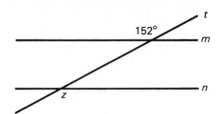

Exercise 26

27. Given $\overline{AB} \parallel \overline{CD}$.

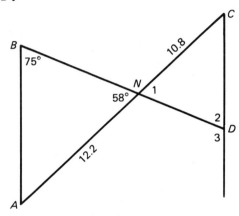

Exercise 27

Find the measure of (a) angle 1, (b) angle 2, (c) angle 3.

Name each polygon:

28.

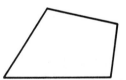

29.

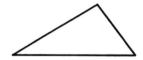

30.

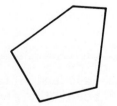

31.

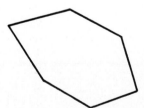

32.

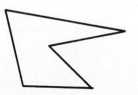

33.

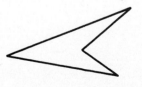

34. 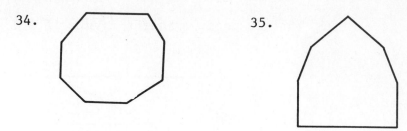   35.

12.2 Quadrilaterals

A *parallelogram* is a quadrilateral with opposite sides parallel. In Fig. 12.23, sides $\overline{AB}$ and $\overline{CD}$ are parallel, and sides $\overline{AD}$ and $\overline{BC}$ are parallel. So, polygon *ABCD* is a parallelogram.

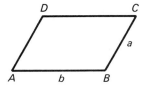

Figure 12.23 Parallelogram

In this parallelogram, Fig. 12.24, draw a line segment from point *D* to side $\overline{AB}$ that forms a 90° angle with $\overline{AB}$. This line segment is an *altitude*.

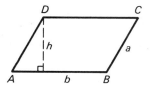

Figure 12.24

Move the part of the parallelogram that is to the left of the altitude to the right so that sides $\overline{AD}$ and $\overline{BC}$ coincide. You now have a rectangle with sides of lengths *b* and *h*. Note that the area of this rectangle, *bh* square units, is the same as the area of the parallelogram. So, the area of a parallelogram is given by the formula $A = bh$, where *b* is the length of the base and *h* is the length of the altitude drawn to that base. The perimeter is $2a + 2b$ or $2(a + b)$. (See Fig. 12.25.)

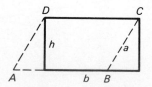

Figure 12.25

A *rectangle* is a parallelogram with four right angles. The area of the rectangle with sides of lengths a and b is given by the formula $A = ab$. (See Fig. 12.26.)

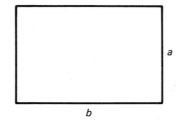

Figure 12.26   Rectangle

Another way to find the area of a rectangle is to count the number of square units in it. In Fig. 12.27, there are 96 squares in the rectangle. So, the area is 96 square units.

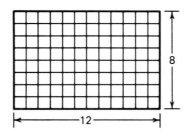

Figure 12.27

The formulas for the areas of each of the following quadrilaterals come from the formula for the area of a rectangle.

A *square* (Fig. 12.28) is a rectangle with the lengths of all four sides equal. Its area is given by the formula $A = b \cdot b = b^2$. The perimeter is $b + b + b + b$, or $4b$. Note that the length of the altitude is also b.

Figure 12.28   Square

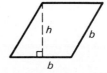

Figure 12.29   Rhombus

A *rhombus* (Fig. 12.29) is a parallelogram with the lengths of all four sides equal. Its area is given by the formula $A = bh$. The perimeter is $b + b + b + b$, or $4b$.

A *trapezoid* (Fig. 12.30) is a quadrilateral with only two sides parallel. Its area is given by the formula $A = \left(\dfrac{a + b}{2}\right)h$. The perimeter is $a + b + c + d$.

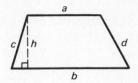

Figure 12.30   Trapezoid

*Summary of Formulas for Area and Perimeter
of Quadrilaterals*

Quadrilateral	Area	Perimeter
Rectangle	$A = ab$	$P = 2(a + b)$
Square	$A = b^2$	$P = 4b$
Parallelogram	$A = bh$	$P = 2(a + b)$
Rhombus	$A = bh$	$P = 4b$
Trapezoid	$A = \left(\dfrac{a + b}{2}\right)h$	$P = a + b + c + d$

<u>Example 1</u>. Find the area and the perimeter of the parallelogram shown in Fig. 12.31.

The formula for the area of a parallelogram is:

$$A = bh$$

So, $A = (27.2\ m)(15.5\ m)$

$$= 422\ m^2$$

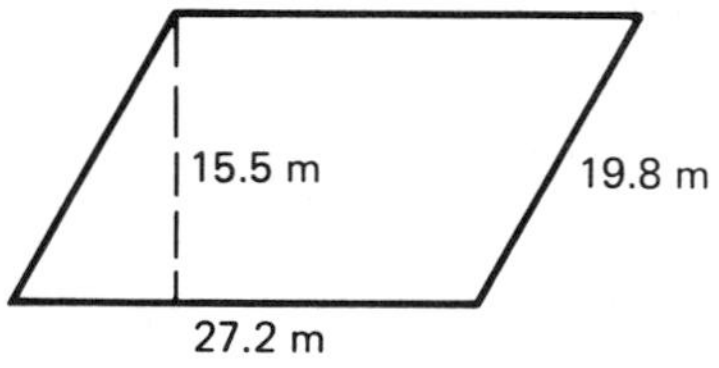

The formula for the perimeter of a parallelogram is:

$$P = 2(a + b)$$

So, $P = 2(27.2\ m + 19.8\ m)$

$$= 2(47.0\ m)$$

$$= 94.0\ m$$

15.5 m   19.8 m

27.2 m

Figure 12.31

<u>Example 2</u>. A rectangular lot that is 121.5 ft by 98.7 ft must be fenced. (See Fig. 12.32.) A fence is installed for $5.50 per running foot. Find the cost of fencing the lot.

The amount of fencing needed is the same as the perimeter of the rectangle. The formula for the perimeter is:

$$P = 2(a + b)$$

So, $P = 2(121.5 \text{ ft} + 98.7 \text{ ft})$

$$= 2(220.2 \text{ ft})$$

$$= 440.4 \text{ ft}$$

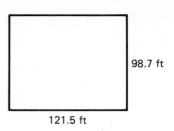

121.5 ft

98.7 ft

Figure 12.32

$$\text{Cost} = \frac{\$5.50}{1 \text{ ft}} \times 440.4 \text{ ft} = \$2422.20$$

<u>Example 3.</u> Find the cost of the fertilizer needed for the lawn in Example 2. One bag covers 2500 ft^2 and costs $6.95.

First, find the area of the rectangle. The formula for the area is $A = bh$

So, $A = (121.5 \text{ ft})(98.7 \text{ ft})$

$$= 12,\overline{0}00 \text{ ft}^2$$

The amount of fertilizer needed is found by dividing the total area by the area covered by one bag: $\frac{12\overline{0}00 \text{ ft}^2}{2500 \text{ ft}^2} = $ 4.8 bags, or 5 bags.

$$\text{Cost} = 5 \text{ bags} \times \frac{\$6.95}{1 \text{ bag}} = \$34.75$$

Use a calculator in the rest of this chapter and follow the rules for working with measurements.

EXERCISES 12.2   *Find the perimeter and the area of each quadrilateral:*

1.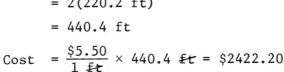

15.0 cm

15.0 cm

2.

10.0 cm

8.00 cm

3.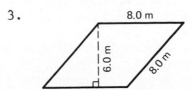

8.0 m

6.0 m

8.0 m

4.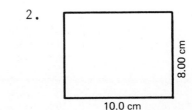

10.0 dm

7.0 dm

5.0 dm

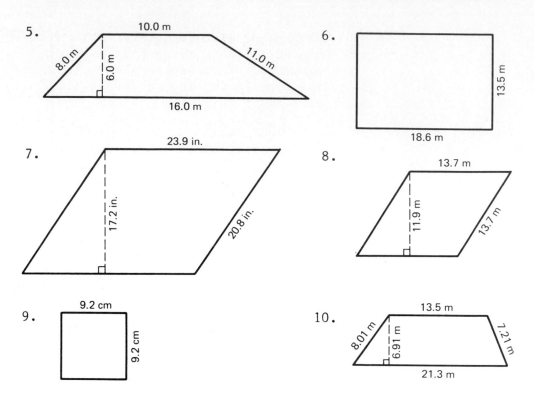

In Exercises 11-15, use the formula A = bh:

11. $A = 24\overline{0}$ cm^2, $b = 10.0$ cm; find h.

12. $A = 792$ m^2, $h = 25.0$ m; find b.

13. $h = 17.8$ mm, $A = 207$ mm^2; find b.

14. $A = 179.3$ cm^2, $b = 13.24$ cm; find h.

15. $A = 29.6$ ft^2, $b = 5.04$ ft; find h.

16. The area of a rectangle is 280 cm^2. Its width is 14 cm. Find its length.

17. The area of a parallelogram is 486 ft^2. The length of its base is 36.2 ft. Find its height.

18. The area of a parallelogram is 273 in^2. The length of its base is 17.6 in. Find its height.

19. The area of a rectangle is 315 m^2. Its height is 38.7 m. Find its width.

20. The area of a rectangle is 1087 cm^2. Its length is 204.2 cm. Find its width.

21. A parallelogram has an area of 2014 cm^2. The length of its base is 52.62 cm. Find its height.

22. A rectangle has a length of 79.2 m. Its area is 1050 m^2. Find its width.

23. The area of a rectangle is 2056.7 km^2. The length of its base is 45.147 km. Find its height.

24. A rectangular X-ray film measures 15 cm by 32 cm. What is its area?

25. Each hospital bed and its accessories take up 96 ft^2 of floor space. How many beds can be placed in a ward 24 ft by 36 ft?

26. A respirator unit needs a rectangular floor space of 0.79 m by 1.2 m. How many units could be placed in a storeroom having 20 m^2 of floor space?

27. A 108 ft^2 roll of fiberglass is 36 in. wide. What is its length in feet?

28. The cost of the fiberglass in Exercise 27 is \$9.16/yd^2. How much would this roll cost?

29. How many pieces of fiberglass, 36 in. wide and 72 in. long, can be cut from the roll in Exercise 27?

30. A mechanic wishes to build a storage garage for 85 cars. Each car will have a space of 15.0 ft by 10.0 ft. What is the floor area of the garage?

31. At a cost of \$14/ft^2, what is the cost of the garage in Exercise 30?

32. A machinist wishes to build a screen around his shop area. The area is rectangular and measures 16.2 ft by 20.7 ft. (a) How many linear feet of screen will he need? The screen is to be 8.0 ft high. (b) How many square feet of screen will he need?

33. A piece of sheet metal in the shape of a parallelogram has a rectangular hole in it, as shown. Find (a) the area of the piece that was punched out and (b) the area of the metal that is left.

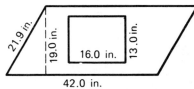

34. A piece of sheet metal has an area of 10,680 in^2. The piece is a rectangle having a length of 72.0 in. How wide is the piece?

35. What is the acreage of a ranch that is a square 25.0 mi on a side?

36. A field of corn is averaging 97 bu/acre. The field measures 1020 yd by 928 yd. How many bushels of corn will there be?

37. Find the amount of sheathing needed for the roof below.

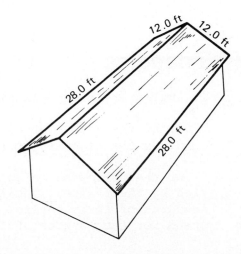

38. A ceiling is 12 ft by 15 ft. How many 1-ft-by-3-ft suspension panels are needed to cover the ceiling?

39. The Smith family is going to paint their home (shown in the drawing). The area of the openings not to be painted is 325 ft^2. The cost per square foot is $0.85. What will be the cost of painting the house?

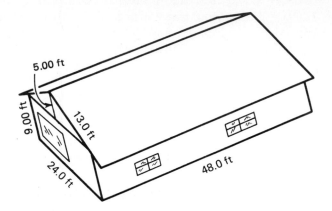

40. An 8-in.-thick wall uses 15 standard bricks (8 in. by $2\frac{1}{4}$ in. by $3\frac{3}{4}$ in.) for one square foot. Find the number of bricks needed for a wall 8 in. thick, 18 ft long, and 8.5 ft high.

41. What is the display floor space of a parallelogram-shaped space that is 29.0 ft long and 8.7 ft deep?

42. Canvas that costs $\frac{3}{4}$¢/in^2 is used in making golf bags. What would be the cost of enough rectangular pieces of canvas, 8 in. by 40 in., to make 200 golf bags?

43. By law, all businesses outside the Parkville city limits must fence their lots. How many feet of fence will be needed to fence the lot shown?

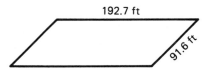

12.3 Triangles

Triangles are often classified in two ways:

1. By the number of equal sides.
2. By the measures of the angles of the triangle.

Of the following triangles, the first three are classified by their sides. The last three are classified by the measures of their angles. In each triangle, *a*, *b*, and *c* stand for the lengths of the sides.

An *equilateral triangle* (Fig. 12.33) has all three sides equal. Also, all three angles are equal.

An *isosceles triangle* (Fig. 12.34) has two sides equal. Also, the angles opposite these two sides are equal. A *scalene triangle* (Fig. 12.35) has no sides equal. Also, no angles are equal.

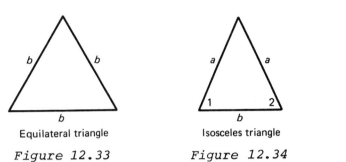

Equilateral triangle	Isosceles triangle	Scalene triangle
Figure 12.33	*Figure 12.34*	*Figure 12.35*

A *right triangle* (Fig. 12.36) has one right angle. The two sides of the right angle are called *legs* of the right triangle. The side opposite the right angle is called the *hypotenuse*.

An *obtuse triangle* (Fig. 12.37) has one obtuse angle.

An *acute triangle* (Fig. 12.38) has three acute angles.

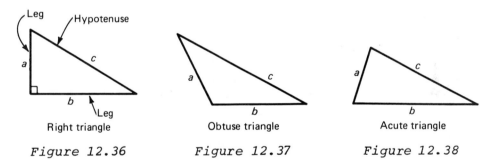

Right triangle	Obtuse triangle	Acute triangle
Figure 12.36	*Figure 12.37*	*Figure 12.38*

The *Pythagorean theorem* is a formula for right triangles only. (See Fig. 12.39.) It relates the lengths of the sides. Ancient Egyptians used this to build the pyramids.

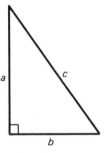

Figure 12.39

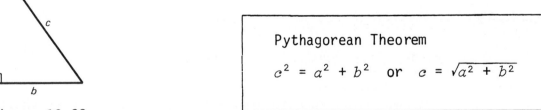

Pythagorean Theorem

$$c^2 = a^2 + b^2 \quad \text{or} \quad c = \sqrt{a^2 + b^2}$$

The Pythagorean theorem states that the *square of the hypotenuse is equal to the sum of the squares of the two legs*. Alternate forms of the Pythagorean theorem are:

$$a^2 = c^2 - b^2 \quad \text{or} \quad a = \sqrt{c^2 - b^2}$$

and
$$b^2 = c^2 - a^2 \quad \text{or} \quad b = \sqrt{c^2 - a^2}$$

Before we use the Pythagorean theorem, you may wish to review square roots in Sec. 10.5. A calculator may be used to find the square root of a number. (See Appendix A.4.) The square root of a nonperfect square, such as $\sqrt{20}$, is usually rounded because its decimal equivalent cannot be found exactly. For example, $\sqrt{20} = 4.4721359$ . . . So, we most often write $\sqrt{20} = 4.47$.

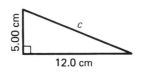

Figure 12.40

Example 1. Find the length of the hypotenuse of the triangle in Fig. 12.40.

Substitute 5.00 cm for a and 12.0 cm for b in the formula:

$$c = \sqrt{a^2 + b^2}$$
$$= \sqrt{(5.00 \text{ cm})^2 + (12.0 \text{ cm})^2}$$
$$= \sqrt{25.0 \text{ cm}^2 + 144 \text{ cm}^2}$$
$$= \sqrt{169 \text{ cm}^2} = 13.0 \text{ cm}$$

Flowchart	Buttons Pushed	Display
Enter 5	5	5
Push square	x^2	25
Push plus	+	25
Enter 12	1 → 2	12
Push square	x^2	144
Push equals	=	169
Push square root	$\sqrt{}$	13

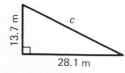

Figure 12.41

Example 2. Find the length of the hypotenuse of the triangle in Fig. 12.41.

$$c = \sqrt{a^2 + b^2}$$
$$= \sqrt{(13.7 \text{ m})^2 + (28.1 \text{ m})^2}$$
$$= 31.3 \text{ m}$$

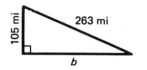

105 mi

263 mi

b

Figure 12.42

Example 3. Find the length of side *b* of the triangle in Fig. 12.42.

$$b = \sqrt{c^2 - a^2}$$

$$= \sqrt{(263 \text{ mi})^2 - (105 \text{ mi})^2}$$

$$= 241 \text{ mi}$$

Example 4. The right triangle in Fig. 12.43 gives the relationship in a circuit among the applied voltage, the voltage across a resistance, and the voltage across a coil.

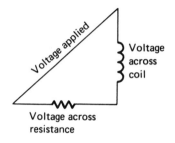

Figure 12.43

In a circuit, the voltage across the resistance is 79 V. The voltage across the coil is 82 V. Find the applied voltage.

Using the Pythagorean theorem:

Voltage applied

$$= \sqrt{(\text{voltage across coil})^2 + (\text{voltage across resistance})^2}$$

$$= \sqrt{(82 \text{ V})^2 + (79 \text{ V})^2}$$

$$= 110 \text{ V}$$

To find the perimeter of a triangle (Fig. 12.44) find the sum of the lengths of the three sides. The formula is $p = a + b + c$, where P is the perimeter and *a*, *b*, and *c* are the lengths of the sides.

An altitude of a triangle is a line segment drawn perpendicular from one vertex to the opposite side. Sometimes this opposite side must be extended.

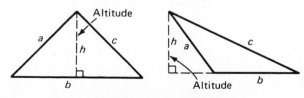

Figure 12.44

Now let us look closely at a parallelogram (Fig. 12.45) to find the formula for the area of a triangle. Remember that the area of a parallelogram with sides of lengths *a* and *b* is given by the formula $A = bh$. In this formula, *b* is the length of the base of the parallelogram and *h* is the length of the altitude drawn to that base.

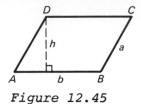

Figure 12.45

If a line segment is drawn from *B* to *D* in the parallelogram, two triangles are formed (Fig. 12.46).

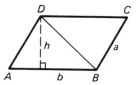

Figure 12.46

We know from geometry that these two triangles have equal areas. Since the area of the parallelogram is *bh* square units, the area of one triangle is one half the area of the parallelogram, or $\frac{1}{2} bh$ square units. So, the formula for the area of a triangle is

$$A = \frac{1}{2} bh$$

where *b* is the length of the base of the triangle (the side to which the altitude is drawn) and *h* is the length of the altitude.

Example 5. The length of the base of a triangle (Fig. 12.47) is $1\overline{0}$ cm. The length of the altitude to that base is 6.0 cm. Find the area of the triangle.

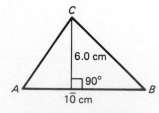

Figure 12.47

Use the formula $A = \frac{1}{2} bh$. Substitute the proper values.

$$A = \frac{1}{2} bh$$

$$= \frac{1}{2}(1\overline{0} \text{ cm})(6.0 \text{ cm})$$

$$= 3\overline{0} \text{ cm}^2$$

The area is $3\overline{0}$ cm^2.

If the lengths only of the three sides are known, the area of a triangle can be found by the formula:

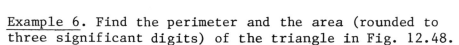

$$A = \sqrt{s(s - a)(s - b)(s - c)}$$

where $s = \frac{1}{2}(a + b + c)$ and a, b, and c are the lengths of the three sides.

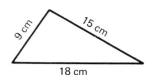

Figure 12.48

Example 6. Find the perimeter and the area (rounded to three significant digits) of the triangle in Fig. 12.48.

$P = a + b + c$

$= 9 \text{ cm} + 15 \text{ cm} + 18 \text{ cm}$

$= 42 \text{ cm}$

$s = \frac{1}{2}(a + b + c)$

$= \frac{1}{2}(9 + 15 + 18)$

$= \frac{1}{2}(42)$

$= 21$

$A = \sqrt{s(s - a)(s - b)(s - c)}$

$= \sqrt{21(21 - 9)(21 - 15)(21 - 18)}$

$= \sqrt{21(12)(6)(3)}$

$= \sqrt{4536}$

$= 67.3 \text{ cm}^2$

Flowchart	Buttons Pushed	Display
Enter 21	2 → 1	21
Push times	×	21
Left parenthesis	(	21
Enter 21	2 → 1	21
Push minus	−	21
Enter 9	9	9
Right parenthesis	)	12
Push times	×	252
Left parenthesis	(	252
Enter 21	2 → 1	21
Push minus	−	21
Enter 15	1 → 5	15
Right parenthesis	)	6
Push times	×	1512
Left parenthesis	(	1512
Enter 21	2 → 1	21
Push minus	−	21

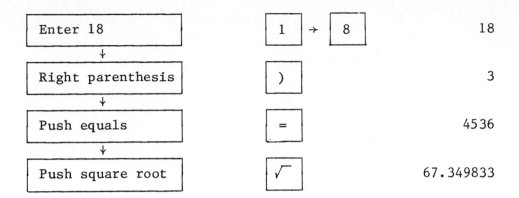

Enter 18	1 → 8	18
Right parenthesis	)	3
Push equals	=	4536
Push square root	√	67.349833

Use a protractor to measure the angles of the triangle below. Then, find the sum of these three angles.

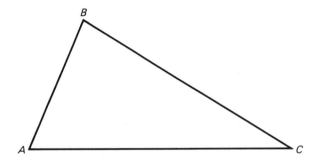

If you were very careful using the protractor, you found that:

> The sum of the measures of the angles of a triangle is 180°.

Now you can use this fact to find missing angles in triangles.

Example 7. Two angles of a triangle have measures 80° and 40°. (See Fig. 12.49.) Find the measure of the third angle of the triangle.

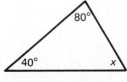

Figure 12.49

Since the sum of the measures of the angles of any triangle is 180°, we know that:

$$40° + 80° + x = 180°$$
$$120° + x = 180°$$
$$x = 60°$$

So, the measure of the missing angle is 60°.

EXERCISES 12.3   *Use the rules for working with measurements.*
                  Find the length of the hypotenuse in each triangle:

1.
6.0 m
c
8.0 m

2.
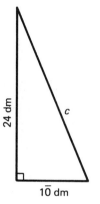
24 dm
c
$1\overline{0}$ dm

3.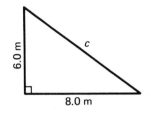
24 m
c
7.0 m

4.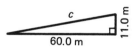
c
60.0 m
11.0 m

5.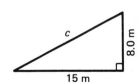
c
8.0 m
15 m

6.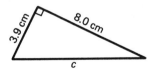
3.9 cm
8.0 cm
c

7.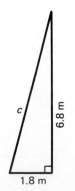
c
6.8 m
1.8 m

8.
3.68 m
3.68 m
c

9.
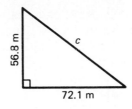

10.
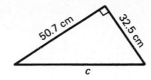

Find the length of the missing side in each triangle:

11.
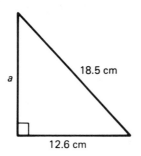

12.
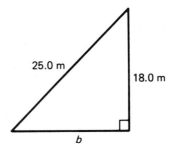

13.
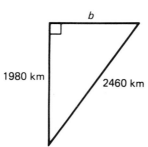

14.

15.
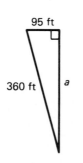

16.
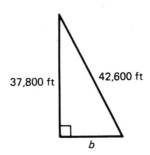

17. Find the length of the braces needed for the supports shown.
18. Find the distance between the two holes.

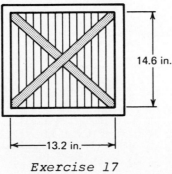

Exercise 17

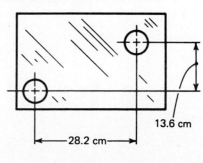

Exercise 18

19. Find the length of the brace needed in the diagram.
20. A piece of flat stock has a diagonal of 4.125 in. and a width of 2.275 in. Find the thickness of the stock.

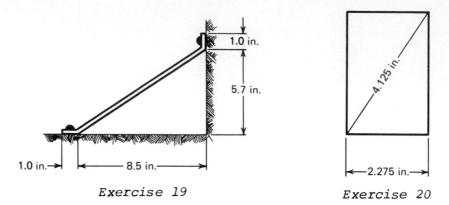

Exercise 19   *Exercise 20*

21. A piece of 4.00-in.-diameter round stock is to be milled into a square piece of stock with the largest dimensions possible. What will be the length of the side of the square?

22. Find the length of the rafter for the roof shown.

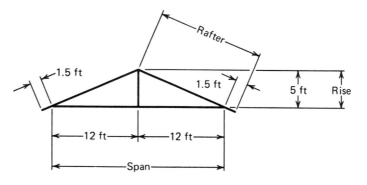

23. Find the offset distance *x* of the 6-ft length of pipe shown.

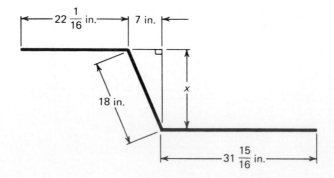

24. A conduit is run in a building as shown. (a) Find the
length of the conduit from point A_1 to A_6. (b) Find
the straight-line distance from A_1 to A_6.

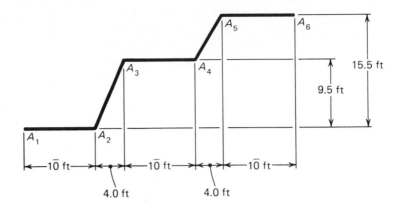

25. The voltage across the resistance is 85.2 V. The
voltage across the coil is 78.4 V. Find the voltage
applied in the circuit.

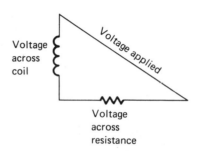

26. The voltage across the coil is 362 V. The voltage applied is 537 V. Find the voltage across the resistance.

27. The resistor current is 24 A. The total current is 32 A.
Find the coil current.

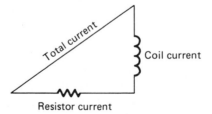

28. The resistor current is 50.2 A. The coil current is
65.3 A. Find the total current.

29. The impedance is 165 Ω. The resistance is 105 Ω. Find
the reactance of the circuit.

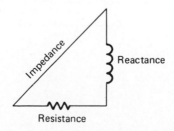

30. The reactance is 20.2 Ω. The resistance is 38.3 Ω. Find the impedance of the circuit.

31. The impedance is 4.5 Ω. The reactance is 3.7 Ω. Find the resistance of the circuit.

32. A window is 7.2 m high. The lower end of the ladder is 3.1 m from the side of the house. How long must a ladder be to reach the window?

Find the area of each triangle:

33.
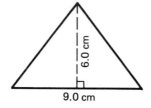
6.0 cm
9.0 cm

34.
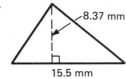
8.37 mm
15.5 mm

35.
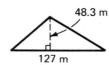
48.3 m
127 m

36.
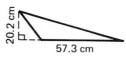
20.2 cm
57.3 cm

37.
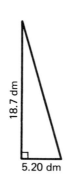
18.7 dm
5.20 dm

38.
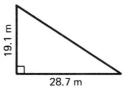
19.1 m
28.7 m

Find the area and perimeter of each triangle:

39.
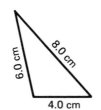
6.0 cm
8.0 cm
4.0 cm

40.

8.0 m
13 m
8.0 m

41.
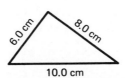
6.0 cm
8.0 cm
10.0 cm

42.
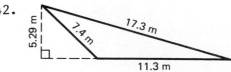
5.29 m
7.4 m
17.3 m
11.3 m

43.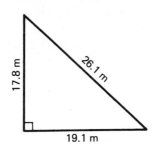

44.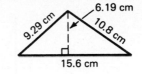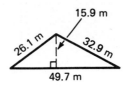

45.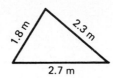

46.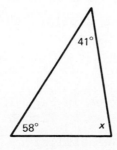

47. A machine guard is to be triangular in shape and is to be made of steel mesh. The machine has a base of 12.5 ft and a height of 8.3 ft. How many square feet of the mesh are needed for the guard?

48. A piece of metal stock is triangular in shape and weighs 38 lb. The length of the base is 4.0 ft and the altitude is 2.5 ft. What is the weight per square foot of the metal?

49. A steel plate is punched with a triangular hole as shown. Find the area of the hole.

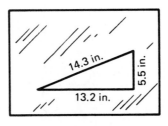

50. A triangular piece of wood with an area of 82 ft² has a base 28 ft in length. Find its height.

51. The sides of a piece of plasterboard are 3.7 ft, 4.8 ft, and 6.5 ft. What is the area of the plasterboard?

Find the measure of the missing angle in each triangle:
(Do not use a protractor.)

52.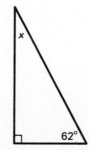

53.

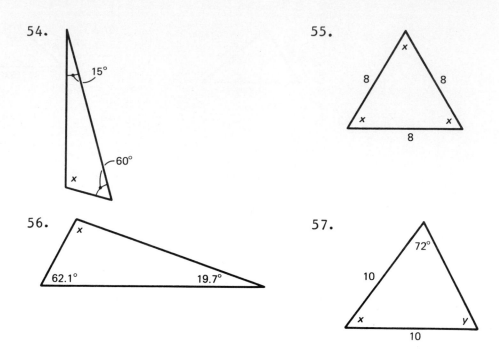

12.4 Similar Polygons

Polygons with the same shape are called *similar* polygons. Polygons are similar when the *corresponding angles are equal*.

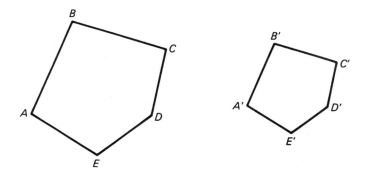

Figure 12.50

Polygon *ABCDE* is similar to polygon *A'B'C'D'E'* because the corresponding angles are equal. (See Fig. 12.50.) That is, $\angle A = \angle A'$, $\angle B = \angle B'$, $\angle C = \angle C'$, $\angle D = \angle D'$, and $\angle E = \angle E'$.

When two polygons are similar, the lengths of the *corresponding sides are proportional*. That is,

$$\frac{AB}{A'B'} = \frac{BC}{B'C'} = \frac{CD}{C'D'} = \frac{DE}{D'E'} = \frac{EA}{E'A'}$$

Example 1. The polygons in Fig. 12.50 are similar and $\overline{AB} = 12$, $DE = 3$, $A'B' = 8$. Find $D'E'$.

Since the polygons are similar,

$$\frac{AB}{A'B'} = \frac{DE}{D'E'}$$

$$\frac{12}{8} = \frac{3}{D'E'}$$

$$12(D'E') = (8)(3)$$

$$D'E' = \frac{24}{12} = 2$$

Two triangles are similar when two pairs of corresponding angles are equal, as in Fig. 12.51. (If two pairs of corresponding angles are equal, then the third pair of corresponding angles must also be equal. Why?)

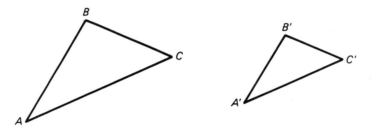

Figure 12.51

Triangle ABC is similar to triangle $A'B'C'$ because $\angle A = \angle A'$, $\angle B = \angle B'$, and $\angle C = \angle C'$.

When two triangles are similar, the lengths of the corresponding sides are proportional. That is, $\frac{AB}{A'B'} = \frac{BC}{B'C'} = \frac{CA}{C'A'}$. Or,

$$\frac{AB}{A'B'} = \frac{BC}{B'C'} \; ; \; \frac{BC}{B'C'} = \frac{CA}{C'A'} \; ; \; \text{and} \; \frac{AB}{A'B'} = \frac{CA}{C'A'}$$

<u>Example 2</u>. Find DE and AE in Fig. 12.52.

Triangles ADE and ABC are similar because $\angle A$ is common to both and each triangle has a right angle. So, the lengths of the corresponding sides are proportional.

$$\frac{AB}{AD} = \frac{BC}{DE}$$

$$\frac{24}{20} = \frac{18}{DE}$$

$$24(DE) = (20)(18)$$

$$DE = \frac{360}{24} = 15$$

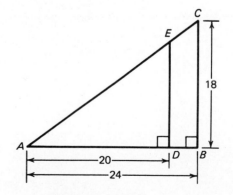

Figure 12.52

Let's use the Pythagorean theorem to find AE.

$$AE = \sqrt{(AD)^2 + (DE)^2}$$
$$= \sqrt{(20)^2 + (15)^2}$$
$$= \sqrt{400 + 225}$$
$$= \sqrt{625}$$
$$= 25$$

EXERCISES 12.4 *Follow the rules for working with measurements beginning with Exercise 6:*

1. Suppose $\overline{DE} \parallel \overline{BC}$. Find DE.

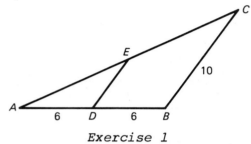

Exercise 1

2. Polygon $ABCD$ is similar to polygon $FGHI$. Find:

   (a) $\angle H$
   (b) FI
   (c) IH
   (d) BC

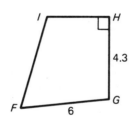

Exercise 2

3. Given $\overline{AB} \parallel \overline{CD}$. Is triangle ABO similar to triangle CDO? Why or why not?
4. Find the lengths of $\overline{AO}$ and $\overline{CD}$.

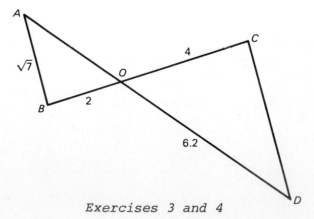

Exercises 3 and 4

5. Quadrilaterals *ABCD* and *XYZW* are similar rectangles. *AB* = 12, *BC* = 8, and *XY* = 8. Find *YZ*.

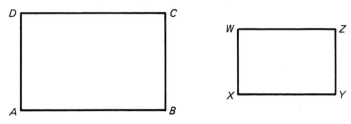

Exercise 5

6. An inclined ramp is to be built so that it reaches a height of 6.00 ft over a 15.00 ft run. Braces are placed every 5.00 ft. Find the height of braces *X* and *Y*.

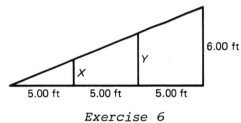

Exercise 6

7. A tree casts a shadow $8\overline{0}$ ft long when a vertical rod 6.0 ft high casts a shadow 4.0 ft long. How tall is the tree?

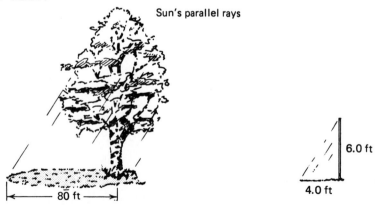

Exercise 7

8. A machinist must follow a blueprint with scale 1 to 16. What must be the dimensions of the finished stock shown below? That is, find lengths *A*, *B*, *C*, and *D*. (The measurements below are the lengths on the blueprint paper.)

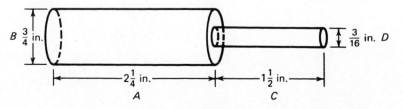

Exercise 8

9. Find the length of DE.
10. Find the length of BC.

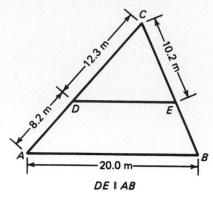

DE ∥ AB

Exercises 9 and 10

11. A 6.00 ft by 8.00 ft bookcase is to be built. It has shelves every foot. A support is to be notched in the shelves diagonally from one corner to the opposite corner. At what point should each of the shelves be notched? That is, find lengths A, B, C, D, and E. How long is the cross-piece?

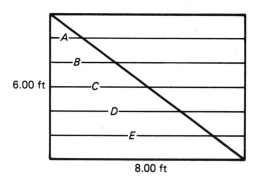

Exercise 11

12. A tower 132 ft high is to be anchored to the ground by wires as shown in the figure below. How long is each of the guy wires?

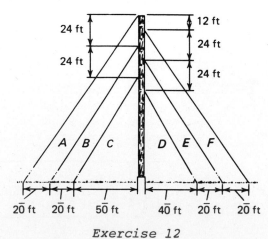

Exercise 12

12.5 Circles

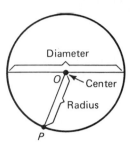

Figure 12.53

A *circle* is a plane closed curve traced by a point, *P*, on the moving side of an angle as the side rotates from 0° to 360°. The vertex of the angle, 0, is called the *center* of the circle. The distance from the vertex to the point *P* is called the *radius*, *r*, of the circle. The *diameter*, *d*, of the circle is a line segment through the center of the circle with endpoints on the circle. Note (in Fig. 12.53) that the length of the diameter is equal to the sum of the lengths of the two radii, that is, $d = 2r$.

The *circumference* of a circle is the distance around the circle. The ratio of the circumference of a circle to the length of the diameter of that same circle is always the same. That is, it is a constant. This constant is π (pi), and its approximate value is usually stated as 3.14 or $3\frac{1}{7}$.

The number π, however, cannot be written exactly as a decimal. So, we must round the value of π to solve problems. When solving problems with π, use the π button on your calculator.

The following formulas for finding the circumference and the area of a circle will be used often. Study them carefully. Let *C* be the circumference of a circle and let *A* be the area of a circle. Remember that *d* stands for the length of the diameter and *r* stands for the radius.

Formulas for finding the circumference of a circle:

$$C = \pi d$$
$$C = 2\pi r$$

Formulas for finding the area of a circle:

$$A = \pi r^2$$
$$A = \frac{\pi d^2}{4}$$

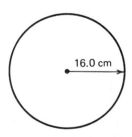

Figure 12.54

Example 1. Find the area and circumference of the circle on the left, in Fig. 12.54.

The formula for finding the area of a circle given the radius is

$$A = \pi r^2$$

So,

$$A = \pi (16.0 \text{ cm})^2$$
$$= 804 \text{ cm}^2$$

The formula for finding the circumference of a circle given the radius is

$$C = 2\pi r$$

So, $$C = 2\pi \ (16.0 \ \text{cm})$$

$$= 101 \ \text{cm}$$

<u>Example 2.</u> The area of a circle is 576 m². Find the radius.

The formula for finding the area of a circle in terms of the radius is

$$A = \pi r^2$$

So, $$576 \ \text{m}^2 = \pi r^2$$

$$\frac{576 \ \text{m}^2}{\pi} = r^2 \qquad \text{Divide both sides by } \pi.$$

$$\sqrt{\frac{576 \ \text{m}^2}{\pi}} = r \qquad \text{Take the square root of both sides.}$$

$$13.5 \ \text{m} = r$$

<u>Example 3.</u> The circumference of a circle is 28.2 cm. Find the radius.

The formula for the circumference of a circle in terms of the radius is

$$C = 2\pi r$$

So, $$28.2 \ \text{cm} = 2\pi r$$

$$\frac{28.2 \ \text{cm}}{2\pi} = r \qquad \text{Divide both sides by } 2\pi.$$

$$4.49 \ \text{cm} = r$$

An angle whose vertex is at the center of a circle is called a *central* angle. Angle *A* in Fig. 12.55 is a central angle.

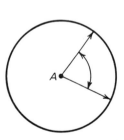

Figure 12.55

<u>Example 4.</u> Measure the central angles in Fig. 12.56 and find their sum.

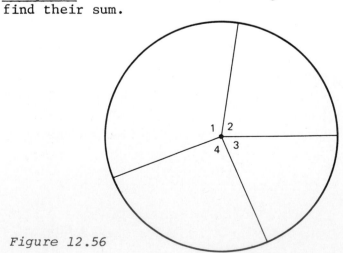

Figure 12.56

```
Measure of angle 1 = 117°
Measure of angle 2 =  83°
Measure of angle 3 =  68°
Measure of angle 4 =  92°
Sum is                360°
```

The sum of the measures of all the central angles of any circle is 360°.

Other common terms and relationships of the circle are:

A *chord* is a line segment that has its endpoints on the circle.

A *secant* is any line which has a chord in it.

A *tangent* is a line (or a line segment) which has only one point in common with a circle and lies totally outside the circle. (See Fig. 12.57.)

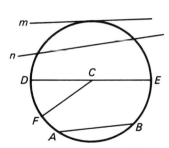

$\overline{AB}$ is a chord. Line n is a secant. Line m is a tangent. $\overline{DE}$ is a diameter. CF is a radius.

Figure 12.57

An *inscribed angle* is an angle whose vertex is on the circle and whose sides are chords.

That part of the circle between the two sides of an inscribed or central angle is called the *intercepted arc*. (See Fig. 12.58.)

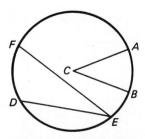

$\angle ACB$ is a central angle.

$\angle DEF$ is an inscribed angle.

$\overset{\frown}{AB}$ is the intercepted arc of $\angle ACB$.

$\overset{\frown}{FD}$ is the intercepted arc of $\angle DEF$.

Figure 12.58

The number of degrees in a central angle is equal to the number of arc degrees in its intercepted arc.

An inscribed angle is equal in degrees to one-half the number of arc degrees in its intercepted arc. (See Fig. 12.59.)

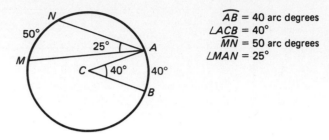

Figure 12.59

An angle formed by two intersecting chords is equal in number to one-half the sum of the measures of the intercepted arcs. (See Fig. 12.60.)

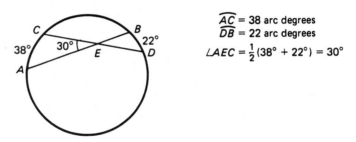

Figure 12.60

A diameter that is perpendicular to a chord bisects the chord. (See Fig. 12.61.)

A line segment from the center of a circle to the point of tangency is perpendicular to the tangent. (See Fig. 12.62.)

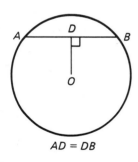

$AD = DB$

Figure 12.61

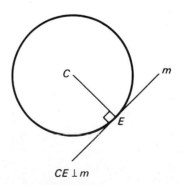

$CE \perp m$

Figure 12.62

Two tangents drawn from a point outside a circle to the circle are equal. And, they make equal angles with the line segment drawn from the center of the circle to the point outside the circle. (See Fig. 12.63.)

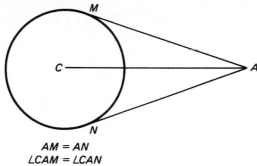

$AM = AN$
$\angle CAM = \angle CAN$

Figure 12.63

EXERCISES 12.5   *Follow the rules for working with measurements.*
Find (a) the circumference and (b) the area of each circle:

1.
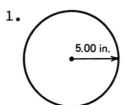

2.
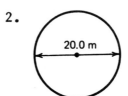

3.
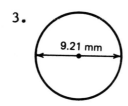

4.
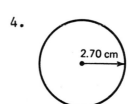

5.
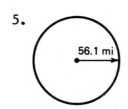

6.
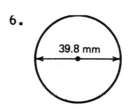

Find the measure of each unknown angle:

7.
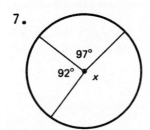

8.
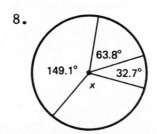

9.

10.

11. The area of a circle is 28.2 cm^2. Find its radius.
12. The area of a circle is 214 ft^2. Find its radius.
13. The area of a circle is 21.2 in^2. Find its radius.
14. The area of a circle is 792 m^2. Find its radius.
15. The circumference of a circle is 62.8 m. Find its radius.
16. The circumference of a circle is 17.2 in. Find its radius.
17. The circumference of a circle is 202 ft. Find its radius.
18. The circumference of a circle is 2.38 mm. Find its radius.
19. How many degrees are in a central angle whose arc is $\frac{1}{4}$ of a circle?
20. How many degrees are in a central angle whose arc is $\frac{2}{3}$ of a circle?
21. How many degrees are in a central angle whose arc is $\frac{2}{5}$ of a circle?
22. What percent of 360° is 45°?
23. What percent of 360° is 90°?
24. What percent of 360° is 135°?
25. What percent of 360° is 37.5°?
26. Find 65% of 360°.
27. Find 62.8% of 360°.
28. Find 3.2% of 360°.
 29. A wheel with radius of 1.80 ft is used to measure a field. The wheel rotates 236 times while going the length of the field. How long is the field?
 30. Find the length of the diameter of a circular silo with circumference of 52.0 ft.
31. A circular hole 4.00 ft in diameter is to receive a heating pipe. The hole is to be cut from a piece of dry wall 16.0 ft by 36.0 ft.
    (a) Find the area of the dry wall that is cut out.
    (b) Find the area of the dry wall that is left to be finished.

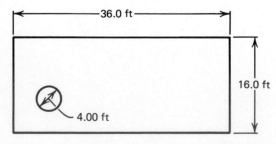

32. A rectangular piece of insulation is to be wrapped around a pipe 4.25 in. in diameter. How wide does the rectangular piece need to be?

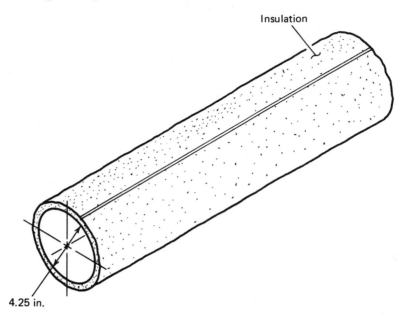

Insulation

4.25 in.

33. How many 1.5-in.-diameter pipes have approximately the same total cross-sectional area as one whose diameter is 5.0 in.?

34. The steel plate below is punched with a hole 5.2 cm in diameter. Find the area of the piece of steel which is left.

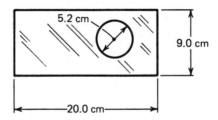

5.2 cm

9.0 cm

20.0 cm

35. A pipe has a 3.50-in. outside diameter and a 3.25-in. inside diameter. Find the area of its cross-section.

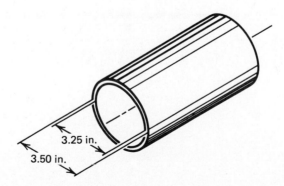

3.25 in.

3.50 in.

36. The outside diameter of a gear is 8.25 in. What is the circumference of the gear?

37. A boiler 5.00 ft in diameter is to be placed in a corner of a room.
    (a) How far from the corner C are points A and B of the boiler?
    (b) How long is a pipe from C to the center of the boiler M?

38. A pulley is connected to a spindle of a wheel by a belt. The distance from the spindle to the center of the pulley is 15.0 in. The diameter of the pulley is 15.0 in. The belt is in contact with the pulley for 31.4 in. What is the length of the belt?

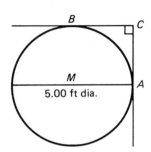

5.00 ft dia.

Exercise 37

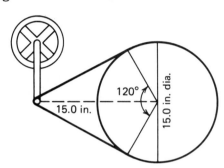

120°

15.0 in.

15.0 in. dia.

Exercise 38

39. Mary needs to punch 5 evenly spaced holes around a circular metal plate as shown. Find the measure of each central angle.

40. Find the measure of $\angle 1$ in the figure below.

41. Find the measure of $\angle 2$ in the figure below.

42. Find the measure of $\angle 3$ in the figure below given that $\overline{AC} \parallel \overline{DB}$.

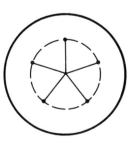

Exercise 39

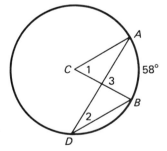

58°

Exercises 40-42

43. Find the measure of $\angle 1$ in the figure below.
44. In the figure below, the length of AC is 5 and the distance between B and C is 13. Find the length of $\overline{AB}$.

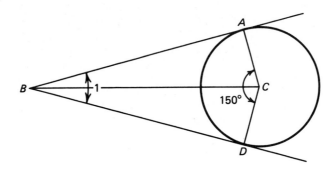

Exercises 43 and 44

45. A satellite is at position P relative to a strange planet of radius 2000 miles. The angle between the tangent lines is 12.4°. The distance from the satellite to Q is 20,500 miles. What is the altitude, $\overline{SP}$, of the satellite above the planet?

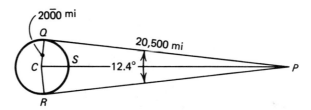

Exercise 45

46. In the figure, $CP = 12.2$ m and $PB = 10.8$ m. Find the radius of the circle.

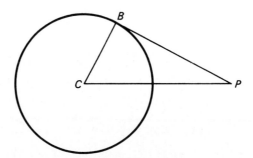

Exercise 46

In Exercises 47—50, $\overleftrightarrow{AB}$ and $\overleftrightarrow{AC}$ are secants; $\overline{CD}$ and $\overline{BF}$ are chords.

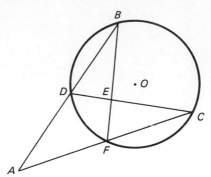

Exercises 47-50

47. Suppose $\overparen{DB}$ = 85 arc degrees and $\angle DEF$ = 52°. Find the number of arc degrees in $\overparen{FC}$.
48. Suppose $\overparen{BC}$ = 100 arc degrees and $\overparen{DF}$ = 40 arc degrees. Find the number of degrees in $\angle BAC$.
49. Suppose $\angle BEC$ = 78° and $\overparen{BC}$ = 142 arc degrees. Find the number of arc degrees in $\overparen{DF}$.
50. Suppose $\angle DCF$ = 30°, $\angle EFC$ = 52°, and $\overparen{FC}$ = 110 arc degrees. Find the number of arc degrees in $\overparen{DB}$.

12.6 Radian Measure

Radian measure, the metric unit of angle measure, is needed in many applications, such as in arc length and rotary motion. The *radian* (rad) unit is defined as the measure of an angle with its vertex at the center of a circle and with an intercepted arc on the circle equal in length to the radius. (In Fig. 12.64, $\angle AOB$ forms the intercepted arc AB on the circle.)

Figure 12.64

In general, the radian is defined to be the ratio of the length of arc an angle intercepts on a circle to the length of its radius. In a complete circle or complete revolution, the circumference $C = 2\pi r$. This means that for any circle the ratio of the circumference to the radius is constant (2π) because $\dfrac{C}{r} = 2\pi$. That is, the radian measure of one complete revolution is 2π rad.

What is the relationship between radians and degrees? Since

$$\text{one complete revolution} = 360°,$$

and $$\text{one complete revolution} = 2\pi \text{ rad},$$

then $$360° = 2\pi \text{ rad}$$

or $$180° = \pi \text{ rad}.$$

This gives us the conversion factors

$$\frac{\pi \text{ rad}}{180°} = 1 \quad \text{and} \quad \frac{180°}{\pi \text{ rad}} = 1$$

For comparison purposes,

$$1 \text{ rad} = \frac{180°}{\pi} = 57.3°$$

$$1° = \frac{\pi \text{ rad}}{180°}$$

$$= 0.01745 \text{ rad}$$

__Example 1.__ How many degrees are in an angle that measures $\frac{\pi}{2}$ rad?

Use the conversion factor $\frac{180°}{\pi \text{ rad}}$.

$$\frac{\pi}{2} \text{ rad} \times \frac{180°}{\pi \text{ rad}} = \frac{180°}{2} = 90°$$

__Example 2.__ How many radians are in an angle that measures 30°?

Use the conversion factor $\frac{\pi \text{ rad}}{180°}$.

$$30° \times \frac{\pi \text{ rad}}{180°} = \frac{\pi}{6} \text{ rad or } 0.524 \text{ rad}$$

As a wheel rolls along a surface, the distance, s, that a point on the wheel travels is equal to the product of the radius, r, and the angle, θ, measured in radians, through which the wheel turns (Fig. 12.65). That is,

$$s = r\theta \quad (\theta \text{ in rad})$$

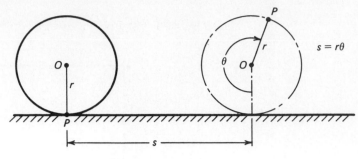

Figure 12.65

Example 3. What is the distance a point on the surface of a pulley travels if its radius is 10.0 cm and the angle of turn is $\frac{5}{4}$ rad?

$$s = r\theta \quad (\theta \text{ in rad})$$

So,

$$s = 10.0 \text{ cm} \times \frac{5}{4}$$

$$= 12.5 \text{ cm}$$

Example 4. What is the distance a point on the surface of a gear travels if its radius is 15 cm and the angle of the turn is 420°?

$$s = r\theta \quad (\theta \text{ in rad})$$

Since the angle is given in degrees, you must change 420° to radians.

Use the conversion factor $\frac{\pi \text{ rad}}{180°}$.

$$\overset{7}{\cancel{420°}} \times \frac{\pi \text{ rad}}{\underset{3}{\cancel{180°}}} = \frac{7\pi}{3} \text{ rad}$$

Then,

$$s = \overset{5}{\cancel{15}} \text{ cm} \times \frac{7\pi}{\cancel{3}} = 35\pi \text{ cm or } 110 \text{ cm}$$

Example 5. A wheel with radius 5.4 cm travels a distance of 21 cm. Find the angle θ (in radians) that the wheel turns.

$$s = r\theta \quad (\theta \text{ in rad})$$

Solve for θ:

$$\frac{s}{r} = \theta$$

$$\frac{21 \text{ cm}}{5.4 \text{ cm}} = \theta$$

$$3.9 \text{ rad} = \theta$$

EXERCISES 12.6

1. π rad = _____ °.   2. 1.7 rad = _____ °.

3. 21° = _____ rad.   4. Change 45° to radians.

5. Change $\frac{\pi}{3}$ rad to degrees.   6. Change 150° to radians.

7. Change 135° to radians.   8. Change $\frac{\pi}{12}$ rad to degrees.

9. How many radians are contained in a central angle which is $\frac{2}{3}$ of a circle?

10. What percent of 2π rad is $\frac{\pi}{2}$ rad?

11. What would be the number of radians in the measure of the central angle whose arc is $\frac{2}{5}$ of a circle?

12. What percent of 2 rad is $\frac{\pi}{12}$ rad?

Complete the table using the formula $s = r\theta$ (θ in radians):

	Radius, r	Angle, θ	Distance, s
13.	25.0 cm	$\frac{2\pi}{5}$ rad	
14.	30.0 cm	$\frac{4\pi}{3}$ rad	
15.	6.0 cm	45°	
16.	172 mm	$\frac{\pi}{4}$ rad	
17.	18.0 cm	330°	
18.	3.0 m	250°	
19.	40.0 cm	rad	112 cm
20.	0.0081 mm	rad	0.011 mm
21.	0.500 m	°	0.860 m
22.	0.027 m	°	0.0283 m

23. A pulley is turning at an angular velocity of 10.0 rad per second. How many revolutions is the pulley making each second? (<u>Hint</u>. One revolution equals 2π rad.)

24. The radius of a wheel is 20.0 in. It turns through an angle of 2.75 rad. What is the distance a point travels on the surface of the wheel?

25. The radius of a gear is 22.0 cm. It turns through an angle of 240°. What is the distance a point travels on the surface of the gear?
26. A wheel of diameter 6.00 m travels a distance of 31.6 m. Find the angle θ (in radians) that the wheel turns.
27. A wheel of diameter 15.2 cm turns through an angle of 3.40 rad. Find the distance a point travels on the surface of the wheel.

12.7 Prisms

Up to now, you studied the geometry of two dimensions. Now you need to study the geometry of three dimensions, called *solid geometry*. The three dimensions are length, width, and depth. There are many different geometric solids.

A *prism* is a solid made up of polygons having at least one pair of these polygons, which are parallel, and having the same size and shape. (Remember that a closed broken line in two dimensions is called a polygon.) The polygons that form a prism are called *faces*. The two parallel polygons (which may be any type polygon) are called the *bases* of the prism. The remaining polygons will be parallelograms and are called *lateral faces*. A *right prism* has lateral faces that are rectangles and are therefore perpendicular to the bases.

The name of the polygon used as the base names the type of prism. For example, a prism with bases that are triangles is called a triangular prism. (Fig. 12.66.) A prism with bases that are pentagons is called a pentagonal prism (Fig. 12.67), and so on.

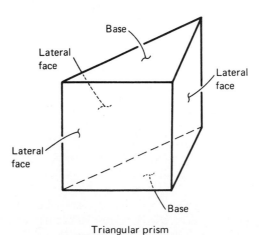

Triangular prism

Figure 12.66

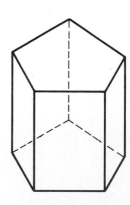

Pentagonal prism

Figure 12.67

> The lateral surface area of a prism is the sum of the areas of the lateral faces of the prism.
>
> The total area of a prism is the sum of the areas of the lateral faces and the areas of the bases.

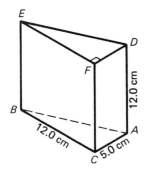

Figure 12.68

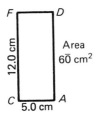

Figure 12.69

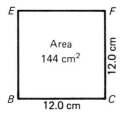

Figure 12.70

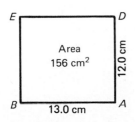

Figure 12.71

<u>Example 1.</u> Find the lateral surface area of the triangular right prism in Fig. 12.68.

To find the lateral surface area, find the area of each lateral face. Then, find the sum of these areas.

The area of the rectangular face located on the right front side (Fig. 12.69) is

$$A = \ell w$$
$$= (12.0 \text{ cm})(5.0 \text{ cm})$$
$$= 6\bar{0} \text{ cm}^2$$

The area of the square face on the left front side (Fig. 12.70) is

$$A = s^2$$
$$= (12.0 \text{ cm})^2$$
$$= 144 \text{ cm}^2$$

To find the area of the third face (back side of the prism), first find length AB. Since AB is also the hypotenuse of the right triangle ABC, we can use the Pythagorean theorem. Use the formula

$$c = \sqrt{a^2 + b^2}$$
$$= \sqrt{(5.0 \text{ m})^2 + (12.0 \text{ m})^2}$$
$$= \sqrt{25 \text{ m}^2 + 144 \text{ m}^2}$$
$$= \sqrt{169 \text{ m}^2}$$
$$= 13.0 \text{ m}$$

The area of the third face (Fig. 12.71) is

$$A = \ell w$$
$$= (13.0 \text{ cm})(12.0 \text{ cm})$$
$$= 156 \text{ cm}^2$$

Therefore, the lateral surface area is

$$6\bar{0} \text{ cm}^2 + 144 \text{ cm}^2 + 156 \text{ cm}^2 = 36\bar{0} \text{ cm}^2$$

Example 2. Find the total area of the prism in Example 1.

To find the total surface area, first find the area of the bases. Then, add this result to the lateral surface area from Example 1. The bases have the same size and shape, so, just find the area of one base and then double it.

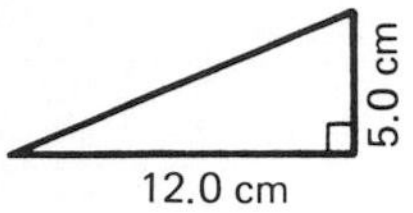

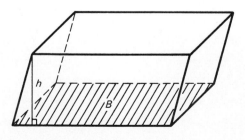

Figure 12.72

The area of one base as shown in Fig. 12.72 is

$$B = A = \frac{1}{2}bh$$

$$= \frac{1}{2}(12.0 \text{ cm})(5.0 \text{ cm})$$

$$= 3\bar{0} \text{ cm}^2$$

Double this to find the area of both the bases.

$$2(3\bar{0} \text{ cm}^2) = 6\bar{0} \text{ cm}^2$$

Add this area to the lateral surface area to find the total area.

$$36\bar{0} \text{ cm}^2 + 6\bar{0} \text{ cm}^2 = 42\bar{0} \text{ cm}^2$$

So, the total surface area is $42\bar{0} \text{ m}^2$.

The volume of a prism is found by the formula:

$$V = Bh$$

where B is the area of one of the bases and h is the altitude of a lateral face. See Fig. 12.73.

Figure 12.73

Example 3. Find the volume of the prism in Example 1.

To find the volume of the prism, use the formula $V = Bh$. B is the area of the base which you found to be $3\bar{0}$ cm^2. The altitude of a lateral face is 12.0 cm.

$$V = Bh$$
$$= (3\bar{0} \text{ cm}^2)(12.0 \text{ cm})$$
$$= 360 \text{ cm}^3$$

Example 4. Find the volume of the prism in Fig. 12.74.

Use the formula $V = Bh$. The base is a parallelogram with sides of length 10.0 cm and 4.0 cm. The altitude is 3.0 cm. First, find B, the area of the base.

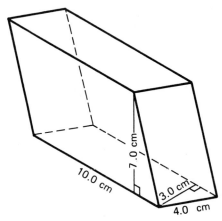

Figure 12.74

$$B = bh$$
$$= (10.0 \text{ cm})(3.0 \text{ cm})$$
$$= 3\bar{0} \text{ cm}^2$$

The altitude of a lateral face of the prism is 7.0 cm. Therefore,

$$V = (3\bar{0} \text{ cm}^2)(7.0 \text{ cm})$$
$$= 210 \text{ cm}^3$$

Example 5. A rectangular piece of steel is 24.1 in. by 13.2 in. by 8.20 in. (See Fig. 12.75.) Steel weighs 0.28 lb/in^3. Find its weight in pounds.

Find the volume using the formula for the volume of a prism, $V = Bh$.

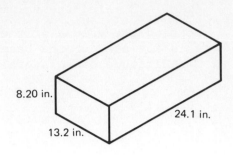

Figure 12.75

First, find B, the area of the base of the prism:

$$B = \ell w$$

$$B = (13.2 \text{ in.})(24.1 \text{ in.})$$

the volume is then

$$V = Bh$$

$$V = [(13.2 \text{ in.})(24.1 \text{ in.})](8.20 \text{ in.})$$

$$= 2610 \text{ in}^3$$

Since steel weighs 0.28 lb/in^3, the total weight is

$$2610 \text{ in}^3 \times \frac{0.28 \text{ lb}}{1 \text{ in}^3} = 730 \text{ lb}$$

EXERCISES 12.7   *Follow the rules for working with measurements.*

    1. (a) Find the lateral surface area of the prism below.
       (b) Find the total area of the prism.
       (c) Find the volume of the prism.
    2. (a) Find the lateral surface area of the prism.
       (b) Find the total area of the prism.
       (c) Find the volume of the prism.
       (d) What is the name given to this geometric solid?

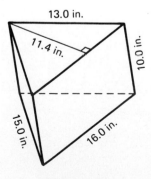

Exercise 1

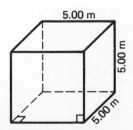

Exercise 2

3. (a) What is the area of the lateral surface to be painted in the figure? (Assume no windows.)
   (b) What is the area of roof to be covered with shingles?
   (c) What is the volume of concrete needed to pour a floor 16 cm deep?
   (d) What is the total area of the figure? (Include painted surface, roof, and floor.)

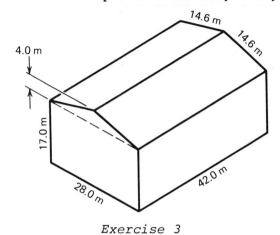

Exercise 3

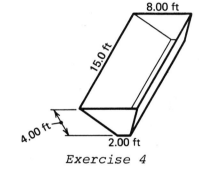

Exercise 4

4. Find the volume of the wagon box.
5. Find the volume of the gravity bin.

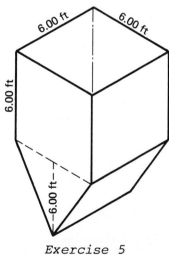

Exercise 5

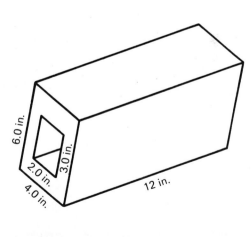

Exercise 8

6. Steel weighs 0.28 lb/in^3. What is the weight of a piece of steel 0.3125 in. by 12.0 in. by 20.0 in.?
7. A steel rod of cross-sectional area 5.0 in^2 weighs 42.0 lb. Find its length. (Steel weighs 0.28 lb/in^3.)
8. The lead sleeve above has a cored hole 2.0 in. by 3.0 in. How many cubic inches of lead are in this sleeve?

12.8 Cylinders

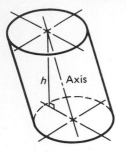

Figure 12.76

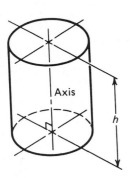

Figure 12.77

A *cylinder* is a geometric solid with a curved lateral surface. Its bases are circles that have the same area. See Fig. 12.76.

The *axis* of the cylinder is the line segment between the centers of the bases. The altitude, h, is the shortest distance between the bases. If the axis is perpendicular to the bases, the cylinder is called a *right circular cylinder* and the axis is the same length as the altitude. See Fig. 12.77.

The volume of a right circular cylinder is found by the formula $V = Bh$, where B is the area of the base. The base is always a circle and its area is found by the formula $B = \pi r^2$. Therefore, the formula for the volume of a cylinder can be written

$$V = \pi r^2 h$$

where r is the radius of the base and h is the altitude.

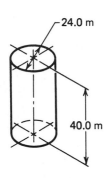

Figure 12.78

Example 1. Find the volume of the right circular cylinder in Fig. 12.78.

Use the formula $V = \pi r^2 h$. Since the diameter is 24.0 m, the radius is 12.0 m and

$$V = \pi \, (12.0 \text{ cm})^2 (40.0 \text{ m})$$
$$= 18,100 \text{ m}^3$$

Example 2. A technician wants to find the diameter of a cylindrical tank. The tank is 23.8 ft high. It has a capacity of 136,000 gallons. (1 ft^3 = 7.48 gal)

First, find the volume of the cylinder in ft^3.

$$136,000 \text{ gal} \times \frac{1 \text{ ft}^3}{7.48 \text{ gal}} = 18,200 \text{ ft}^3$$

The volume of a cylinder is found by using the formula $V = \pi r^2 h$.

Since V and h are known, we can find r.

$$V = \pi r^2 h$$

$$r^2 = \frac{V}{\pi h}$$

$$r = \sqrt{\frac{V}{\pi h}}$$

$$r = \sqrt{\frac{18,200 \text{ ft}^3}{\pi (23.8 \text{ ft})}}$$

$$= 15.6 \text{ ft}$$

Diameter is $2r$. So, the diameter is $2(15.6 \text{ ft}) = 31.2 \text{ ft}$.

The lateral surface area of a right circular cylinder can be shown as follows: Think of the cylinder as a can without ends. Cut through the side of the can and then flatten it out, as shown in Fig. 12.79.

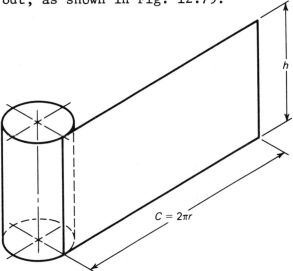

Figure 12.79

The lateral surface area of a right cylinder may then be thought of as a rectangle with the base $2\pi r$ and the altitude h. The formula for the lateral surface area is

$$A = 2\pi rh$$

The total area of a cylinder can be found by finding the area of the bases and adding it to the lateral surface area.

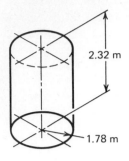

2.32 m

1.78 m

Figure 12.80

Example 3. Find the total area of the cylinder in Fig. 12.80.

The area of one base is

$$A = \pi r^2 = \pi(1.78 \text{ m})^2$$
$$= 9.95 \text{ m}^2$$

The area of both bases, then, is

$$2(9.95 \text{ m}^2) = 19.9 \text{ m}^2$$

The lateral surface area $= 2\pi rh$

$$= 2\pi(1.78 \text{ m})(2.32 \text{ m})$$
$$= 25.9 \text{ m}^2$$

The total area $\quad = (19.9 \text{ m}^2) + (25.9 \text{ m}^2)$

$$= 45.8 \text{ m}^2$$

EXERCISES 12.8   *Follow the rules for working with measurements. Find the volume of each cylinder:*

1.

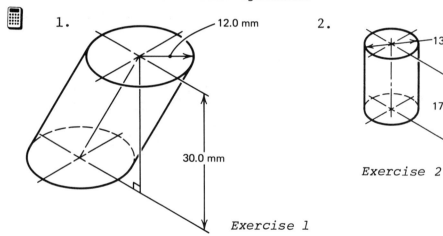

12.0 mm

30.0 mm

Exercise 1

2.

13.2 m

17.9 m

Exercise 2

3. A technician needs to know how many litres a certain cylindrical tank will hold. The height is 39.2 m. The radius of the base is 8.20 m. How many litres does the tank hold? ($1 \text{ m}^3 = 1000 \text{ L}$)

4. How many bushels of wheat may be stored in the cylindrical bin shown below? ($0.804 \text{ bu} = 1 \text{ ft}^3$)

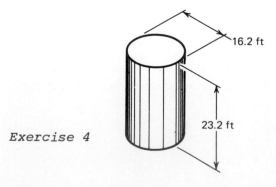

16.2 ft

23.2 ft

Exercise 4

5. A technician is asked to draw up plans for a $40\bar{0},000-$ gallon cylindrical tank with a radius of 20.0 ft. What should the height be? (1 ft³ = 7.48 gal)

 6. The compression ratio of a cylinder C is the ratio of the volume of a cylinder when the piston is at the bottom of its stroke (V_B) to the volume of the cylinder when the piston is at the top of the stroke (V_T), i.e., $C = \dfrac{V_B}{V_T}$. The diameter of the cylinder is 5.500 in. The height of the cylinder with the piston at the bottom of the stroke is 5.250 in. The height at the top of the stroke is 0.750 in. Find C.

 7. Use the formula in Exercise 6 to find the compression ratio of a cylinder of diameter 8.250 in., a height of 7.00 in. with piston at the bottom of the stroke, and a height of 0.500 in. with piston at the top of the stroke.

 8. An engine has 8 cylinders. Each cylinder has a bore of 4.70 in. diameter and a stroke of 5.25 in. Find its piston displacement.

 9. A cylindrical tank is 25 ft 8 in. long and 7 ft 8 in. in diameter. How many cubic feet does it hold?

 10. A 3.0-in.-diameter cylindrical rod is 16 in. long. Find its volume.

 11. A cylindrical piece of steel is 10.0 in. long. Its volume is 25.3 in³. Find its diameter.

12. A 2.50-in.-diameter cylindrical rod has a volume of 15.6 in³. Find its length.

13. Find the lateral surface area of the cylinder.

14. Find its total area.

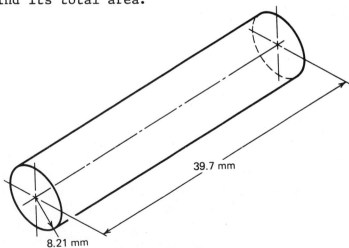

Exercises 13 and 14

15. Find the lateral surface area of the cylinder below.
16. Find its total area.
17. Find the total amount (area) of paper used for labels for 1000 cans like the one shown below.

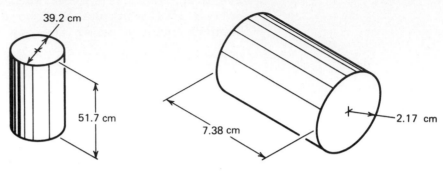

Exercises 15 and 16   *Exercise 17*

18. How many square feet of sheet metal are needed for the sides of the cylindrical tank shown below? (Allow 4.0 in. for seam overlap.)

19. A cylindrical piece of stock is turned on a lathe from 3.10 in. down to 2.24 in. in diameter. The cut is 5.00 in. long. What is the volume of the metal removed?

20. What is the volume of lead in the "pig" shown below? What is the volume of the mold?

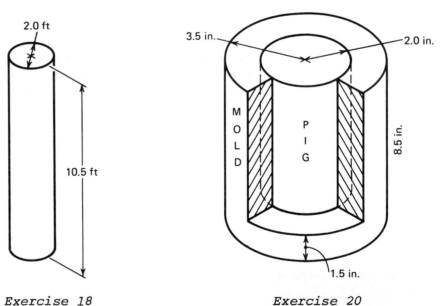

Exercise 18   *Exercise 20*

21. A cylinder bore is increased in diameter from 2.78 in. to 2.86 in. The cylinder is 5.50 in. high. How much has the surface area of the walls been increased?

22. Each cylinder bore of a 6-cylinder engine has a diameter of 2.50 in. and a height of 4.90 in. What is the lateral surface area of the six cylinder bores?

23. The sides of a silo 15 ft in diameter and 26 ft high are to be painted. Each gallon of paint will cover $2\bar{0}0$ ft^2. How many gallons of paint will be needed?

 24. How many square feet of sheet metal are needed to form the trough shown below?

25. Find the number of kilograms of metal needed for 2,700,000 cans of the type shown in the diagram. The metal has a density of 0.000147 g/cm^2.

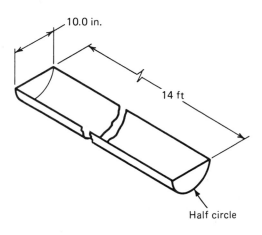

Half circle

Exercise 24

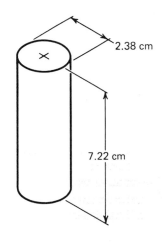

Exercise 25

12.9 Pyramids and Cones

A *pyramid* is a geometric solid whose base is a polygon and whose lateral faces are triangles with a common vertex. The common vertex is called the *apex* of the pyramid. If a pyramid has a base that is a triangle, then it is called a *triangular pyramid*. If a pyramid has a base that is a hexagon, then it is called a *hexagonal pyramid*. In general, a pyramid is named by the shape of its base. Two types of pyramids are shown in Fig. 12.81.

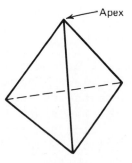
Triangular pyramid

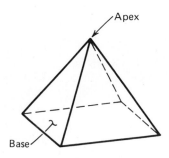
Square pyramid

Figure 12.81

The volume of a pyramid can be found using the formula

$$V = \frac{1}{3}Bh$$

where B is the area of the base and h is the height of the
pyramid. The height of a pyramid is the shortest distance
between the apex and the base of the pyramid and so is
perpendicular to the base.

Example 1. Find the volume of the pyramid in Fig. 12.82.

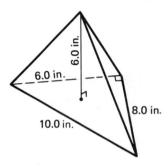

6.0 in.

6.0 in.

8.0 in.

10.0 in.

Figure 12.82

The base is a right triangle with legs 6.0 in. and
8.0 in. Therefore,

$$B = \frac{1}{2}(6.0 \text{ in.})(8.0 \text{ in.}) = 24 \text{ in}^2$$

The height is 6.0 in. Therefore,

$$V = \frac{1}{3}Bh = \frac{1}{3}(24 \text{ in}^2)(6.0 \text{ in.}) = 48 \text{ in}^3$$

A *cone* (Fig. 12.83) is a geometric solid whose base is a
circle. It has a curved lateral surface that comes to a
point called a vertex. The *axis* of a cone is a line segment
from the vertex to the center of the base.

The height, h, of a cone is the shortest distance between
the vertex and the base. A *right circular cone* is a cone in
which the height is the distance from the vertex to the
center of the base. The *slant height* of a right circular cone
is the length of a line segment which joins the vertex to
any point on the circle which forms the base of the cone.

The volume of a circular cone is given by the formula
$V = \frac{1}{3}Bh$. Since the base, B, is always a circle, its area is
πr^2, where r is the radius of the base. Thus the formula for
the volume of a right circular cone is usually written as
follows:

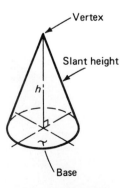

Vertex

Slant height

h

Base

Figure 12.83

$$V = \frac{1}{3}\pi r^2 h$$

The lateral surface area of a right circular cone is found by using the formula

$$A = \pi rs$$

where s is the slant height of the cone.

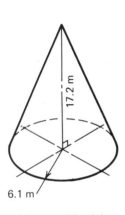

6.1 m

17.2 m

Figure 12.84

Example 2. Find the volume of the right circular cone in Fig. 12.84.

$$V = \frac{1}{3}\pi r^2 h$$

$$= \frac{1}{3}\pi (6.1 \text{ m})^2 (17.2 \text{ m})$$

$$= 670 \text{ m}^3$$

Example 3. Find the lateral surface area of the right circular cone in Fig. 12.84.

The formula for the lateral surface area is $A = \pi rs$. The slant height is not given. However, a right triangle is formed by the axis, the radius, and the slant height. Therefore, to find the slant height, s, use the formula

$$c = \sqrt{a^2 + b^2}$$

Then, $$s = \sqrt{(6.1 \text{ m})^2 + (17.2 \text{ m})^2}$$

$$= 18.2 \text{ m}$$

The lateral surface area can then be found as follows.

$$A = \pi rs$$

$$= \pi (6.1 \text{ m})(18.2 \text{ m})$$

$$= 350 \text{ m}^2$$

EXERCISES 12.9   *Follow the rules for working with measurements.*
Find the volume of each figure:

1.

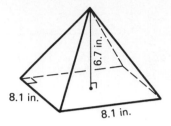

6.7 in.

8.1 in.

8.1 in.

2.   22.6 cm

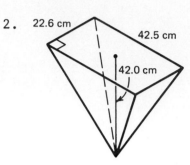

42.5 cm

42.0 cm

3.

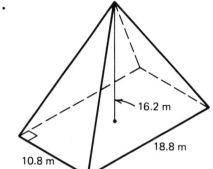

16.2 m

18.8 m

10.8 m

4.

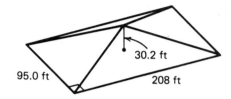

30.2 ft

95.0 ft

208 ft

5.

36.0 mm

115 mm

101 mm

6.

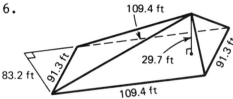

109.4 ft

29.7 ft

83.2 ft   91.3 ft

91.3 ft

109.4 ft

7.

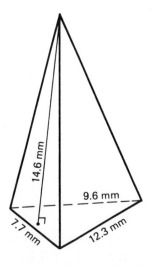

14.6 mm

9.6 mm

7.7 mm

12.3 mm

Hint. To find *B* use

$$B = A = \sqrt{s(s - a)(s - b)(s - c)}$$

8.

7.9 mm

4.6 mm

5.9 mm

6.8 mm

4.6 mm

2.7 mm

5.9 mm

Hint. To find the area of
the base of the pyramid,
find the sum of the areas
of the rectangle and the
triangle.

9.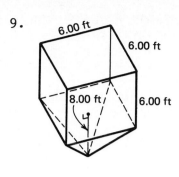

10.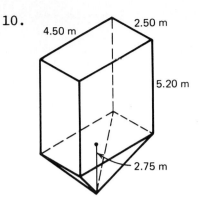

11. Find the volume of the right circular cone.
12. Find its lateral surface area.
13. Find the lateral surface area of the right circular cone.
14. Find its volume.

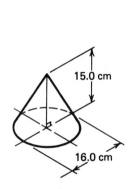

Exercises 11 and 12

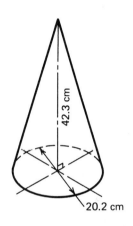

Exercises 13 and 14

15. A loading chute in a flour mill goes directly into a feeding bin. The feeding bin is in the shape of an inverted right circular cone, as shown. How many bushels of wheat can be placed in the feeding bin? (0.804 bu = 1 ft³)

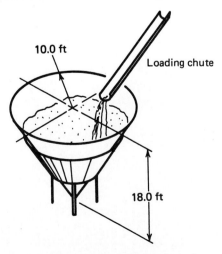

Exercise 15

16. An engineer is asked to give specifications for the top of a 24.2-ft-diameter tank. The top is to be in the shape of a cone and made of 1-in. steel. The 1-in. steel weighs 39.6 lb/ft^2. What is the total weight of the top?

 17. What is the total weight of the top, sides, and bottom of the tank in Exercise 16?

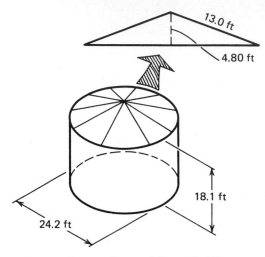

Exercises 16 and 17

18. The volume of a right circular cone is 238 cm^3. The radius of its base is 5.82 cm. Find its height.
19. Find the volume of the right circular cone below.

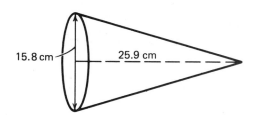

Exercises 19 and 20

20. Find the lateral surface area of the right circular cone in Exercise 19.
21. Find the weight of the display model shown below. The model is made of pine. Pine weighs 31.2 lb/ft^3.

Exercise 21

T 22. Gravel is piled in the shape of a cone. The circumference of the base is 224 ft. The slant height is 45 ft. Find the volume of gravel. If gravel weighs 3200 lb/yd^3, how many 22-ton truckloads are needed to transport the gravel?

12.10 Spheres

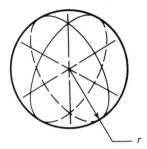

Figure 12.85

A *sphere* (Fig. 12.85) is a geometric solid formed by a closed curved surface with all points on the surface the same distance from a given point, the center. The given distance from a point on the surface to the center is called the *radius*.

The volume of a sphere can be found by using the formula

$$V = \frac{4}{3}\pi r^3$$

where r is the radius of the sphere. The surface area can be found by using the formula

$$A = 4\pi r^2$$

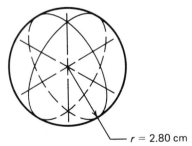

Figure 12.86

Example 1. Find the surface area of a sphere of radius 2.80 cm. (See Fig. 12.86.)

$$A = 4\pi r^2$$
$$= 4\pi (2.80 \text{ cm})^2$$
$$= 4\pi (2.80)^2 \text{ cm}^2$$
$$= 98.5 \text{ cm}^2$$

Example 2. Find the volume of the sphere in Example 1. The formula for the volume of a sphere is

$$V = \frac{4}{3}\pi r^3$$
$$= \frac{4}{3}\pi (2.80 \text{ cm})^3$$
$$= \frac{4}{3}\pi (2.80)^3 \text{ cm}^3$$
$$= 92.0 \text{ cm}^3$$

EXERCISES 12.10   *Follow the rules for working with measurements.*
Find (a) the surface area and (b) the volume of each sphere.

1.

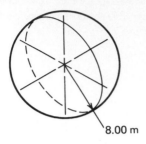

8.00 m

2.

18.7 cm

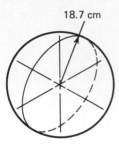

3.

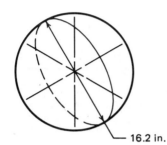

16.2 in.

4.

20.6 ft

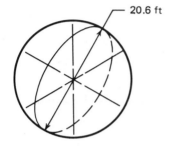

5. A balloon 30.1 m in radius is to be filled with helium. How many m³ of helium are needed to fill it?

6. An experimental balloon is to have a diameter of 5.72 m. How much material is needed for this balloon?

7. How many gallons of water can be stored in the spherical portion of the water tank below? (7.48 gal = 1 ft³)

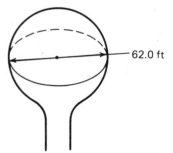

62.0 ft

8. A city drains 150,000 gal of water from a full spherical tank with a radius of 26 ft. How many gallons of water are left in the tank? (1 ft³ = 7.48 gal)

Chapter 12 Review

 Use a protractor to measure each angle below:

1.

2.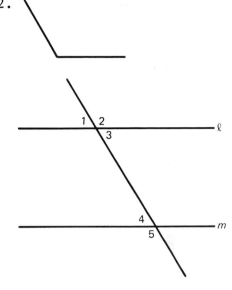

3. In the figure at the
   right, ℓ is parallel to
   m and $\angle 5 = 121°$.
   Find the measure of all
   the angles.

4. In figure at the right,
   $\angle 4$ and $\angle 5$ are called
   ? angles.

5. Suppose $\angle 1 = 4x + 5$
   and $\angle 2 = 2x + 55$. Find
   the value of x.

Exercises 3-5

6. Name the polygon which has
   (a) 4 sides   (b) 5 sides   (c) 6 sides   (d) 3 sides
   (e) 8 sides

Find the perimeter and the area of each quadrilateral:

7.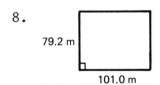

   6.00 cm | 5.00 cm

   12.00 cm

8. 79.2 m

   101.0 m

9.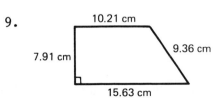

   10.21 cm

   7.91 cm   9.36 cm

   15.63 cm

10. The area of a rectangle is 79.6 m². The length is
    10.3 m. What is the width?

11. The area of a parallelogram is 2.53 cm². Find its
    height if the base is 1.76 cm.

Find the area and the perimeter of each triangle:

12.

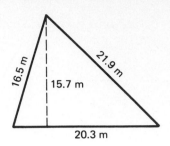

13.

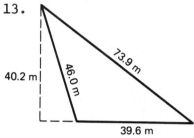

14.

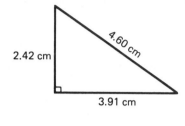

Find the length of the hypotenuse of each triangle:

15.

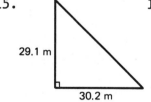

16.

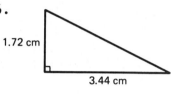

17.

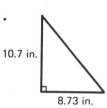

18. **Find the measure of the missing angle.**

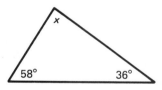

Exercise 18

19. **Suppose $\overline{DE} \parallel \overline{BC}$. Find length BC.**

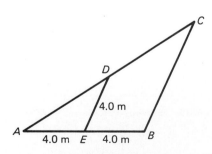

Exercise 19

20. Find the area and circumference of the circle below.

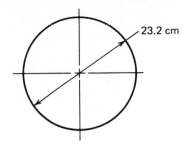

Exercise 20

21. The area of a circle is 462 cm². Find its radius.
22. How many degrees are in a central angle whose arc
    is $\frac{2}{5}$ of a circle?
23. What percent of 360° is 150°?
24. Find 21% of 360°.
25. Change 24° to radians.
26. Change $\frac{\pi}{18}$ rad to degrees.
27. The radius of a wheel is 75.3 cm. The angle of turn is
    0.561 rad. Find the distance the wheel travels.
28. A wheel of diameter 25.8 cm travels a distance of 20.0
    cm. Find the angle θ (in radians) that the wheel turns.
29. A wheel of radius 16.2 cm turns an angle of 1028°. Find
    the distance a point travels on the surface of the
    wheel.
30. Find the lateral surface area of the figure.
31. Find the total surface area of the figure.
32. Find the volume of the figure.
33. Find the volume of the cylinder.

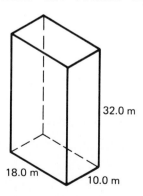

Exercises 30-32

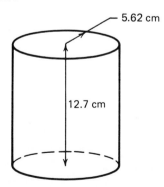

Exercise 33

34. A metallurgist needs to cast a molten alloy in the shape of a cylinder. The dimensions of the mold are as shown in the figure. Find the amount (volume) of molten alloy that he will need.

35. Find the lateral surface area of the cylinder which was cast in Exercise 34.

36. Find the total surface area of the cylinder which was cast in Exercise 34.

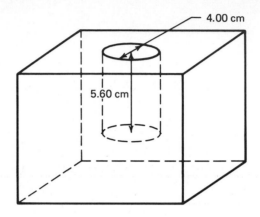

Exercises 34–36

37. Find the volume of the pyramid.

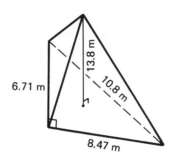

38. Find the volume of the right circular cone.

39. Find the lateral surface area of the right circular cone.

40. Find the volume of the sphere.

41. Find the surface area of the sphere.

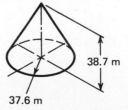

Exercises 38 and 39

Exercises 40 and 41

13

TRIGONOMETRY*

13.1 Trigonometric Ratios

Trigonometry is the study of triangles. In the first part of this chapter, we are going to consider only right triangles. A right triangle has one right angle, two acute angles, a hypotenuse, and two legs.

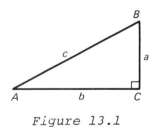

Figure 13.1

The right angle, as shown in Fig. 13.1, is usually labeled with the capital letter C. The vertices of the two acute angles are usually labeled with the capital letters A and B. The hypotenuse is the side opposite the right angle. The hypotenuse is usually labeled with the lowercase letter c. The legs are the sides opposite the acute angles. The leg (side) opposite angle A is usually labeled a and the leg opposite angle B is usually labeled b. Note that each side of the triangle is labeled with the lower case of the letter of the angle of which it is opposite.

The two legs are also named as the side opposite angle A and the side adjacent (or next to) angle A (Fig. 13.2); or as the side opposite angle B and the side adjacent to angle B (Fig. 13.3).

 *A calculator is needed for the work in this chapter.

463

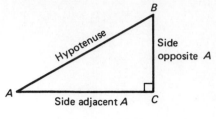

Figure 13.2

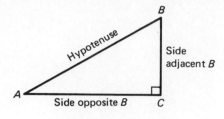

Figure 13.3

In any right triangle (Fig. 13.4) the following formula may be used:

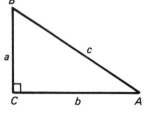

Figure 13.4

Pythagorean Theorem

$$c^2 = a^2 + b^2$$

That is, the square of the length of the hypotenuse is equal to the sum of the squares of the lengths of the legs. The following equivalent formulas are often more useful:

$c = \sqrt{a^2 + b^2}$ Used to find the length of the hypotenuse

$a = \sqrt{c^2 - b^2}$ Used to find the length of leg a

$b = \sqrt{c^2 - a^2}$ Used to find the length of leg b

Recall that the Pythagorean theorem was developed in detail in Section 12.3.

Example 1. Find the length of side b in Fig. 13.5.

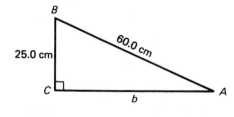

Figure 13.5

Using the formula to find the length of the leg b, we have

$$b = \sqrt{c^2 - a^2}$$

$$b = \sqrt{(60.0 \text{ cm})^2 - (25.0 \text{ cm})^2}$$

$$= 54.5 \text{ cm}$$

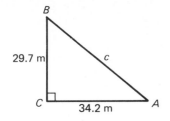

Figure 13.6

<u>Example 2</u>. Find the length of side c in Fig. 13.6.
Using the formula to find the hypotenuse c, we have

$$c = \sqrt{a^2 + b^2}$$
$$c = \sqrt{(29.7 \text{ m})^2 + (34.2 \text{ m})^2}$$
$$= 45.3 \text{ m}$$

A *ratio* is the comparison of two quantities by division.
We will name and use ratios of the sides of a right triangle to find an unknown part—or parts—of that right
triangle. Such a ratio is called a *trigonometric ratio*
and expresses the relationship between an acute angle and
the lengths of two of the sides of a right triangle.

The *sine* of the angle A, abbreviated "sin A," is equal
to the ratio of the length of the side opposite angle A,
which is a, to the length of the hypotenuse, c.

The *cosine* of the angle A, abbreviated "cos A," is the
ratio of the length of the side adjacent to angle A,
which is b, to the length of the hypotenuse, c.

The *tangent* of the angle A, abbreviated "tan A," is
equal to the ratio of the length of the side opposite
angle A, which is a, to the length of the side adjacent
to angle A, which is b.

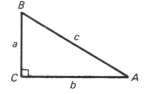

Figure 13.7

That is, in a right triangle (Fig. 13.7) we have:

Trigonometric Ratios

$$\sin A = \frac{\text{length of side opposite angle } A}{\text{length of hypotenuse}} = \frac{a}{c}$$

$$\cos A = \frac{\text{length of side adjacent to angle } A}{\text{length of hypotenuse}} = \frac{b}{c}$$

$$\tan A = \frac{\text{length of side opposite angle } A}{\text{length of side adjacent to angle } A} = \frac{a}{b}$$

Similarly, the ratios can be defined for angle B.

$$\sin B = \frac{\text{length of side opposite angle } B}{\text{length of hypotenuse}} = \frac{b}{c}$$

$$\cos B = \frac{\text{length of side adjacent to angle } B}{\text{length of hypotenuse}} = \frac{a}{c}$$

$$\tan B = \frac{\text{length of side opposite angle } B}{\text{length of side adjacent to angle } B} = \frac{b}{a}$$

<u>Example 3</u>. Find the three trigonometric ratios for angle A in the triangle in Fig. 13.8.

$$\sin A = \frac{\text{length of side opposite angle } A}{\text{length of hypotenuse}} = \frac{a}{c} = \frac{144 \text{ m}}{156 \text{ m}}$$

$$= 0.9231$$

$$\cos A = \frac{\text{length of side adjacent to angle } A}{\text{length of hypotenuse}} = \frac{b}{c} = \frac{60.0 \text{ m}}{156 \text{ m}}$$

$$= 0.3846$$

$$\tan A = \frac{\text{length of side opposite angle } A}{\text{length of side adjacent to angle } A} = \frac{a}{b} = \frac{144 \text{ m}}{60.0 \text{ m}}$$

$$= 2.400$$

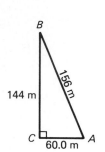

Figure 13.8

Values of the trigonometric ratios of various angles are found in tables and from calculators. For our purposes, we shall use calculators. You will need a calculator that has sin, cos, and tan buttons.

<u>Example 4</u>. Find sin 37.5° rounded to four significant digits.

Flowchart	Buttons Pushed	Display
Enter 37.5	3 → 7 → . → 5	37.5
Push sin	sin	0.608761

Then, sin 37.5° = 0.6088 rounded to four significant digits.

<u>Example 5</u>. Find cos 18.63° rounded to four significant digits.

Flowchart	Buttons Pushed	Display
Enter 18.63	1 → 8 → . → 6 → 3	18.63
Push cos	cos	0.947601

Thus, cos 18.63° = 0.9476 rounded to four significant digits.

<u>Example 6</u>. Find tan 81.7° rounded to four significant digits.

Flowchart	Buttons Pushed	Display

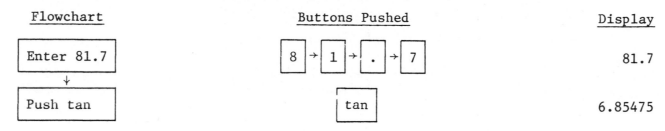

Enter 81.7	$8 \to 1 \to . \to 7$	81.7
Push tan	tan	6.85475

So, tan 81.7° = 6.855 rounded to four significant digits.

A calculator may also be used to find the *angle* when the value of the trigonometric ratio is known. The procedure is shown in the examples below.

<u>Example 7</u>. Find angle A to the nearest tenth of a degree when sin A = 0.6372.

Flowchart	Buttons Pushed	Display

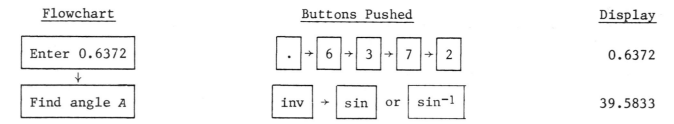

Enter 0.6372	$. \to 6 \to 3 \to 7 \to 2$	0.6372
Find angle A	inv → sin or $\sin^{-1}$	39.5833

Thus, angle A = 39.6° rounded to the nearest tenth of a degree.

<u>Example 8</u>. Find angle B to the nearest tenth of a degree when tan B = 0.3106.

Flowchart	Buttons Pushed	Display
Enter 0.3106	$. \to 3 \to 1 \to 0 \to 6$	0.3106
Find angle B	inv → tan or $\tan^{-1}$	17.2548

So, angle B = 17.3° rounded to the nearest tenth of a degree.

Example 9. Find angle A to the nearest hundredth of a degree when cos A = 0.4165.

Flowchart	Buttons Pushed	Display

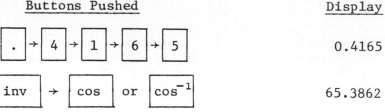

Enter 0.4165	$. \to 4 \to 1 \to 6 \to 5$	0.4165
↓		
Find angle A	inv $\to$ cos or cos^{-1}	65.3862

Then, angle A = 65.39° rounded to the nearest hundredth of a degree.

EXERCISES 13.1

Use the illustration of right triangle ABC below for Exercises 1-10:

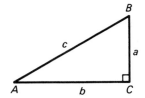

1. The side opposite angle A is ____?____.

2. The side opposite angle B is ____?____.

3. The hypotenuse is ____?____.

4. The side adjacent to angle A is ____?____.

5. The side adjacent to angle B is ____?____.

6. The angle opposite side a is ____?____,

7. The angle opposite side b is ____?____.

8. The angle opposite side c is ____?____.

9. The angle adjacent to side a is ____?____.

10. The angle adjacent to side b is ____?____.

Use the illustration of the right triangle above and the Pythagorean theorem to find each unknown side, rounded to three significant digits:

11. c = 75.0 m, a = 45.0 m

12. a = 25.0 cm, b = 60.0 cm

13. a = 29.0 mi, b = 47.0 mi

14. a = 12.0 km, c = 61.0 km

15. c = 18.9 cm, a = 6.71 cm

16. a = 20.2 mi, b = 19.3 mi

17. a = 171 ft, b = 203 ft

18. c = 35.3 m, b = 25.0 m

19. a = 202 m, c = 404 m

20. a = 1.91 km, b = 3.32 km

21. b = 1520 km, c = 2160 km

22. a = 203,000 ft, c = 521,000 ft

23. a = 45,800 m, b = 38,600 m

24. c = 3960 m, b = 3540 m

Use the triangle below for Exercises 25-30:

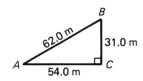

25. Find sin A.   26. Find cos A.   27. Find tan A.
28. Find sin B.   29. Find cos B.   30. Find tan B.

Find the value of each trigonometric ratio rounded to four significant digits:

31. sin 49.6°   32. cos 55.2°   33. tan 65.3°

34. sin 69.7°   35. cos 29.7°   36. tan 14.6°

37. sin 31.64°   38. tan 13.25°   39. cos 75.31°

40. cos 84.83°   41. tan 3.05°   42. sin 6.74°

43. sin 37.62°   44. cos 18.94°   45. tan 21.45°

46. sin 11.31°   47. cos 47.16°   48. tan 81.85°

Find each angle rounded to the nearest tenth of a degree:

49. sin A = 0.7941   50. tan A = 0.2962   51. cos B = 0.4602

52. cos A = 0.1876   53. tan B = 1.386   54. sin B = 0.3040

55. sin B = 0.1592   56. tan B = 2.316   57. cos A = 0.8592

58. cos B = 0.3666   59. tan A = 0.8644   60. sin A = 0.5831

Find each angle rounded to the nearest hundredth of a degree:

61. tan A = 0.1941   62. sin B = 0.9324   63. cos A = 0.3572

64. cos B = 0.2597   65. sin A = 0.1506   66. tan B = 2.500

67. tan B = 3.806   68. sin A = 0.4232   69. cos B = 0.7311

70. cos A = 0.6427   71. sin B = 0.3441   72. tan A = 0.5536

73. In Exercises 25-30 there are two pairs of ratios that are equal. Name them.

13.2 Using Trigonometric Ratios to Find Angles

A useful and time-saving fact about right triangles (Fig. 13.9) is that the sum of the acute angles of a right triangle is always 90°.

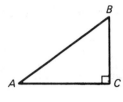

Figure 13.9

$$A + B = 90°$$

Why? We know that the sum of the interior angles of any triangle is 180°. A right triangle by definition contains a right angle, whose measure is 90°. That leaves 90° to be divided between the two acute angles.

Note, then, that if one acute angle is given or known, the other acute angle may be found by subtracting the known angle from 90°. That is,

$$A = 90° - B$$
$$B = 90° - A$$

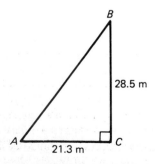

Exercise 13.10

Example 1. Find angle A in the triangle in Fig. 13.10.

We know the sides opposite and adjacent to angle A. So, we use the tangent ratio:

$$\tan A = \frac{\text{length of side opposite angle } A}{\text{length of side adjacent to angle } A}$$

$$= \frac{28.5 \text{ m}}{21.3 \text{ m}} = 1.338$$

Next, find angle A to the nearest tenth of a degree when tan A = 1.338. The complete set of operations on a calculator is:

Flowchart	Buttons Pushed	Display

Enter 28.5 $2 \rightarrow 8 \rightarrow . \rightarrow 5$ 28.5

Push divide $\div$ 28.5

Enter 21.3 $2 \rightarrow 1 \rightarrow . \rightarrow 3$ 21.3

Push equals $=$ 1.33803

Find angle A inv $\rightarrow$ tan or $\tan^{-1}$ 53.2267

Thus, angle A = 53.2° rounded to the nearest tenth of a degree.

When calculations involve a trigonometric ratio, we shall use the following rule:

Angles expressed to the nearest	Length of sides of triangle will contain
1°	Two significant digits
0.1°	Three significant digits
0.01°	Four significant digits

Example 2. Find angle B in the triangle in Fig. 13.11.

We know the hypotenuse and the side adjacent to angle B. So, let's use the cosine ratio.

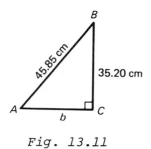

$$\cos B = \frac{\text{length of side adjacent to angle } B}{\text{length of hypotenuse}}$$

$$= \frac{35.20 \text{ cm}}{45.85 \text{ cm}}$$

Fig. 13.11

Find angle B using a calculator as follows:

Flowchart	Buttons Pushed	Display
Enter 35.2	3 → 5 → . → 2	35.2
Push divide	÷	35.2
Enter 45.85	4 → 5 → . → 8 → 5	45.85
Push equals	=	0.767721
Find angle *B*	inv → cos or cos⁻¹	39.8503

So, angle *B* = 39.85° rounded to the nearest hundredth of a degree.

The question is often raised, "Which of the three trig ratios do I use?" First, notice that each trigonometric ratio consists of two sides and one angle, or three quantities in all. To find the solution to such an equation, two of the quantities must be known. Let's answer the question in two parts:

1. If you are finding an angle, then two sides must be known. Label these two known sides as *side opposite* the angle you are finding, *side adjacent* to the angle you are finding, or *hypotenuse*. Then, choose the trig ratio which has these two sides.
2. If you are finding a side, then one side and one angle must be known. Label the known side and the unknown side as *side opposite* the known angle, *side adjacent* to the known angle, or *hypotenuse*. Then, choose the trig ratio which has these two sides.

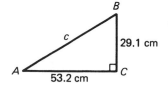

Figure 13.12

Example 3. Find angle *A* and angle *B* in the triangle in Fig. 13.12.

$$\tan A = \frac{\text{length of side opposite angle } A}{\text{length of side adjacent to angle } A}$$

$$= \frac{29.1 \text{ cm}}{53.2 \text{ cm}} = 0.5470$$

Therefore, angle *A* = 28.7°

Then, angle *B* = 90° − 28.7° = 61.3°

EXERCISES 13.2   *Using the illustration below, find the measure of each acute angle for each right triangle:*

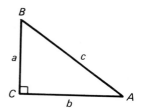

1. a = 36.0 m, b = 50.9 m   2. a = 72.0 cm, c = 144 cm

3. a = 39.7 cm, b = 43.6 cm

4. a = 171 km, b = 695 km

5. b = 13.6 m, c = 18.7 m   6. b = 409 km, c = 612 km

7. a = 29.7 m, b = 29.7 m, c = 42.0 m

8. a = 36.2 mm, b = 62.7 mm, c = 72.4 mm

9. a = 2902 km, b = 1412 km

10. b = 21.34 m, c = 47.65 m

11. a = 0.6341 cm, c = 0.7982 cm

12. b = 4.372 m, c = 5.806 m

13. b = 1455 ft, c = 1895 ft

14. a = 25.45 in., c = 41.25 in.

15. a = 243.2 km, b = 271.5 km

16. a = 351.6 m, b = 493.0 m

17. a = 16.7 m, c = 81.4 m

18. a = 847 m, b = 105 m

19. b = 1185 ft, c = 1384 ft

20. a = 48.7 cm, c = 59.5 cm

21. a = 845 km, b = 2960 km

22. b = 2450 km, c = 3570 km

23. a = 8897 m, c = 9845 m

24. a = 58.44 mi, b = 98.86 mi

13.3 Using Trigonometric Ratios to Find Sides

We may also use a trigonometric ratio to find a side of a right triangle, given one side and the measure of one of the acute angles.

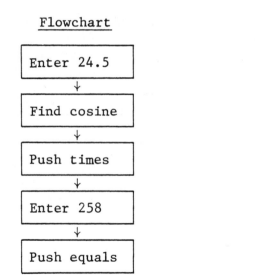

Figure 13.13

Example 1. Find side a in the triangle in Fig. 13.13.

With respect to the known angle B, we know the hypotenuse and are finding the adjacent side. So, we use the cosine ratio.

$$\cos B = \frac{\text{length of side adjacent to angle } B}{\text{length of hypotenuse}}$$

$$\cos 24.5° = \frac{a}{258 \text{ ft}}$$

$$a = (\cos 24.5°)(258 \text{ ft}) \quad \text{(Multiply both sides by 258 ft.)}$$

Side a can be found using a calculator as follows:

Flowchart	Buttons Pushed	Display
Enter 24.5	2 → 4 → . → 5	24.5
Find cosine	cos	0.909961
Push times	×	0.909961
Enter 258	2 → 5 → 8	258
Push equals	=	234.770

Then, side a = 235 ft rounded to three significant digits.

Example 2. Find the sides b and c in the triangle of Fig. 13.14.

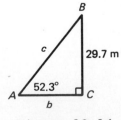

Figure 13.14

If we find side b first, we are looking for the side adjacent to angle A, the known angle. We are given the side opposite angle A. Thus we should use the tangent ratio.

$$\tan A = \frac{\text{length of side opposite angle } A}{\text{length of side adjacent to angle } A}$$

$$\tan 52.3° = \frac{29.7 \text{ m}}{b}$$

$b(\tan 52.3°) = 29.7 \text{ m}$ \qquad (Multiply both sides by b.)

$$b = \frac{29.7 \text{ m}}{\tan 52.3°}$$ \qquad (Divide both sides by $\tan 52.3°$.)

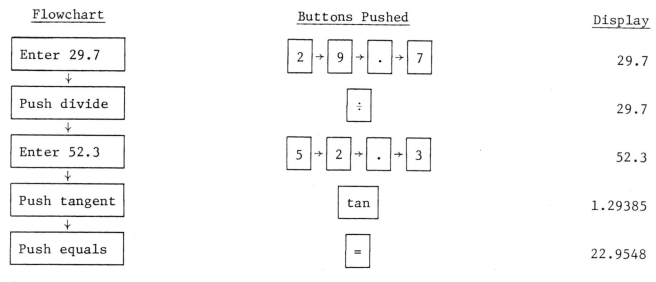

Flowchart	Buttons Pushed	Display
Enter 29.7	2 → 9 → . → 7	29.7
Push divide	÷	29.7
Enter 52.3	5 → 2 → . → 3	52.3
Push tangent	tan	1.29385
Push equals	=	22.9548

Then, side b = 23.0 m rounded to three significant digits.

To find side c we are looking for the hypotenuse and we have the opposite side given. Thus we should use the sine ratio.

$$\sin A = \frac{\text{length of side opposite angle } A}{\text{length of hypotenuse}}$$

$$\sin 52.3° = \frac{29.7 \text{ m}}{c}$$

$c(\sin 52.3°) = 29.7 \text{ m}$ \qquad (Multiply both sides by c.)

$$c = \frac{29.7 \text{ m}}{\sin 52.3°}$$ \qquad (Divide both sides by $\sin 52.3°$.)

$$= 37.5 \text{ m}$$

EXERCISES 13.3   *Find the unknown sides of each right triangle:*

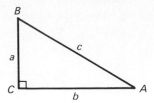

1. $a = 36.7$ m, $A = 42.1°$ 2. $b = 73.6$ cm, $B = 19.0°$

3. $a = 236$ km, $B = 49.7°$ 4. $b = 28.9$ ft, $A = 65.2°$

5. $c = 49.1$ cm, $A = 36.7°$ 6. $c = 236$ m, $A = 12.9°$

7. $b = 23.7$ cm, $A = 23.7°$ 8. $a = 19.2$ km, $B = 63.2°$

9. $b = 29,200$ km, $A = 12.9°$

10. $c = 36.7$ mi, $B = 68.3°$

11. $a = 19.72$ m, $A = 19.75°$

12. $b = 125.3$ cm, $B = 23.34°$

13. $c = 255.6$ mi, $A = 39.25°$

14. $c = 7.363$ km, $B = 14.80°$

15. $b = 12,350$ m, $B = 69.72°$

16. $a = 3678$ m, $B = 10.04°$

17. $a = 1980$ m, $A = 18.4°$

18. $a = 9820$ ft, $B = 35.7°$

19. $b = 841.6$ km, $A = 18.91°$

20. $c = 289.5$ cm, $A = 24.63°$

21. $c = 185.6$ m, $B = 61.45°$

22. $b = 21.63$ km, $B = 82.06°$

23. $c = 256$ cm, $A = 25.6°$

24. $a = 18.3$ mi, $A = 71.2°$

13.4   Solving Right Triangles

In. Sec. 13.1 we defined *trigonometry* as the study of triangles. To "solve" a triangle means to find the measures of the various parts of a triangle that are not given. We proceed as we did in the last two sections.

    Example 1. Solve the right triangle in Fig. 13.15.

Since $A + B = 90°$

                 $A = 90° - B$

                 $= 90° - 36.7°$

                 $= 53.3°$

Figure 13.15

We then can use either the sine or the cosine ratio to find side c.

$$\sin B = \frac{\text{length of side opposite angle } B}{\text{length of hypotenuse}}$$

$$\sin 36.7° = \frac{19.2 \text{ m}}{c}$$

$\sin 36.7° = 19.2$ m   (Multiply both sides by c.)

$$c = \frac{19.2 \ m}{\sin 36.7°}$$ (Divide both sides by $\sin 36.7°$.)

$$= 32.1 \text{ m}$$

Now we may use a trigonometric ratio or the Pythagorean theorem to find side a.

Solution by a Trigonometric Ratio

$$\tan B = \frac{\text{length of side opposite angle } B}{\text{length of side adjacent to angle } B}$$

$$\tan 36.7° = \frac{19.2 \text{ m}}{a}$$

$a(\tan 36.7°) = 19.2$ m   (Multiply both sides by a.)

$$a = \frac{19.2 \text{ m}}{\tan 36.7°}$$ (Divide both sides by $\tan 36.7°$.)

$$= 25.8 \text{ m}$$

Solution by the Pythagorean Theorem

$$a = \sqrt{c^2 - b^2}$$

$$= \sqrt{(32.1 \text{ m})^2 - (19.2 \text{ m})^2}$$

$$= 25.7 \text{ m}$$

Can you explain the difference in these two results?

<u>Example 2</u>. Solve the right triangle in Fig. 13.16.

We are given the measure of one of the acute angles and the length of the hypotenuse.

$$A = 90° - B$$

$$= 90° - 45.7°$$

$$= 44.3°$$

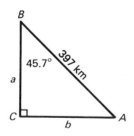

Figure 13.16

To find side b, we must use the sine or the cosine ratio since the hypotenuse is given.

$$\sin B = \frac{\text{length of side opposite angle } B}{\text{length of hypotenuse}}$$

$$\sin 45.7° = \frac{b}{397 \text{ km}}$$

$(\sin 45.7°)(397 \text{ km}) = b$ (Multiply both sides by

$284 \text{ km} = b$ 397 km.)

We again can use either a trigonometric ratio or the Pythagorean theorem to find side a.

Solution by a Trigonometric Ratio

$$\cos B = \frac{\text{length of side adjacent to angle } B}{\text{length of hypotenuse}}$$

$$\cos 45.7° = \frac{a}{397 \text{ km}}$$

$(\cos 45.7°)(397 \text{ km}) = a$ (Multiply both sides by

                        397 km.)

$277 \text{ km} = a$

Solution by the Pythagorean Theorem

$$a = \sqrt{c^2 - b^2}$$
$$= \sqrt{(397 \text{ km})^2 - (284 \text{ km})^2}$$
$$= 277 \text{ km}$$

The third type of solving a right triangle is when two of the sides of the right triangle are given.

<u>Example 3</u>. Solve the right triangle in Fig. 13.17.

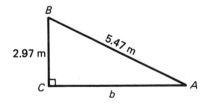

Figure 13.17

To find either angle A or angle B, we could use either sine or cosine since the hypotenuse is given.

$$\sin A = \frac{\text{length of side opposite angle } A}{\text{length of hypotenuse}}$$

$$\sin A = \frac{2.97 \text{ m}}{5.47 \text{ m}} = 0.5430$$

$$A = 32.9°$$

Then, $B = 90° - A$

$= 90° - 32.9°$

$= 57.1°$

The unknown side b can be found using the Pythagorean
theorem.

$$b = \sqrt{c^2 - a^2}$$

$$= \sqrt{(5.47 \text{ m})^2 - (2.97 \text{ m})^2}$$

$$= 4.59 \text{ m}$$

EXERCISES 13.4   *Using the illustration, solve each right triangle:*

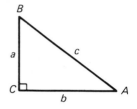

1. $A = 50.6°$, $c = 49.0$ m   2. $a = 30.0$ cm, $b = 40.0$ cm

3. $B = 41.2°$, $a = 267$ ft   4. $A = 39.7°$, $b = 49.6$ km

5. $b = 72.0$ mi, $c = 78.0$ mi 6. $B = 22.4°$, $c = 46.0$ mi

7. $A = 52.1°$, $a = 72.0$ mm   8. $B = 42.3°$, $b = 637$ m

9. $A = 68.8°$, $c = 39.4$ m   10. $a = 13.6$ cm, $b = 13.6$ cm

11. $a = 12.00$ m, $b = 24.55$ m

12. $B = 38.52°$, $a = 4315$ m

13. $A = 29.19°$, $c = 2975$ ft

14. $B = 29.86°$, $a = 72.62$ m

15. $a = 46.72$ m, $b = 19.26$ m

16. $a = 2436$ ft, $c = 4195$ ft

17. $A = 41.1°$, $c = 485$ m

18. $a = 1250$ km, $b = 1650$ km

19. $B = 9.45°$, $a = 1585$ ft

20. $A = 14.60°$, $b = 135.7$ cm

21. $b = 269.5$ m, $c = 380.5$ m

22. $B = 75.65°$, $c = 92.75$ km

23. $B = 81.5°$, $b = 9370$ ft

24. $a = 14.6$ mi, $c = 31.2$ mi

13.5 Applications Involving Trigonometric Ratios

Trigonometric ratios can be used to solve many applications similar to those problems we solved in the last sections. However, instead of having to find all the parts of a right triangle, we usually need to find only one.

Example 1. The roof in Fig. 13.18 has a rise of 7.50 ft and a run of 18.0 ft. Find angle A.

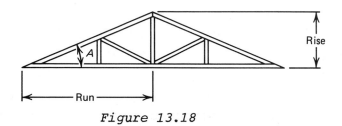

Figure 13.18

In this problem we know the length of the side opposite angle A and the length of the side adjacent to angle A. So, we use the tangent ratio.

$$\tan A = \frac{\text{length of side opposite angle } A}{\text{length of side adjacent to angle } A}$$

$$\tan A = \frac{7.50 \text{ ft}}{18.0 \text{ ft}} = 0.4167$$

Then, $A = 22.6°$

The *angle of depression* is the angle between the horizontal and the line of sight to an object that is *below* the horizontal. The *angle of elevation* is the angle between the horizontal and the line of sight to an object that is *above* the horizontal.

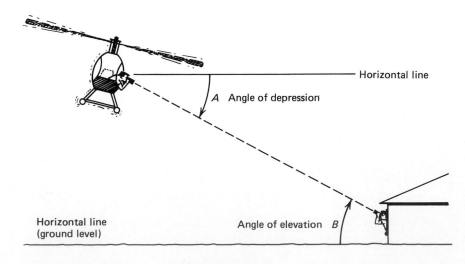

Figure 13.19

In Fig. 13.19, angle *A* is the angle of depression for an observer in the helicopter sighting down to the building on the ground, and angle *B* is the angle of elevation for an observer in the building sighting up to the helicopter.

Example 2. A ship's navigator measures the angle of elevation to the beacon of a lighthouse to be 10.1°. He knows that this particular beacon is 232 m above sea level. How far is the ship from the lighthouse?

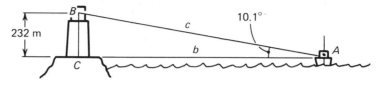

Figure 13.20

First, you should sketch the problem, as shown in Fig. 13.20. Since this problem involves finding the length of the side adjacent to an angle when the opposite side is known, we use the tangent ratio.

$$\tan A = \frac{\text{length of side opposite angle } A}{\text{length of side adjacent to angle } A}$$

$$\tan 10.1° = \frac{232 \text{ m}}{b}$$

$b(\tan 10.1°) = 232 \text{ m} \qquad \text{(Multiply both sides by } b.\text{)}$

$b = \dfrac{232 \text{ m}}{\tan 10.1°} \quad \text{(Divide both sides by } \tan 10.1°.\text{)}$

$\qquad = 13\overline{0}0 \text{ m}$

EXERCISES 13.5

1. A conveyor is used to lift paper to a shredder. The most efficient operating angle of elevation for the conveyor is 35.8°. The paper is to be elevated 11.0 m. What length of conveyor is needed?
2. A millwright is to weld a support for a 23-m conveyor so that it will operate at a $2\overline{0}°$ angle. What is the length of the support? See the figure below.

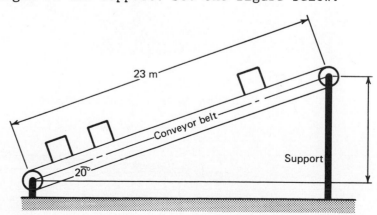

3. A bullet is found embedded in the wall of a room 2.3 m above the floor. The bullet entered the wall going upward at an angle of 12°. How far from the wall was the gun fired if the gun was held at a distance of 1.2 m from the floor?

4. The recommended safety angle of a ladder against a building is 78°. A 10-m ladder will be used. How high up on the side of the building will the ladder reach?

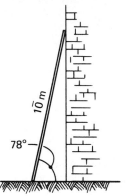

5. A piece of conduit 38.0 ft long cuts across the corner of a room as shown. Find the length x and the angle A.

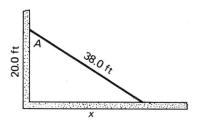

6. Find the width of the river in the figure below.

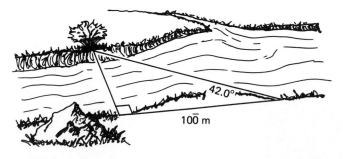

7. A roadbed rises 220 ft for each 2300 ft of road. Find the angle of inclination of the roadbed.

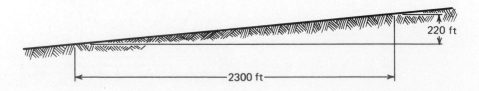

8. A smokestack is 18$\overline{0}$ ft high. A guy wire must be fastened to the stack 2$\overline{0}$ ft from the top. This makes an angle of 40.0° with the ground. Find the length of the guy wire.

9. A railroad track has an angle of elevation of 1.0°. What is the difference in height (in feet) of two points of the track which are 1.00 mi apart?

10. A contractor builds a roof with a pitch of 20.0°. The width of the building is 40.0 ft and the peak of the roof is at the center of the building. How high above the outside walls will the peak of the roof be?

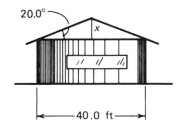

11. A gauge is used to check the diameter of a crankshaft journal. It is constructed to make measurements on the basis of a right triangle with a 60.0° angle. The distance *AB*, in the figure below, is 11.4 cm. Find the radius, *BC*, of the journal.

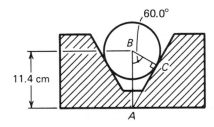

12. A television station has cables cut in 18-m lengths. These are to be fastened to the top of their relay antennae and to the ground. The cables make an angle of 25° with level ground. Allow 1 m for fastening. What is the height of the relay antennae they are using?

13. The cables attached to a TV relay tower are 11$\overline{0}$ m long. They meet level ground at an angle of 60.0°. Find the height of the tower.

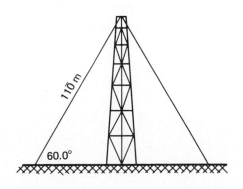

14. A lunar module is resting on the moon's surface directly below a spaceship in its orbit, 12.0 km above the moon. Two lunar explorers find that the angle from their position to that of the spaceship is 82.9°. What distance are they from the lunar module?

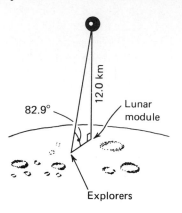

15. In ac (alternating current) circuits, the relationship between the impedance Z (in ohms), the resistance R (in ohms), the phase angle θ, and the reactance X (in ohms) is shown by the right triangle below.

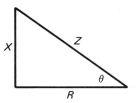

(a) Find the impedance if the resistance is 350 Ω and the phase angle is 35°.

(b) Suppose the resistance is 550 Ω and the impedance is 700 Ω. What is the phase angle?

(c) Suppose the reactance is 182 Ω and the resistance is 240 Ω. Find the impedance and the phase angle.

16. A conical tank with its point down has a height of 4.0 m and a radius of 1.2 m. The tank is filled to a height of 3.7 m with liquid. How many litres of liquid are in the tank? (1000 litres = 1 m^3)

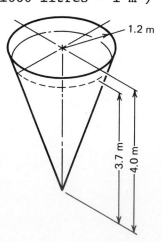

17. Use the right triangle below.
    (a) Find the voltage applied if the voltage across the
        coil is 35.6 V and the voltage across the resis-
        tance is 40.2 V.
    (b) Find the voltage across the resistance if the volt-
        age applied is 378 V and the voltage across the coil
        is 268 V.
    (c) Find the voltage across the coil if the voltage
        applied is 448 V and the voltage across the resis-
        tance is 381 V.

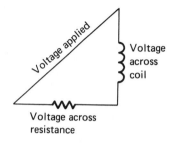

18. The centers of eight holes are equally spaced around
    a 4.0-in.-diameter circle, as shown below. Find the
    distance between the centers of holes A and C.

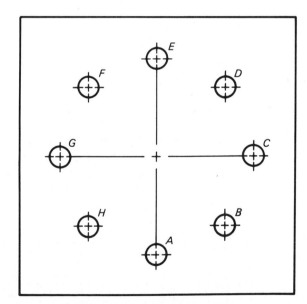

19. Find angle θ of the taper drawn below.

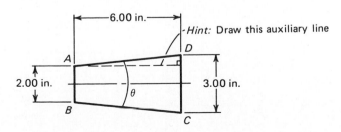

20. In the figure below, angle θ is 3$\overline{0}$° and AB has length
3.0 in. The taper has a height of 5.0 in. What is the
length of BC?

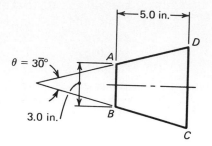

21. Find (a) distance x and (b) distance BD in the figure
below. Length BC = 5.50 in.

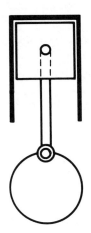 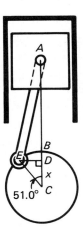

22. Find the length AB along the roofline of the building
below. The slope of the roof is 45.0°.

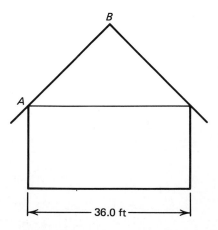

23. Find length x and angle A in the figure below.

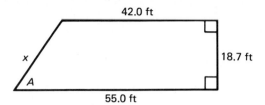

24. From the base of a building, measure out a horizontal
distance of 215 ft along the ground. The angle of elevation to the top of the building is 63.0°. Find the
height of the building.

13.6 Sine and Cosine Graphs

Up to this point, we have considered only angles of trigonometric ratios between 0° and 90° because we were dealing
only with right triangles. For many applications, we need
to consider values greater than 90°. In this section, we
will use a calculator to find values of the sine and cosine
ratios of angles greater than 90°. Then, we will use these
values to construct various sine and cosine graphs.

The procedure for finding the value of sine or cosine of
an angle greater than 90° is the same as for angles between
0° and 90°, as shown in Sec. 13.1.

Example 1. Find sin 255° rounded to four significant
digits.

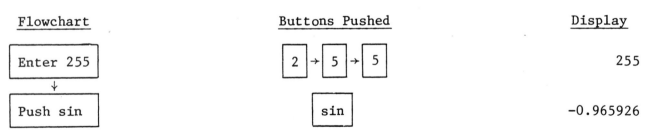

Flowchart	Buttons Pushed	Display
Enter 255	2 → 5 → 5	255
Push sin	sin	−0.965926

Then, sin 255° = −0.9659 rounded to four significant
digits.

Next, let's graph $y = \sin x$ for values of x between 0°
and 360°. First, find a large number of values of x and y
that satisfy the equation and plot them in the xy plane.
For convenience, we will choose values of x in multiples of
30° and round values of y to two significant digits.

x	0°	30°	60°	90°	120°	150°	180°	210°	240°	270°	300°	330°	360°
y	0	0.50	0.87	1.0	0.87	0.50	0	−0.50	−0.87	−1.0	−0.87	−0.50	0

Now, choose a convenient scale for the x axis so that one unit equals 30° and mark off on the x axis between 0° and 360°. Then, choose a convenient scale for the y axis so that one unit equals 0.1 and mark the y axis between +1.0 to −1.0. Plot the points corresponding to the ordered pairs (x, y) from the table above. Then, connect the points with a smooth, continuous curve. The graph is shown below in Fig. 13.21.

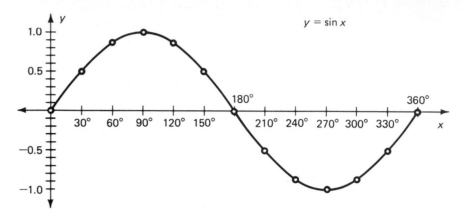

Figure 13.21

The graph of y = cos x for values of x between 0° and 360° can be found in a similar manner and is shown below.

x	0°	30°	60°	90°	120°	150°	180°	210°	240°	270°	300°	330°	360°
y	1.0	0.87	0.50	0	−0.50	−0.87	−1	−0.87	−0.50	0	0.50	0.87	1.0

Plot the points corresponding to these ordered pairs, and connect them with a smooth curve as shown below in Fig. 13.22.

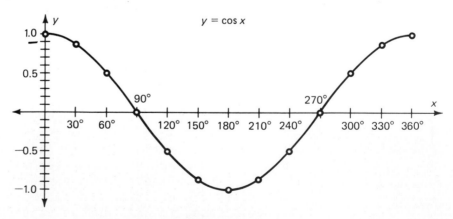

Figure 13.22

Example 2. Graph $y = 4 \sin x$ for values of x between $0°$ and $360°$.

Prepare a table listing values of x in multiples of $30°$. To find each value of y, find the sine of the angle, multiply this value by 4, and round to two significant digits.

x	$0°$	$30°$	$60°$	$90°$	$120°$	$150°$	$180°$	$210°$	$240°$	$270°$	$300°$	$330°$	$360°$
y	0	2.0	3.5	4	3.5	2.0	0	-2.0	-3.5	-4	-3.5	-2.0	0

Here, let's choose the scale for the y axis so that one unit equals 0.5, and mark off the y axis between +4.0 and -4.0. Plot the points corresponding to these ordered pairs in the table and then connect them with a smooth curve as shown below in Fig. 13.23.

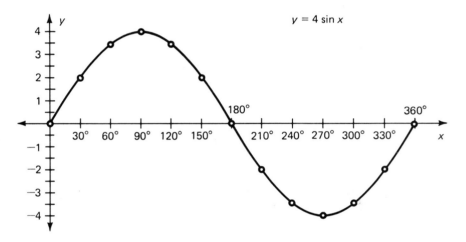

Figure 13.23

Note that the graphs for $y = \sin x$ and $y = 4 \sin x$ are similar. That is, each starts at $(0°, 0)$, reaches its maximum at $x = 90°$, crosses the x axis at $(180°, 0)$, reaches its minimum at $x = 270°$, and meets the x axis at $(360°, 0)$. In general, the graphs of equations in the form

$$y = A \sin x \quad \text{and} \quad y = A \cos x, \text{ where } A > 0$$

reach a maximum value of A and a minimum value of $-A$. The value of A is usually called the *amplitude*.

One of the most common applications of waves is in alternating current. In a generator, a coil of wire is rotated in a magnetic field, which produces an electric current. See Fig. 13.24.

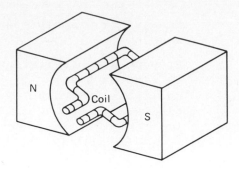

Figure 13.24

In a simple generator, the current i changes as the coil rotates according to the equation

$$i = I \sin x$$

where I is the maximum current and x is the angle through which the coil rotates.

Similarly, the voltage v also changes as the coil rotates according to the equation

$$v = V \sin x$$

where V is the maximum voltage and x is the angle through which the coil rotates.

Example 3. The maximum voltage V in a simple generator is 25 V. The changing voltage v as the coil rotates is given by

$$v = 25 \sin x$$

Graph the equation in multiples of 30° for one complete revolution of the coil.

First, prepare a table. To find each value of y, find the sine of the angle, multiply this value by 25, and round to two significant digits.

x	0°	30°	60°	90°	120°	150°	180°	210°	240°	270°	300°	330°	360°
y	0	13	22	25	22	13	0	−13	−22	−25	−22	−13	0

Let's choose the scale for the y axis so that one unit equals 5 V. Mark off the y axis between +25 V and −25 V. Plot the points corresponding to the ordered pairs in the table. Then connect these with a smooth curve as shown in Fig. 13.25.

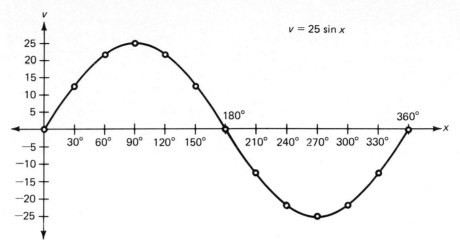

Figure 13.25

    If you were to continue finding ordered pairs in the
previous tables by choosing values of x greater than 360°
and less than 0°, you would find that the y values repeat
themselves and that the graphs form *waves* as shown below
in Fig. 13.26.

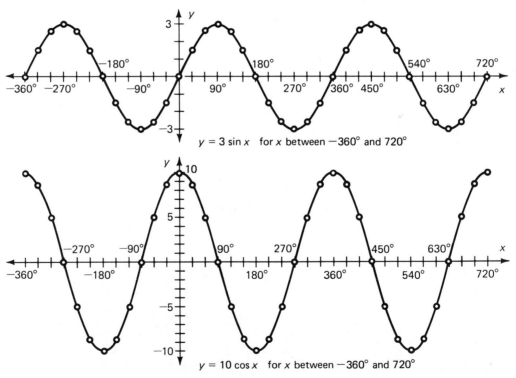

Figure 13.26

In general, if the *x*-axis variable is *distance*, the length of one complete wave is called the *wavelength* and is given by the symbol λ, the Greek letter "lambda." See Fig. 13.27.

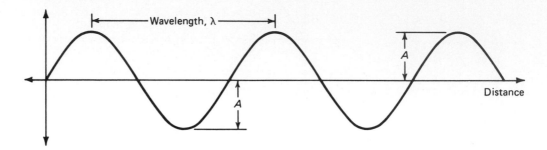

Figure 13.27

If the *x*-axis variable is *time*, the time required for one complete wave to pass a given point is called the *period*, *T*. See Fig. 13.28.

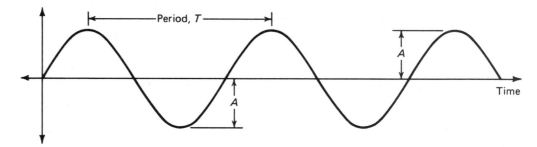

Figure 13.28

The frequency *f* is the number of waves that pass a given point each second. That is,

$$f = \frac{1}{T}$$

The unit of frequency is the hertz (Hz), where

$$1 \text{ Hz} = 1 \text{ wave/s} \quad \text{or} \quad 1 \text{ cycle/s}$$

Common multiples of the hertz are the kilohertz (kHz, 10^3 Hz) and the megahertz (MHz, 10^6 Hz).

Frequency and wavelength are related to wave velocity by the formula

$$v = \lambda f$$

where *v* is wave velocity, λ is the wavelength, and *f* is the frequency.

EXERCISES 13.6

Find each value rounded to four significant digits:

1. sin 137°   2. sin 318°   3. cos 246°

4. cos 295°   5. sin 205.8°   6. sin 106.3°

7. cos 166.5°   8. cos 348.2°   9. tan 217.6°

10. tan 125.5°   11. tan 156.3°   12. tan 418.5°

Graph each equation for values of x between 0° and 360° in multiples of 30°:

13. $y = 6 \sin x$ 14. $y = 2 \sin x$

15. $y = 5 \cos x$ 16. $y = 4 \cos x$

Graph each equation for values of x between 0° and 360° in multiples of 15°:

17. $y = \sin 2x$ 18. $y = \cos 2x$

Graph each equation for values of x between 0° and 360° in multiples of 10°:

19. $y = 4 \cos 3x$ 20. $y = 2 \sin 3x$

The maximum voltage in a simple generator is V_{max}. The changing voltage v as the coil rotates is given by

$$v = V_{max} \sin x$$

Graph this equation in multiples of 30° for one complete revolution of the coil for each value of V_{max}:

21. $V_{max} = 36$ V   22. $V_{max} = 48$ V

The maximum current in a simple generator is I. The changing current i as the coil rotates is given by

$$i = I \sin x$$

Graph this equation in multiples of 30° for one complete revolution of the coil for each value of I:

23. I = 5.0 A   24. I = 7.5 A

25. From the graph in Exercise 21, estimate the value of *v* at x = 45° and x = 295°.
26. From the graph in Exercise 22, estimate the value of *v* at x = 135° and x = 225°.
27. From the graph in Exercise 23, estimate the value of *i* at x = 135° and x = 225°.
28. From the graph in Exercise 24, estimate the value of *i* at x = 45° and x = 190°.

29. A radar unit operates at a wavelength of 3.4 cm. Radar waves travel at the speed of light, which is 3.0×10^8 m/s. What is the frequency of the radar waves?

30. A local AM radio station broadcasts at 1400 kHz. What is the wavelength of its radio waves? (They travel at the speed of light, which is 3.0×10^8 m/s.)

13.7   Solving Oblique Triangles: Law of Sines

An *oblique triangle* is a triangle with no right angle. We will use the common notation of labeling vertices of a triangle by the capital letters A, B, and C and using the small letters a, b, and c as the sides opposite angles A, B, and C, respectively. See Fig. 13.29.

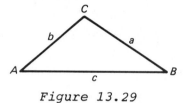

Figure 13.29

Recall that an *acute* angle is an angle that has a measure less than 90°. An *obtuse* angle is an angle that has a measure greater than 90° but less than 180°.

The trigonometric ratios used earlier in this chapter apply *only* to right triangles. So, we must use other ways to solve oblique triangles. One law that we will use to solve oblique triangles is the Law of Sines.

Law of Sines

$$\frac{a}{\sin A} = \frac{b}{\sin B} = \frac{c}{\sin C}$$

In words: *for any triangle, the ratio of any side to the sine of the opposite angle equals the ratio of any other side to the sine of its opposite angle.*

When using this law, you must form a proportion by choosing two of the three ratios in which three of the four terms are known. In order to use the Law of Sines, you must know:

(a) two angles and a side opposite one of them (actually knowing two angles and any side is enough, because in knowing two angles, the third is easily found), or

(b) two sides and an angle opposite one of them.

<u>Example 1</u>. If $C = 28.0°$, $c = 46.8$ cm, and $B = 101.5°$, solve the triangle.*

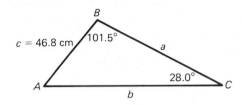

Figure 13.30

First, let's find side b, in Fig. 13.30.

$$\frac{c}{\sin C} = \frac{b}{\sin B}$$

$$\frac{46.8 \text{ cm}}{\sin 28.0°} = \frac{b}{\sin 101.5°} \qquad \text{(Multiply both sides by LCD.)}$$

$$b(\sin 28.0°) = (\sin 101.5°)(46.8 \text{ cm})$$

(Divide both sides by $\sin 28.0°$.)

$$b = \frac{(\sin 101.5°)(46.8 \text{ cm})}{\sin 28.0°}$$

$$b = 97.7 \text{ cm}$$

You may use a calculator to do this calculation as follows:

Flowchart	Buttons Pushed	Display
Enter 101.5	$1 \to 0 \to 1 \to . \to 5$	101.5
Find sin	sin	0.979925
Push times	×	0.979925
Enter 46.8	$4 \to 6 \to . \to 8$	46.8
Push divide	÷	45.8605
Enter 28	$2 \to 8$	28

*As in previous sections, we will round sides to three significant digits and angles to the nearest tenth of a degree.

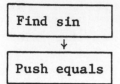

sin	0.469472
=	97.6853

$A = 180° - B - C = 180° - 101.5° - 28.0° = 50.5°$.

To find side a,

$$\frac{c}{\sin C} = \frac{a}{\sin A}$$

$$\frac{46.8 \text{ cm}}{\sin 28.0°} = \frac{a}{\sin 50.5°}$$

$a(\sin 28.0°) = (\sin 50.5°)(46.8 \text{ cm})$ (Multiply both sides by the LCD.)

$$a = \frac{(\sin 50.5°)(46.8 \text{ cm})}{\sin 28.0°}$$ (Divide both sides by $\sin 28.0°$.)

$a = 76.9 \text{ cm}$

The solution is $a = 76.9$ cm, $b = 97.7$ cm, and $A = 50.5°$.

A wide variety of applications may be solved using the Law of Sines.

<u>Example 2</u>. Find the lengths of rafters AC and BC for the roofline in Fig. 13.31.

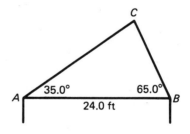

Figure 13.31

First, find angle C:

$C = 180° - A - B = 180° - 35.0° - 65.0° = 80.0°$

To find side AC,

$$\frac{AC}{\sin B} = \frac{AB}{\sin C}$$

$$\frac{AC}{\sin 65.0°} = \frac{24.0 \text{ ft}}{\sin 80.0°}$$

$AC(\sin 80.0°) = (\sin 65.0°)(24.0 \text{ ft})$ (Multiply both sides by the LCD.)

$$AC = \frac{(\sin 65.0°)(24.0 \text{ ft})}{\sin 80.0°}$$ (Divide both sides by $\sin 80.0°$.)

$AC = 22.1 \text{ ft}$

To find side BC,

$$\frac{BC}{\sin A} = \frac{AB}{\sin C}$$

$$\frac{BC}{\sin 35.0°} = \frac{24.0 \text{ ft}}{\sin 80.0°}$$

$$BC(\sin 80.0°) = (\sin 35.0°)(24.0 \text{ ft})$$

$$BC = \frac{(\sin 35.0°)(24.0 \text{ ft})}{\sin 80.0°}$$

$$BC = 14.0 \text{ ft}$$

EXERCISES 13.7

Solve each triangle using the labels as shown at the right: (Round lengths of sides to three significant digits and angles to the nearest tenth of a degree.)

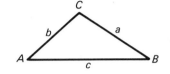

1. $A = 68.0°$, $a = 24.5$ m, $b = 17.5$ m
2. $C = 56.3°$, $c = 142$ cm, $b = 155$ cm
3. $A = 61.5°$, $B = 75.6°$, $b = 255$ ft
4. $B = 41.8°$, $C = 59.3°$, $c = 24.7$ km
5. $A = 14.6°$, $B = 35.1°$, $c = 43.7$ cm
6. $B = 24.7°$, $C = 136.1°$, $a = 342$ m
7. $A = 54.0°$, $C = 43.1°$, $a = 26.5$ m
8. $B = 64.3°$, $b = 135$ m, $c = 118$ m
9. $A = 20.1°$, $a = 47.5$ mi, $c = 35.6$ mi
10. $B = 75.2°$, $A = 65.1°$, $b = 305$ ft
11. $C = 48.7°$, $B = 56.4°$, $b = 5960$ m
12. $A = 118.0°$, $a = 5750$ m, $b = 4750$ m
13. $B = 105.5°$, $c = 11.3$ km, $b = 31.4$ km
14. $A = 58.2°$, $a = 39.7$ mi, $c = 27.5$ mi
15. $A = 16.5°$, $a = 206$ ft, $b = 189$ ft
16. $A = 35.0°$, $B = 49.3°$, $a = 48.7$ m
17. Find the distance AC across the river.

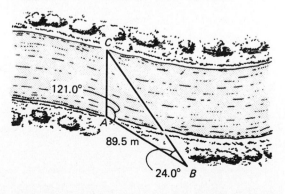

18. Find the lengths of rafters *AC* and *BC* of the roof below.

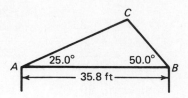

19. Find the distance *AB* between the ships below.

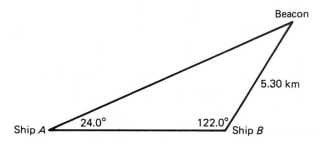

20. Find the height of the cliff below.

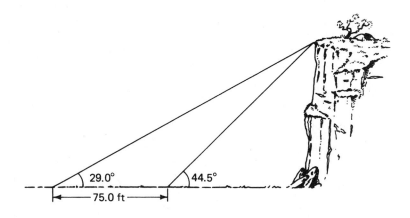

21. A contractor needs to correct the slope of a subdivision lot to place a house on level ground. The present slope of the lot is 12.5°. The contractor needs a level lot that is 105 ft deep. To control erosion, the back of the lot must be cut to a slope of 24.0°. How far from the street, measured along the present slope, will the excavation extend?

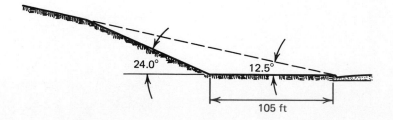

22. A weather balloon is sighted from points A and B, which are 4.00 km apart on level ground. The angle of elevation of the balloon from point A is 29.0°. Its angle of elevation from point B is 48.0°. (a) Find the height (in m) of the balloon if it is between A and B. (b) Find its height (in m) if point B is between point A and the weather balloon.

13.8   Law of Sines: The Ambiguous Case

The solution of a triangle in which two sides and an angle opposite one of the sides are given needs special care. In this case, there may be one, two, or no triangles formed from the given information. By construction, let's discuss the possibilities.

Example 1. Construct a triangle given that $A = 32°$, $a = 18$ cm, and $b = 24$ cm.

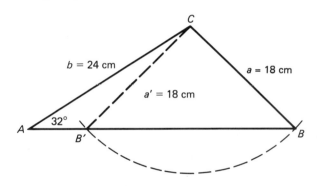

Figure 13.32

As you can see from Fig. 13.32, there are two triangles that satisfy these given conditions: triangles ABC and $AB'C$. In one case, angle B is acute. In the other, angle B' is obtuse.

Example 2. Construct a triangle given that $A = 40°$, $a = 12$ cm, and $b = 24$ cm.

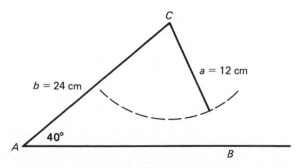

Figure 13.33

As you can see from Fig. 13.33, there is no triangle that satisfies these conditions. Side *a* is just not long enough to reach *AB*.

Example 3. Construct a triangle given that A = 50°, *a* = 12 cm, and *b* = 8 cm.

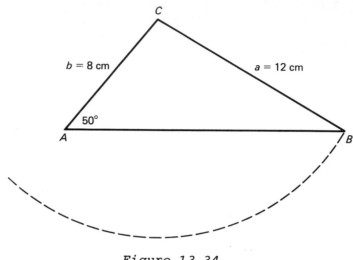

Figure 13.34

As you can see from Fig. 13.34, there is only one triangle that satisfies these conditions. Side *a* is too long for two solutions.

Let's summarize the possible cases when two sides and an angle opposite one of the sides are given. Assume that *acute* angle A and adjacent side *b* are given. From these two parts, the altitude (*h* = *b* sin A) is determined and fixed. Then, depending on the length of the opposite side *a*, we have four possible cases, as shown in Fig. 13.35.

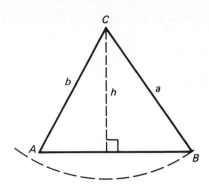

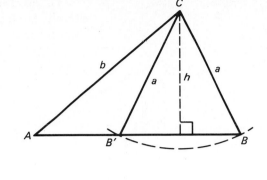

(a) When $h < b < a$, we have
   only one possible
   triangle. That is, when
   the side opposite the
   given acute angle is
   greater than the known
   adjacent side, there is
   only one possible
   triangle.

(b) When $h < a < b$, we have
   two possible triangles.
   That is, when the side
   opposite the given
   acute angle is less than
   the known adjacent side
   but greater than the altitude, there are two
   possible triangles.

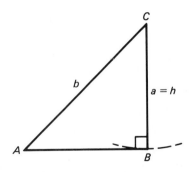

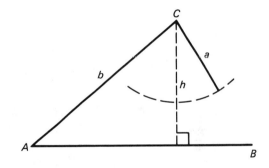

(c) When $a = h$, we have
   one possible (right)
   triangle. That is,
   when the side opposite
   the given acute angle
   equals the altitude,
   there is only one pos-
   sible (right) triangle.

(d) When $a < h$, there is no
   possible triangle. That
   is, when the side opposite the given acute
   angle is less than the
   altitude, there is no
   possible triangle.

Figure 13.35

If angle A is *obtuse*, we have two possible cases as shown in Fig. 13.36.

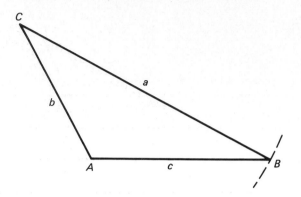
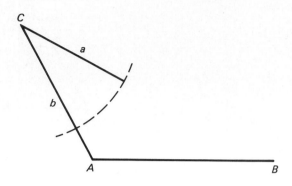

(e) When $a > b$, we have one possible triangle. That is, when the side opposite the given obtuse angle is greater than the known adjacent side, there is only one possible triangle.

(f) When $a \leq b$, there is no possible triangle. That is, when the side opposite the given obtuse angle is less than or equal to the known adjacent side, there is no possible triangle.

Figure 13.36

The table below summarizes these possibilities for the ambiguous case of the law of sines. Assume that the given parts are angle A, side opposite a, and side adjacent b.

Angle A					
Acute				Obtuse	
$h < b < a$ (a)	$h < a < b$ (b)	$a = h$ (c)	$a < h$ (d)	$a > b$ (e)	$a \leq b$ (f)
One triangle	Two triangles	One right triangle	No triangle	One triangle	No triangle

Note: If the given parts are not angle A, side opposite a, and side adjacent b, then you must substitute the given angle for A, the side opposite for a, and the side adjacent for b in this table. This is why it is important to understand the word description given in each case (a) through (f).

<u>Example 4</u>. If $A = 25.0°$, $a = 50.0$ m, and $b = 80.0$ m, solve the triangle.

First, find h.

$$h = b \sin A = (80.0 \text{ m})(\sin 25.0°) = 33.8 \text{ m}$$

Since $h < a < b$, we have two solutions. First, let's find *acute* angle B in Fig. 13.37.

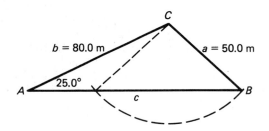

Figure 13.37

$$\frac{a}{\sin A} = \frac{b}{\sin B}$$

$$\frac{50.0 \text{ m}}{\sin 25.0°} = \frac{80.0 \text{ m}}{\sin B}$$

$(\sin B)(50.0 \text{ m}) = (\sin 25.0°)(80.0 \text{ m})$ (Multiply both sides by the LCD.)

$$\sin B = \frac{(\sin 25.0°)(80.0 \text{ m})}{50.0 \text{ m}}$$ (Divide both sides by 50.0 m.)

$$\sin B = 0.6762$$

$$B = 42.5°$$

You may use a calculator to do this calculation as follows:

Flowchart	Buttons Pushed	Display

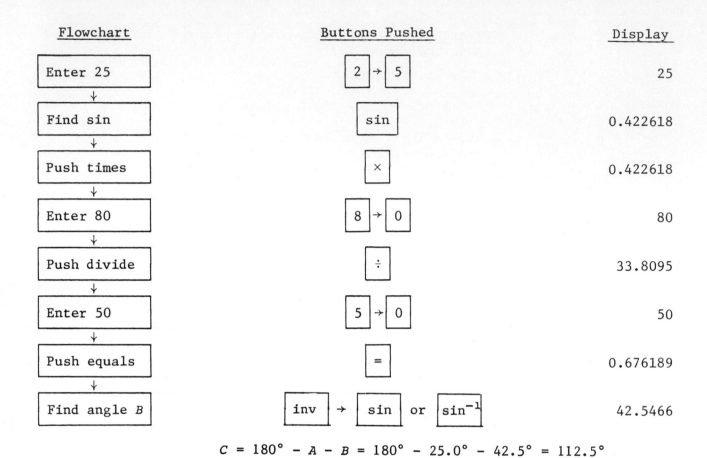

Flowchart	Buttons Pushed	Display
Enter 25	2 → 5	25
Find sin	sin	0.422618
Push times	×	0.422618
Enter 80	8 → 0	80
Push divide	÷	33.8095
Enter 50	5 → 0	50
Push equals	=	0.676189
Find angle B	inv → sin or sin^{-1}	42.5466

$C = 180° - A - B = 180° - 25.0° - 42.5° = 112.5°$

To find side c,

$$\frac{c}{\sin C} = \frac{a}{\sin A}$$

$$\frac{c}{\sin 112.5°} = \frac{50.0 \text{ m}}{\sin 25.0°}$$

$c(\sin 25.0°) = (\sin 112.5°)(50.0 \text{ m})$ (Multiply both sides by the LCD.)

$$c = \frac{(\sin 112.5°)(50.0 \text{ m})}{\sin 25.0°}$$ (Divide both sides by sin 25.0°.)

$c = 109 \text{ m}$

Next, let's find *obtuse* angle B in Fig. 13.38.

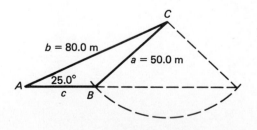

Figure 13.38

$$\frac{a}{\sin A} = \frac{b}{\sin B}$$

$$\frac{50.0 \text{ m}}{\sin 25.0°} = \frac{80.0 \text{ m}}{\sin B}$$

$$(\sin B)(50.0 \text{ m}) = (\sin 25.0°)(80.0 \text{ m})$$

$$\sin B = \frac{(\sin 25.0°)(80.0 \text{ m})}{50.0 \text{ m}}$$

$$\sin B = 0.6762$$

$$B = 180° - 42.5° = 137.5°$$

Note: If B is acute, B is the $\boxed{\text{inv}}$ → $\boxed{\text{sin}}$ of 0.6762.
If B is obtuse, B is 180° − ($\boxed{\text{inv}}$ → $\boxed{\text{sin}}$) of 0.6762.
$C = 180° - A - B = 180° - 25.0° - 137.5° = 17.5°$

To find side c,

$$\frac{c}{\sin C} = \frac{a}{\sin A}$$

$$\frac{c}{\sin 17.5°} = \frac{50.0 \text{ m}}{\sin 25.0°}$$

$$c(\sin 25.0°) = (\sin 17.5°)(50.0 \text{ m})$$

$$c = \frac{(\sin 17.5°)(50.0 \text{ m})}{\sin 25.0°}$$

$$c = 35.6 \text{ m}$$

The two solutions are c = 109 m, B = 42.5°, C = 112.5°
and c = 35.6 m, B = 137.5°, C = 17.5°.

Example 5. If A = 59.0°, a = 205 m, and b = 465 m, solve
the triangle.

    First, find h.

$$h = b \sin A = (465 \text{ m})(\sin 59.0°) = 399 \text{ m}$$

Since $a < h$, there is no possible triangle.

What would happen if you tried to apply the law of
sines anyway?

$$\frac{a}{\sin A} = \frac{b}{\sin B}$$

$$\frac{205 \text{ m}}{\sin 59.0°} = \frac{465 \text{ m}}{\sin B}$$

$$\sin B = \frac{(\sin 59.0°)(465 \text{ m})}{205 \text{ m}} = 1.944$$

Note: sin B = 1.944 is impossible because $-1 \leq \sin B \leq 1$. Recall that the graph of $y = \sin x$ has an amplitude of 1, which means that the values of sin x vary between 1 and -1. Your calculator will also indicate an error when you try to find angle B.

As a final check to make certain that your solution is correct, check to see that the following geometric triangle property is satisfied: In any triangle, the largest side is opposite the largest angle and the smallest side is opposite the smallest angle.

EXERCISES 13.8

For each general triangle, (a) determine the number of solutions, and (b) solve the triangle, if possible, using the labels as shown at the right: (Round lengths to three significant digits and angles to the nearest tenth of a degree.)

1. A = 38.0°, a = 42.3 m, b = 32.5 m

2. C = 47.6°, a = 85.2 cm, c = 96.1 cm

3. A = 25.6°, b = 306 m, a = 275 m

4. B = 41.2°, c = 1860 ft, b = 1540 ft

5. A = 71.6°, b = 48.5 m, a = 15.7 m

6. B = 40.3°, b = 161 cm, c = 288 cm

7. C = 71.2°, a = 245 cm, c = 238 cm

8. A = 36.1°, b = 14.5 m, a = 12.5 m

9. B = 105.0°, b = 33.0 mi, a = 24.0 mi

10. A = 98.3°, a = 1420 ft, b = 1170 ft

11. A = 31.5°, a = 376 m, c = 406 m

12. B = 50.0°, b = 4130 ft, c = 4560 ft

13. C = 60.0°, c = 151 m, b = 181 m

14. A = 30.0°, a = 4850 mi, c = 3650 mi

15. B = 8.0°, b = 451 m, c = 855 m

16. C = 8.7°, c = 89.3 mi, b = 61.9 mi

17. An owner of a triangular lot wishes to fence it in along the lot lines. Lot markers at A and B have been located, but the lot marker at C cannot be found. The owner's attorney gives the following information by phone: AB = 355 ft, BC = 295 ft, and A = 36.0°. What is the length of AC?

18. The average distance from the sun to Earth is 1.5×10^8 km and from the sun to Venus is 1.1×10^8 km. Find the distance between Earth and Venus when the angle between Earth and the sun and Earth and Venus is 24.7°. (Assume that Earth and Venus have circular orbits around the sun.)

13.9 Solving Oblique Triangles: Law of Cosines

A second law used to solve oblique triangles is the Law of Cosines.

LAW OF COSINES

$$a^2 = b^2 + c^2 - 2bc \cos A$$
$$b^2 = a^2 + c^2 - 2ac \cos B$$
$$c^2 = a^2 + b^2 - 2ab \cos C$$

In words: *for any triangle, the square of any side equals the sum of the squares of the other two sides minus twice the product of these two sides and the cosine of their included angle.* (See Fig. 13.39.)

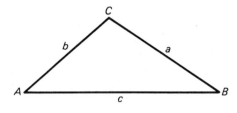

Figure 13.39

In order to use the Law of Cosines, you must know:
(a) two sides and the included angle or
(b) all three sides.

Example 1. If A = 115.2°, b = 18.5 m, and c = 21.7 m, solve the triangle. (See Fig. 13.40.)

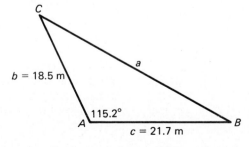

Figure 13.40

To find side a,

$$a^2 = b^2 + c^2 - 2bc \cos A$$

$$a^2 = (18.5 \text{ m})^2 + (21.7 \text{ m})^2 - 2(18.5 \text{ m})(21.7 \text{ m})(\cos 115.2°)$$

$$a = 34.0 \text{ m}$$

You may use a calculator to do this calculation as follows:

Flowchart	Buttons Pushed	Display
Enter 115.2	$1 \to 1 \to 5 \to . \to 2$	115.2
Find cos	cos	−0.425779
Push times	×	−0.425779
Enter 21.7	$2 \to 1 \to . \to 7$	21.7
Push times	×	−9.23941
Enter 18.5	$1 \to 8 \to . \to 5$	18.5
Push times	×	−170.929
Enter −2	$2 \to +/-$	−2
Push plus	+	341.858
Enter 21.7	$2 \to 1 \to . \to 7$	21.7
Find square	x^2	470.89
Push plus	+	812.748
Enter 18.5	1 8 . 5	18.5

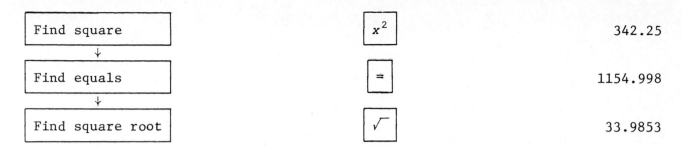

Find square	x^2	342.25
Find equals	$=$	1154.998
Find square root	$\sqrt{}$	33.9853

To find angle B, use the Law of Sines as it requires less computation.

$$\frac{a}{\sin A} = \frac{b}{\sin B}$$

$$\frac{34.0 \text{ m}}{\sin 115.2°} = \frac{18.5 \text{ m}}{\sin B}$$

$(\sin B)(34.0 \text{ m}) = (\sin 115.2°)(18.5 \text{ m})$ (Multiply both sides by the LCD.)

$\sin B = \dfrac{(\sin 115.2°)(18.5 \text{ m})}{34.0 \text{ m}}$ (Divide both sides by 34.0 m.)

$$\sin B = 0.4923$$

$$B = 29.5°$$

$C = 180° - A - B = 180° - 115.2° - 29.5° = 35.3°$

Example 2. If $a = 125$ cm, $b = 285$ cm, and $c = 382$ cm, solve the triangle in Fig. 13.41.

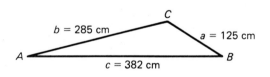

Figure 13.41

When three sides are given, you are advised to find the angle opposite the largest side first. Why?

To find angle C,

$$c^2 = a^2 + b^2 - 2ab \cos C$$

$(382 \text{ cm})^2 = (125 \text{ cm})^2 + (285 \text{ cm})^2 - 2(125 \text{ cm})(285 \text{ cm}) \cos C$

$(382 \text{ cm})^2 - (125 \text{ cm})^2 - (285 \text{ cm})^2 = -2(125 \text{ cm})(285 \text{ cm}) \cos C$

$$\frac{(382 \text{ cm})^2 - (125 \text{ cm})^2 - (285 \text{ cm})^2}{-2(125 \text{ cm})(285 \text{ cm})} = \cos C$$

$$-0.6888 = \cos C$$

$$133.5° = C$$

You may use a calculator to do this calculation as follows:

Flowchart	Buttons Pushed	Display
Enter 382	$3 \rightarrow 8 \rightarrow 2$	382
Find square	x^2	145924
Push minus	$-$	145924
Enter 125	$1 \rightarrow 2 \rightarrow 5$	125
Find square	x^2	15625
Push minus	$-$	130299
Enter 285	$2 \rightarrow 8 \rightarrow 5$	285
Find square	x^2	81225
Push equals	$=$	49074
Push divide	$\div$	49074
Enter –2	$2 \rightarrow +/-$	–2
Push divide	$\div$	–24537
Enter 125	$1 \rightarrow 2 \rightarrow 5$	125
Push divide	$\div$	–196.296
Enter 285	$2 \rightarrow 8 \rightarrow 5$	285
Push equals	$=$	–0.688758
Find angle C	$\text{inv} \rightarrow \text{cos}$ or $\cos^{-1}$	133.532

To find angle A, let's use the Law of Sines.

$$\frac{c}{\sin C} = \frac{a}{\sin A}$$

$$\frac{382 \text{ cm}}{\sin 133.5°} = \frac{125 \text{ cm}}{\sin A}$$

$$(382 \text{ cm})(\sin A) = (\sin 133.5°)(125 \text{ cm})$$

$$\sin A = \frac{(\sin 133.5°)(125 \text{ cm})}{382 \text{ cm}}$$

$$\sin A = 0.2374$$

$$A = 13.7°$$

$$B = 180° - A - C = 180° - 13.7° - 133.5° = 32.8°$$

<u>Example 3</u>. Find the lengths of guy wires AC and BC for the tower in Fig. 13.42(a). The height of the tower is 50.0 m. Angle $ADC = 120.0°$. $AD = 20.0$ m; $BD = 15.0$ m.

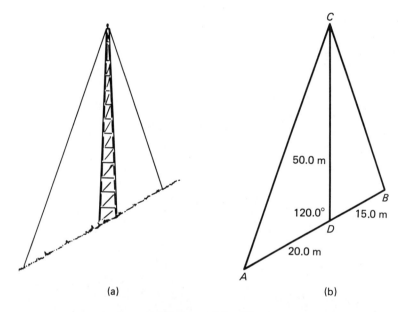

(a)   (b)

Figure 13.42

First, let's use triangle ACD in Fig. 13.42(b) to find length AC.

Using the Law of Cosines,

$$(AC)^2 = (AD)^2 + (DC)^2 - 2(AD)(DC) \cos ADC$$
$$(AC)^2 = (20.0 \text{ m})^2 + (50.0 \text{ m})^2 - 2(20.0 \text{ m})(50.0 \text{ m}) \cos 120.0°$$
$$AC = 62.4 \text{ m}$$

Next, use triangle CDB and the Law of Cosines to find length BC. Note that angle $CDB = 180° - 120.0° = 60.0°$.

$$(BC)^2 = (BD)^2 + (DC)^2 - 2(BD)(DC) \cos CDB$$

$$(BC)^2 = (15.0 \text{ m})^2 + (50.0 \text{ m})^2 - 2(15.0 \text{ m})(50.0 \text{ m}) \cos 60.0°$$

$$BC = 44.4 \text{ m}$$

EXERCISES 13.9

Solve each triangle using the labels as shown at the right: (Round lengths of sides to three significant digits and angles to nearest tenth of a degree.)

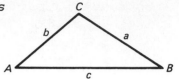

1. $A = 55.0°$, $b = 21.2$ m, $c = 24.0$ m

2. $B = 14.5°$, $a = 37.6$ cm, $c = 48.2$ cm

3. $C = 115.0°$, $a = 247$ ft, $b = 316$ ft

4. $A = 130.0°$, $b = 15.2$ km, $c = 9.50$ km

5. $a = 136$ m, $b = 155$ m, $c = 168$ m

6. $a = 3.96$ in., $b = 4.81$ in., $c = 6.45$ in.

7. $C = 71.6°$, $a = 5.42$ km, $b = 6.03$ km

8. $B = 96.1°$, $a = 452$ yd, $c = 365$ yd

9. $a = 38,500$ mi, $b = 67,500$ mi, $c = 47,200$ mi

10. $a = 146$ cm, $b = 271$ cm, $c = 205$ cm

11. $B = 19.3°$, $a = 4820$ ft, $c = 1930$ ft

12. $C = 108.5°$, $a = 415$ m, $b = 325$ m

13. $a = 19.5$ m, $b = 36.5$ m, $c = 25.6$ m

14. $a = 207$ mi, $b = 106$ mi, $c = 142$ mi

15. **Find the distance** a **across the pond in the figure below.**

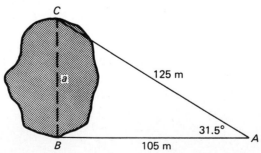

16. **Find the length of rafter** AC **in the figure below.**

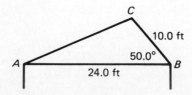

17. **Find angles** A **and** C **in the roof at top of next page.**

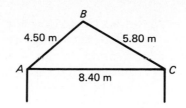

18. (a) Find angles A and C in the roof below. AC = BC
    (b) Find length CD.

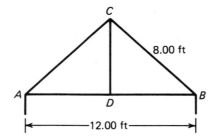

19. In the framework below, we know that AE = CD, AB = BC,
    BD = BE, and $\overline{AC} \parallel \overline{ED}$. Find

    (a) angle BEA
    (b) angle A
    (c) length BE
    (d) length DE

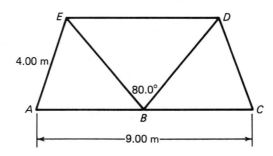

20. In the framework below, we know that
    AB = DE, BC = CD, AH = FE, HG = GF. Find

    (a) length HB
    (b) angle AHB
    (c) length GC
    (d) length AG

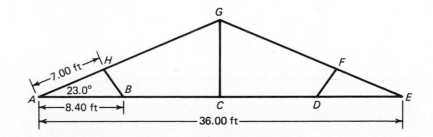

21. A triangular lot has sides 1580 ft, 2860 ft, and 1820 ft. Find its largest angle.
22. A ship travels 125 mi northeast and then 150 mi due east. What is the minimum distance that the ship could travel?

13.10   The Other Trigonometric Ratios

In Section 13.1, we studied three trigonometric ratios: sine, cosine, and tangent. While these three are the most widely used and applied trigonometric ratios, three other ratios are also used. These are defined as follows:

The *cotangent* of the angle A, abbreviated "cot A," is equal to the ratio of the length of the side adjacent to angle A, which is b, to the length of the side opposite angle A, which is a.

The *secant* of the angle A, abbreviated "sec A," is equal to the ratio of the length of the hypotenuse, c, to the length of the side adjacent to angle A, which is b.

The *cosecant* of the angle A, abbreviated "csc A," is equal to the ratio of the length of the hypotenuse, c, to the length of the side opposite angle A, which is a.

That is, in a right triangle, such as triangle ABC in Fig. 13.43, we have:

Other Trigonometric Ratios

$$\cot A = \frac{\text{length of side adjacent to angle } A}{\text{length of side opposite angle } A} = \frac{b}{a}$$

$$\sec A = \frac{\text{length of hypotenuse}}{\text{length of side adjacent to angle } A} = \frac{c}{b}$$

$$\csc A = \frac{\text{length of hypotenuse}}{\text{length of side opposite angle } A} = \frac{c}{a}$$

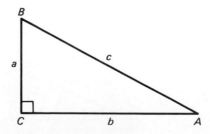

Figure 13.43

Note that these ratios are reciprocals of the first three ratios. That is,

1. The values of csc A and sin A are reciprocals of each other.
2. The values of sec A and cos A are reciprocals of each other.
3. The values of cot A and tan A are reciprocals of each other.

Or,

$$csc\ A = \frac{1}{sin\ A}, \quad sec\ A = \frac{1}{cos\ A}, \quad and \quad cot\ A = \frac{1}{tan\ A}$$

We use this relationship to find the values of these last three trigonometric ratios on a calculator since calculators do not have buttons for cosecant, secant, and cotangent.

Example 1. Find sec 34.5° rounded to four significant digits.

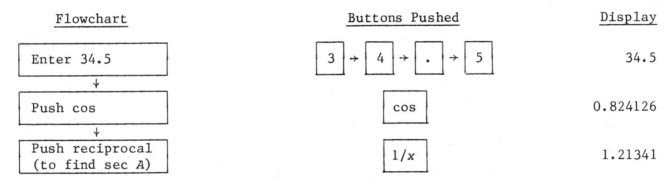

Flowchart	Buttons Pushed	Display
Enter 34.5	3 → 4 → . → 5	34.5
Push cos	cos	0.824126
Push reciprocal (to find sec A)	1/x	1.21341

Then, sec 34.5° = 1.213 rounded to four significant digits.

Example 2. Find cot 72.3° rounded to four significant digits.

Flowchart	Buttons Pushed	Display
Enter 72.3	7 → 2 → . → 3	72.3
Push tan	tan	3.13341
Push reciprocal (to find cot A)	1/x	0.319141

Then, cot 72.3° = 0.3191 rounded to four significant digits.

A calculator may also be used to find the angle when the value of the trigonometric ratio is given. Here, we limit angle A to $0° \le A \le 90°$. The process is shown by example.

<u>Example 3.</u> Find angle A to the nearest tenth of a degree when csc $A = 1.290$.

Flowchart	Buttons Pushed	Display

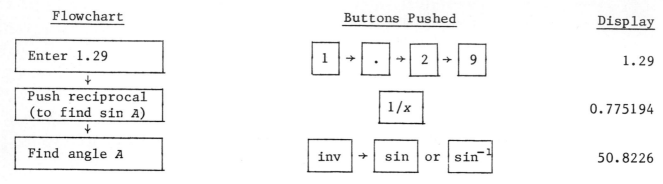

		1.29
		0.775194
		50.8226

Then, angle $A = 50.8°$ rounded to the nearest tenth of a degree.

<u>Example 4.</u> Find angle A to the nearest hundredth of a degree when cot $A = 2.500$.

Flowchart	Buttons Pushed	Display

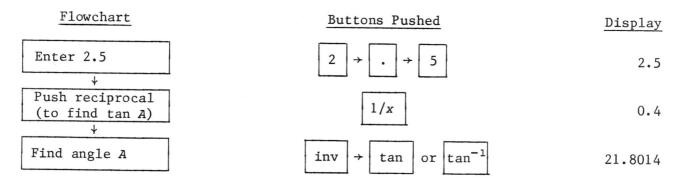

		2.5
		0.4
		21.8014

So, angle $A = 21.80°$ rounded to the nearest hundredth of a degree.

EXERCISES 13.10   *Find the value of each trigonometric ratio rounded to four significant digits:*

1. sec 30°	2. csc 45°	3. cot 45°
4. cot 60°	5. csc 21°	6. sec 67°
7. csc 43.5°	8. cot 4.9°	9. sec 28.7°
10. sec 18.34°	11. csc 23.39°	12. cot 33.45°
13. cot 77.25°	14. csc 85.55°	15. sec 57.88°
16. cot 48.75°	17. csc 50.333°	18. cot 24.307°

Find each angle rounded to the nearest tenth of a degree:

19. sec A = 1.223   20. csc A = 1.980   21. cot A = 0.7681

22. cot A = 2.350   23. csc B = 2.398   24. sec B = 2.117

25. cot B = 3.451   26. sec B = 4.670   27. csc A = 1.008

28. csc B = 2.061   29. sec A = 1.902   30. cot A = 0.0331

Find each angle rounded to the nearest hundredth of a degree:

31. sec B = 2.031   32. cot A = 0.4475   33. csc A = 1.909

34. csc B = 4.500   35. cot B = 2.505   36. sec A = 1.889

Chapter 13 Review

In Exercises 1-7, refer to the triangle below:

1. What is the length of the side opposite angle A in the right triangle?

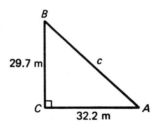

2. What is the angle adjacent to the side whose length is 29.7 m?

3. The side of the triangle denoted by c is known as the ___?___ .

4. What is the length of the side denoted by c?

5. $\dfrac{\text{length of side opposite angle } A}{\text{length of hypotenuse}}$ is what ratio?

6. $\cos A = \dfrac{?}{\text{length of hypotenuse}}$

7. $\tan B = \dfrac{?}{?}$

Find the value of each trigonometric ratio rounded to four significant digits:

8.   cos 36.2°   9. tan 48.7°   10. sin 23.72°

Find each angle rounded to the nearest tenth of a degree:

11. sin A = 0.7136   12. tan B = 0.1835   13. cos A = 0.4104

14. Find angle A.   16. Find side b.
15. Find angle B.   17. Find side c.

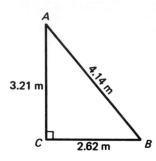

Exercises 14 and 15

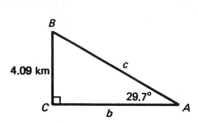

Exercises 16 and 17

Solve each right triangle:

18.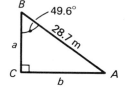

19.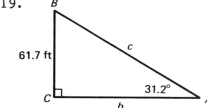

20.

21. A satellite is directly overhead
    one observer station when it is
    at an angle of 68.0° from another
    observer station. The distance
    between the two stations is
    2000 m. What is the height of
    the satellite?

Exercise 21

Find each value rounded to four significant digits:

22. tan 143°   23. sin 209.8°   24. cos 317.4°

Graph each equation for values of x between 0° and 360° in multiples of 15°:

25. $y = 6 \cos x$ 26. $y = 3 \sin 2x$

Solve each triangle using the labels as shown at the right: (*Round length of sides to three significant digits and angles to the nearest tenth of a degree.*)

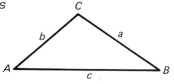

27. $B = 52.7°$, $b = 206$ m, $a = 175$ m

28. $A = 61.2°$, $C = 75.6°$, $c = 88.0$ cm

29. $B = 17.5°$, $a = 345$ m, $c = 405$ m

30. $a = 48.6$ cm, $b = 31.2$ cm, $c = 51.5$ cm

31. $A = 29.5°$, $b = 20.5$ m, $a = 18.5$ m

32. $B = 18.5°$, $a = 1680$ m, $b = 1520$ m

33. $a = 575$ ft, $b = 1080$ ft, $c = 1250$ ft

34. $C = 73.5°$, $c = 58.2$ ft, $b = 81.2$ ft

35. Find angle B in the figure below.
36. Find length x in the figure below.

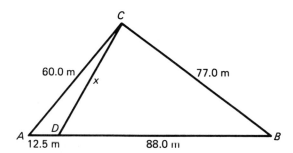

37. The centers of five holes are equally spaced around a circle of diameter 16.00 in. Find the distance between two successive holes.

38. In the roof truss below, $AB = DE$, $BC = CD$, and
    $AE = 36.0$ m. Find each length: (a) AF (b) BF (c) CF
    (d) BC.

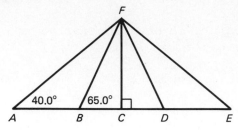

*Find the value of each trigonometric ratio rounded to four
significant digits:*

39. csc 54.6°   40. cot 77.2°   41. sec 25.1°

Find each angle rounded to the nearest tenth of a degree:

42. sec A = 2.035   43. cot B = 0.5454   44. csc A = 1.505

BASIC STATISTICS

14.1 Bar Graphs

What Is Statistics?

Statistics is a branch of mathematics dealing with the collection, analysis, interpretation, and presentation of masses of numerical data. In this chapter, we will first study the different ways that data can be presented using graphs. Then, in the later sections, we will study some of the different ways to describe rather small sets of data.

An in-depth study of statistics would take years. We will study only the most basic parts. These will help you read and better understand newspapers, magazines, and some of the technical reports in your fields of interest.

A graph is a picture that shows the relationship between several types of collected information. A graph is very useful when there are large quantities of information to analyze. There are many ways of graphing. One common way is the *bar graph*. It is made of several parallel bars of lengths proportional to the quantities shown by each bar. Each bar stands for a fixed category, and the length of the bar shows the amount or value of that category. Look closely at the bar graph in Fig. 14.1.

Example. What are the monthly earnings of a data-processing technician with two years of experience?

Find the data-processing technician in the "classifica-tion of workers" column. Read the right end of the bar on the "monthly salary" scale as $1800.

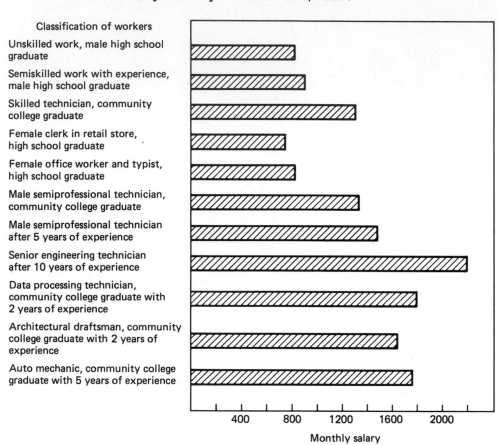

Fig. 14.1 Typical Earnings of Unskilled, Technical, and Semiprofessional Workers in the United States in the Mid-1980s.

EXERCISES 14.1   Find the monthly earnings of the workers listed below from the bar graph in Fig. 14.1:

1. Auto mechanic
2. Office worker and typist
3. Clerk in retail store
4. Skilled technician, community college graduate
5. Unskilled worker
6. Semiskilled worker
7. Architectural draftsman
8. Semiprofessional technician, community college graduate
9. Semiprofessional technician after 5 years' experience
10. Senior engineering technician after 10 years' experience

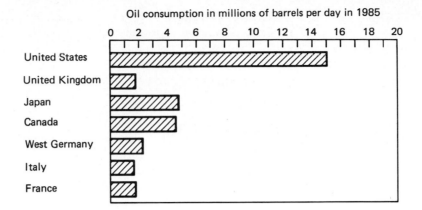

Oil consumption in millions of barrels per day in 1985

Find the following information from the given bar graph:

11. How many barrels per day were used by France?
12. How many barrels per day were used by Japan?
13. What country used the most barrels per day?
14. What country used the least barrels per day?
15. How many barrels per day were used by Canada?
16. How many barrels per day were used by Italy?
17. How many barrels per day were used by the United States?
18. How many barrels per day were used by the United Kingdom?
19. How many barrels per day were used by West Germany?
20. What was the total number of barrels per day used by all the countries listed?

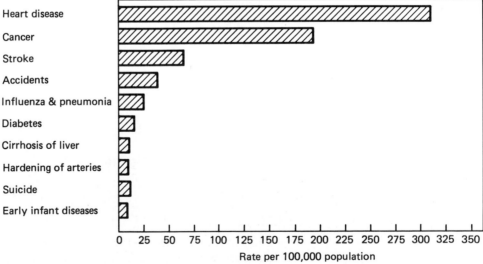

Rate per 100,000 population

Ten leading causes of death in United States in 1985

Using the given bar graph, find the number of deaths per 100,000 for each cause listed below:

21. Heart disease   22. Accidents
23. Influenza and pneumonia   24. Suicide
25. Diabetes   26. Hardening of arteries
27. Cancer   28. Early infant diseases
29. Cirrhosis of liver   30. Stroke

14.2 Circle Graphs

Another type of graph used quite often to give results of surveys is the *circle graph* (see Fig. 14.2). The circle graph is used to show the relationship between the parts and the whole.

To make a circle graph with data given in percents, first draw the circle. Since there are 360 degrees in a circle, multiply the percent of an item by 360 to find what part of the circle is used by that item.

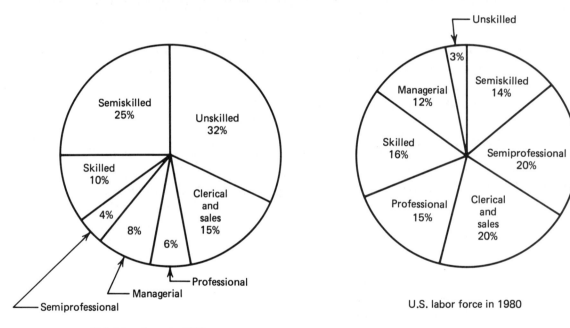

U.S. labor force in 1930

U.S. labor force in 1980

Figure 14.2

Example 1. Draw a circle graph with the following data.

In 1930 58% of the people working had a grade-school education or less, 32% had a high-school education, and 10% had a college education.

58% of 360° = 0.58 × 360° = 208.8°, or about 209°
32% of 360° = 0.32 × 360° = 115.2°, or about 115°
10% of 360° = 0.10 × 360° = 36°

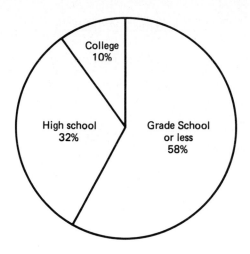

Fig. 14.3 Educational training of those working in 1930

With a protractor draw central angles of 209°, 115°, and 36°. Then, label the sections (see Fig. 14.3).

Sometimes data are not written in percent form. When a circle graph is to be drawn from data not in percent form, the data must first be converted to percents. Once the data are in this form, the steps in drawing the graph are the same as those already given.

<u>Example 2.</u> Draw a circle graph with the following data.

Suggested credit-hour requirements for a community college curriculum in engineering technology are as follows:

Course	*Hours*
Mathematics (technical)	10
Applied science	10
Technical specialty courses in major	30
Supporting technical courses	6
General education courses	$\underline{18}$
	74

Write each area of study as a percent of the whole program (see Fig. 14.4).

Mathematics:
$$\frac{10}{74} = \frac{r}{100}$$

$$74r = 1000$$

$$r = 13.5\%$$

$$13.5\% \times 360° = 0.135 \times 360°$$

$$= 49° \quad \text{(rounded to nearest whole degree)}$$

Science: Same as above (49°)

Technical courses: $\dfrac{30}{74} = \dfrac{r}{100}$

$$74r = 3000$$

$$r = 40.5\%$$

$$40.5\% \times 360° = 0.405 \times 360° = 146°$$

Supporting courses: $\dfrac{6}{74} = \dfrac{r}{100}$

$$74r = 600$$

$$r = 8.1\%$$

$$8.1\% \times 360° = 0.081 \times 360° = 29°$$

General education: $\dfrac{18}{74} = \dfrac{r}{100}$

$$74r = 1800$$

$$r = 24.3\%$$

$$24.3\% \times 360° = 0.243 \times 360° = 87°$$

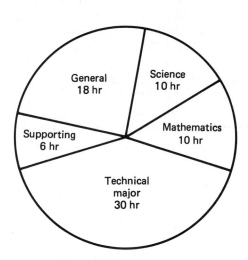

Fig. 14.4 Breakdown of Areas of Study in an Engineering-Technology Program

EXERCISES 14.2   1. What is 26% of 360°?
2. What is 52% of 360°?
3. What is 15.2% of 360°?
4. What is 37.1% of 360°?
5. What is 75% of 360°?
6. What is 47.7% of 360°?

7. Of 744 students, 52 are taking mathematics. What angle of a circle would show the percent of students taking mathematics?

8. Of 2017 students, 89 are taking technical physics. What angle of a circle would show the percent of students taking technical physics?

9. Of 5020 TV sets, 208 are found to be defective. What angle of a circle would show the percent of defective TV sets?

10. Of 29,106 votes cast, 4060 were for Candidate A. (Candidate A was one of four candidates in an election.) What angle of a circle would show the percent of votes not cast for Candidate A?

11. A school spends $16,192 of its $182,100 budget for supplies. What angle of a circle would show the percent of money the school spends on things other than supplies?

12. In one month, the sales of calculators were as follows:

    Brand A: 29   Brand D: 75
    Brand B: 52   Brand E: 43
    Brand C: 15

What central angle of a circle graph would show what percent of the total sold during the month were Brand B?

13. Draw a circle graph with the following data. In 1980, 11% of the expenditures for education came from the federal government, 36% came from state governments, 28% came from local governments, and 25% came from other sources.

14. Draw a circle graph with the following data. Suggested credit-hour requirements for a community college curriculum in industrial technology are as follows.

Course	Credits
Mathematics	6
Applied science	8
Technical specialties	36
Supporting technical courses	9
General education courses	15
	74

15. Draw a circle graph with the following data. In 1980, federal budget receipts, in millions of dollars, were as follows: individual income taxes, 298,000; corporate income taxes, 227,322; social security taxes, 155,000; excise taxes, 18,455; miscellaneous taxes, 14,600.

16. Draw a circle graph with the following data. In 1985, total cotton production, in millions of 480-lb bales, was as follows: United States, 13.8; Canada, 5.8; Europe, 1.2; USSR, 12.3; China, 24; Africa, 9.1; Brazil, 3.0; Mexico, 0.8.

17. Draw a circle graph with the following data. Wholesale sales, in millions of dollars, for durable goods in 1982 were as follows: automobiles, 39,460; electrical goods, 29,170; furniture, 12,498; hardware, 20,815; lumber, 17,041; machinery, 99,250; metals, 10,121; other, 27,748.

14.3   Broken-Line Graphs

The broken-line graph is used to show changing conditions, often over a certain time interval.

Example 1. An industrial technician must keep a chemical at a temperature below 60°F. He must also keep an hourly record of its temperature and record each day's temperatures on a broken-line graph. The following table shows the data he has collected.

Time	8	9	10	11	12	1	2	3	4	5
Temp. (°F)	60°	58°	54°	51°	52°	57°	54°	52°	54°	58°

When drawing line graphs, (a) use graph paper, because it is already subdivided both vertically and horizontally; (b) choose horizontal and vertical scales so that the broken line uses up most of the space allowed for the graph; (c) name and label each scale so that all marks on the scale are the same distance apart and show equal intervals; (d) plot the points from the given data; (e) connect each pair of points in order by a straight line. When you have taken all these steps, you will have a broken-line graph (see Fig. 14.5).

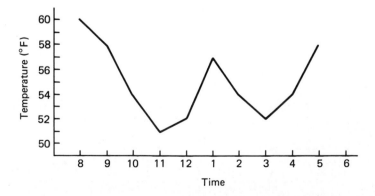

Figure 14.5

EXERCISES 14.3   1. The data below are from the records of the industrial technician in Example 1. These data were recorded on the following day. Draw a broken-line graph for them.

Time	8	9	10	11	12	1	2	3	4	5
Temperature in °F	59°	57°	55°	54°	53°	57°	55°	53°	56°	59°

2. An inspector recorded the number of faulty calculators and the hour in which they passed by his station.

Time	7-8	8-9	9-10	10-11	11-12	1-2	2-3	3-4	4-5	5-6
Number of faulty calculators	1	2	2	3	6	2	4	4	7	10

Draw a broken-line graph for these data.

3. A survey of 100 families was taken to find the number of times the families had gone out to eat in the past month. The data are given in the table below:

Times out in past month	Number of families
0	2
1	15
2	51
3	17
4	10
5 or more	5

Draw a broken-line graph for this survey.

4. The following is the winning batting average recorded for the years 1978 through 1985 for the major leagues:

Year	1978	1979	1980	1981	1982	1983	1984	1985
Batting average	0.334	0.344	0.390	0.341	0.332	0.361	0.351	0.368

Draw a broken-line graph for these data.

5. The following are the average test scores of chapter tests given in a mathematics class:

Chapter	1	2	3	4	5	6	7	8	9	10	11	12	13	14
Score	78	81	75	77	84	81	79	70	72	73	75	69	81	72

Draw a broken-line graph for these scores.

6. The following table gives the average wage per week in 1985 for workers in certain industries:

Industry	Metal mining	Coal mining	Oil and gas	Contract construc-tion	Transporta-tion	Electric and gas service	Retail trade	Hotel
Wage (in $)	556.93	581.39	484.66	463.04	310.31	529.98	180.58	176.18

Draw a broken-line graph for these data.

7. The following table gives the male life expectancy for certain countries:

Country	Egypt	Nigeria	Israel	Japan	France	Sweden	Canada	Mexico	U.S.	USSR
Years	55.9	46.9	72.8	74.2	70.2	73.1	73.0	63.9	71.8	61.9

Draw a broken-line graph for these data.

8. The following table gives the female life expectancy for certain countries:

Country	Egypt	Nigeria	Israel	Japan	France	Sweden	Canada	Mexico	U.S.	USSR
Years	58.4	50.2	76.2	79.8	78.5	79.1	79.0	68.2	78.8	72.0

Draw a broken-line graph for these data.

A technician is often asked to read graphs drawn by a
machine. The machine records measurements by inking them
on a graph. Any field in which quality control or continuous information is needed might use this way of recording measurements. The figure below shows a microbarograph used by the weather service to record atmospheric
pressure in inches. For example, the reading on Monday at
2:00 p.m. was 29.16 in.

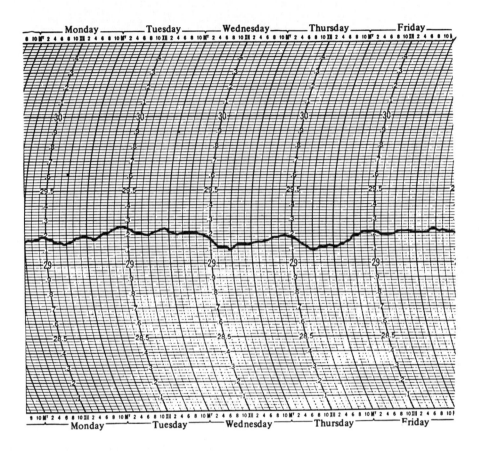

*Use the microbarograph above to answer the following
questions:*

9. What was the atmospheric pressure recorded for
   Tuesday at 4:00 p.m.?
10. What was the highest atmospheric pressure recorded?
    When was it recorded?
11. What was the lowest atmospheric pressure recorded?
12. What was the atmospheric pressure recorded for Thursday
    at 9:00 a.m.?
13. What was the atmospheric pressure recorded for Monday
    at noon?

Shown below is a hygrothermograph used by the weather services to record temperature and relative humidity. The lower part of the graph is used to measure relative humidity from 0% to 100%. The top part of the graph is used to measure temperature from 10°F to 110°F. For example, at 8:00 p.m., the temperature was 81°F and the relative humidity 83%.

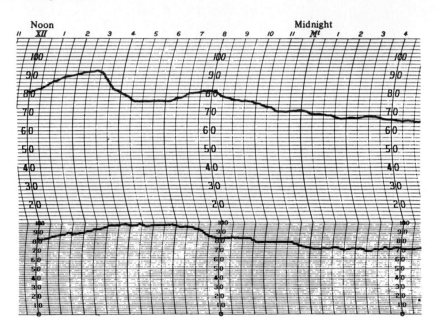

Use the hygrothermograph above to answer the following questions:

14. What was the relative humidity at 12:00 midnight?
15. What was the temperature at 3:30 a.m.?
16. What was the highest temperature recorded? When was it recorded?
17. What was the lowest temperature recorded? When was it recorded?
18. What was the relative humidity at 8:00 a.m.?
19. What was the lowest relative humidity recorded? When was it recorded?

14.4 Other Graphs

A graph can be a curved line, as shown by Graph A (Fig. 14.6). This graph shows typical power gain for class B push-pull amplifiers with 9-volt power supply.

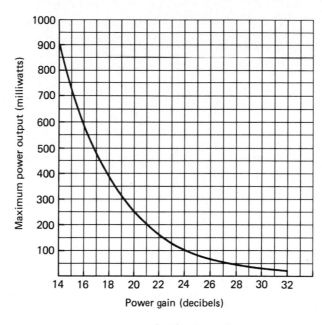

Graph A

Figure 14.6

<u>Example</u>. What is the power output when the gain is 22 decibels (db)?

Find 22 on the horizontal axis and read up until you meet the graph. Read left to the vertical axis and read 160 milliwatts (mW).

EXERCISES 14.4   *Use Graph A to find answers for Exercises 1-5:*

1. What is the highest power output? What is the power gain when this happens?
2. What is the power gain when the power output is 600 mW?
3. What is the power output when the power gain is 25 db?
4. Between what 2-db readings is the greatest change in power output found?
5. Between what 2-db readings is the least change in power output found?

One way not to have to use a curved line for a graph is to use *semilogarithmic* graph paper. It has a logarithmic scale for one axis and a uniform scale for the other axis. Graph B shows the data from Graph A on semilogarithmic graph paper.

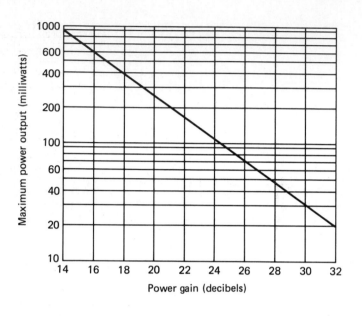

Graph B

Use Graph B to find answers to Exercises 6-10:

6. What is the highest power output? What is the power gain when this happens?
7. What is the power gain when the power output is 600 mW?
8. What is the power output when the power gain is 25 db?
9. Between what 2-db readings is the greatest change in power output found?
10. Between what 2-db readings is the least change in power output found?

We can see that each type of graph has advantages. The least and greatest changes are easier to read from the curved-line graph, but it is easier to read specific values from the straight-line graph.

14.5 Mean Measurement

We have already seen in other chapters that with each technical measurement, a certain amount of error is made. One way that a technician can offset this error is to use what is called the *mean measurement* or the *mean average* of the measurements. To find the mean measurement, the technician takes several measurements. The mean measurement is then found by taking the sum of these measurements and dividing it by the number of measurements taken. That is,

$$\text{mean measurement} = \frac{\text{sum of the measurements}}{\text{number of measurements}}$$

Example. A machinist measured the thickness of a metal disc with a micrometer at four different places. He found the following values:

2.147 in., 2.143 in., 2.151 in., 2.148 in.

Find the mean measurement.

Step 1: Add the measurements.

2.147 in.
2.143 in.
2.151 in.
2.148 in.
8.589 in.

Step 2: Divide the sum of the measurements by the number of measurements.

That is,

$$\text{mean measurement} = \frac{\text{sum of measurements}}{\text{number of measurements}}$$

$$= \frac{8.589 \text{ in.}}{4}$$

$$= 2.14725 \text{ in.}$$

So, the mean measurement is 2.147 in.

Note that the mean measurement is written so that it has the same precision as each of the measurements.

EXERCISES 14.5   *Find the mean measurement for each of the following sets of measurements:*

1. 47.61 cm, 48.23 cm, 47.92 cm, 47.81 cm

2. 9234 m, 9228 m, 9237 m, 9235 m, 9231 m

3. 0.2617 in., 0.2614 in., 0.2624 in., 0.2620 in., 0.2619 in., 0.2617 in.

4. 6.643 mm, 6.644 mm, 6.647 mm, 6.645 mm, 6.650 mm

5. 25,740 mi, 25,780 mi, 25,790 mi, 25,810 mi, 25,720 mi, 25,760 mi

6. 3414 mg, 3433 mg, 3431 mg, 3419 mg, 3441 mg, 3417 mg, 3427 mg, 3434 mg, 3435 mg, 3432 mg

7. 2018 km, 210$\overline{0}$ km, 2005 km, 2025 km, 203$\overline{0}$ km

8. 69°, 81°, 74°, 83°, 67°, 71°, 75°, 63°

9. 728 lb, 475 lb, 803 lb, 915 lb, 1002 lb, 256 lb, 781 lb

10. 108 kW, 21$\overline{0}$ kW, 175 kW, 16$\overline{0}$ kW, 19$\overline{0}$ kW, 12$\overline{0}$ kW

11. 6091; 505$\overline{0}$; 7102; 4111; 606$\overline{0}$; 591$\overline{0}$; 7112; 5855; 628$\overline{0}$; 10,171; 902$\overline{0}$; 10,172

12. 2.7; 8.1; 9.3; 7.2; 10.6; 11.4; 12.9; 13.5; 16.1; 10.9; 12.7; 15.9; 20.7; 21.9; 30.6; 42.9

14.6 Other Average Measurements and Percentiles

There are other procedures to determine an average measurement beside finding the mean measurement. The *median measurement* is the value that falls in the middle of a group of measurements that are arranged in order of size. Notice that one-half of the measurements will be larger than or equal to the median and one-half of the measurements will be less than or equal to the median.

Example 1. Find the median of the following set of measurements.

2.151 mm, 2.148 mm, 2.146 mm, 2.143 mm, 2.149 mm

Step 1: Arrange the measurements in order of size.

2.151 mm
2.149 mm
2.148 mm
2.146 mm
2.143 mm

Step 2: Find the middle measurement.

Since there are five measurements, the third measurement, 2.148 mm, is the median.

In the above example, there was an odd number of measurements. When there is an even number of measurements, there is no one middle measurement. In this case, the

median is found by taking the mean average of the two
middle measurements.

Example 2. Find the median measurement of the following
set of measurements.

$$54°, 57°, 59°, 55°, 53°, 57°, 50°, 56°$$

Step 1: Arrange the measurements in order of size.

$$59°$$
$$57°$$
$$57°$$
$$56°$$
$$55°$$
$$54°$$
$$53°$$
$$50°$$

Step 2: Since there are eight measurements, find the
mean of the two middle measurements.

$$\frac{55° + 56°}{2} = \frac{111°}{2} = 55.5°$$

So, the median measurement is 55.5°.

Another kind of average often used is the mode. The *mode*
is the measurement that appears most often. In Example 2
above, 57° is the mode. However, the mode can present problems. There can be more than one mode, and the mode may or
may not be near the "middle."

Example 3. Find the mode of the following set of measurements:

3.8 cm, 3.2 cm, 3.7 cm, 3.5 cm, 3.8 cm, 3.9 cm,
3.5 cm, 3.1 cm

The measurements 3.5 cm and 3.8 cm are both modes because each appears most often—twice.

Related to averages is the idea of measuring the position of a piece of data relative to the rest of the data.
Percentiles are numbers that divide the data into 100 equal
parts. The *nth percentile* is the number P_n such that
n percent of the data (ranked from smallest to largest)
is smaller than P_n. For example, if you score in the 64th
percentile on some standardized test, this means that you
scored higher than 64% of those who took the test and you
scored lower than 36%.

Example 4. The table below gives ranked data (from smallest to largest). Note that there are 50 pieces of data.

16	49	82	121	147
19	50	88	125	148
23	51	89	126	150
27	52	99	129	155
31	57	101	130	156
32	64	103	131	161
32	71	104	138	163
39	72	107	142	169
43	78	118	143	172
47	79	120	145	179

Ranked Data

(a) Find the 98th percentile.
(b) Find the 75th percentile.
(c) Find the 26th percentile.

Solution:

(a) The 98th percentile is 172 (the 49th piece of data: $0.98 \times 50 = 49$). 98 percent of the data is smaller in value than 172.
(b) The 75th percentile is 142 (the 38th piece of data: $0.75 \times 50 = 37.5$ or 38). 75 percent of the data is smaller in value than 142.
(c) The 26th percentile is 51 (the 13th piece of data: $0.26 \times 50 = 13$). 26 percent of the data is smaller in value than 51.

EXERCISES 14.6

1-12. Find the median measurement for each set of measurements in Exercises 1-12 on pages 535—536 (Sec. 14.5).

Find the following percentiles for the data listed in the table with Example 4:

13. 94th percentile
14. 80th percentile
15. 55th percentile
16. 12th percentile
17. 5th percentile
18. 50th percentile

14.7   Grouped Data

Finding the mean average for a large number of measurements can take a great deal of time and can be subject to mistakes. Grouping the measurements (the data) can make the work in finding the mean much easier.

Grouping data means arranging the data in groups. The groups are determined by setting up intervals. An *interval* is made up of all numbers between two given numbers a and b. We will show such an interval here by $a - b$. For example, 2 - 8 stands for all numbers between 2 and 8.

The number a is called the *lower limit* and b is called the *upper limit* of the interval. The number midway between a and b, $\dfrac{a + b}{2}$, is called the *midpoint* of the interval. In the above example, the lower limit is 2, the upper limit is 8, and the midpoint is $\dfrac{2 + 8}{2} = 5$.

While there are no given rules for choosing these intervals, the following general rules are helpful.

1. The number of group intervals chosen should be between six and twenty.
2. The length of each interval should be the same and should always be an odd number.
3. The midpoint of each interval should have the same number of digits as the raw data. This will mean that the lower limit and the upper limit of each interval will have an extra digit.
4. The lower limit of the first interval should be chosen to include the lowest measurement value.
5. No actual measurement will then fall on these limits so that it will be clear to which interval each measurement belongs.

Once the intervals have been chosen, form a frequency distribution. A *frequency distribution* lists each interval, its midpoint, and the number of measurements that lie in that interval (frequency).

<u>Example 1.</u> Make a frequency distribution for the recorded high temperatures for the days from November 1 to January 31 as given below:

November	High temperature (°F)	December	High temperature (°F)	January	High temperature (°F)
1	42	1	20	1	29
2	45	2	27	2	29
3	36	3	32	3	30
4	41	4	45	4	26
5	29	5	26	5	2
6	40	6	24	6	45
7	29	7	28	7	41
8	18	8	45	8	12
9	45	9	13	9	31
10	49	10	32	10	26
11	30	11	41	11	25
12	38	12	49	12	15
13	20	13	32	13	52
14	41	14	23	14	42
15	26	15	46	15	22
16	15	16	31	16	30
17	46	17	12	17	19
18	50	18	31	18	19
19	31	19	40	19	19
20	36	20	9	20	55
21	31	21	42	21	23
22	38	22	40	22	17
23	22	23	15	23	26
24	29	24	24	24	12
25	39	25	28	25	16
26	52	26	27	26	21
27	29	27	29	27	39
28	25	28	8	28	20
29	30	29	36	29	23
30	36	30	45	30	22
		31	12	31	9

First, choose the number and size of group intervals to be used. We must have enough group intervals to cover the range of the data (the difference between the highest and the lowest values). Here, the range is 55° – 2° = 53°. Since 53 is close to 54, let us choose the odd number 9 as the interval length. This means that we will need 54 ÷ 9 = 6 group intervals. This satisfies our general rule for the number of intervals.

We will choose 6 as our first midpoint, with 1.5 – 10.5 as our first interval. Here, 1.5 is the lower limit and 10.5 is the upper limit of the interval. We then make the frequency distribution as follows:

Temperature °F	Midpoint x	Tally	Frequency f
1.5 – 10.5	6	////	4
10.5 – 19.5	15	𝓣𝓗𝓛 𝓣𝓗𝓛 ////	14
19.5 – 28.5	24	𝓣𝓗𝓛 𝓣𝓗𝓛 𝓣𝓗𝓛 𝓣𝓗𝓛 ///	23
28.5 – 37.5	33	𝓣𝓗𝓛 𝓣𝓗𝓛 𝓣𝓗𝓛 𝓣𝓗𝓛 ///	23
37.5 – 46.5	42	𝓣𝓗𝓛 𝓣𝓗𝓛 𝓣𝓗𝓛 𝓣𝓗𝓛 //	22
46.5 – 55.5	51	𝓣𝓗𝓛 /	6
			92

To find the mean from the frequency distribution: (a) multiply the frequency of each interval by the midpoint of that interval, xf; (b) add the products, xf; and (c) divide by the number of data, sum of f.

$$\text{mean} = \frac{\text{sum of } xf}{\text{sum of } f}$$

The following frequency-distribution table gives the information for finding the mean of Example 1:

Temperature °F	Midpoint x	Frequency f	Product xf
1.5 – 10.5	6	4	24
10.5 – 19.5	15	14	210
19.5 – 28.5	24	23	552
28.5 – 37.5	33	23	759
37.5 – 46.5	42	22	924
46.5 – 55.5	51	6	306
		92	2775

The mean temperature is found as follows:

$$\text{mean} = \frac{\text{sum of } xf}{\text{sum of } f}$$

$$= \frac{2775}{92} = 30.2°F$$

Note that if the mean of the data in this example were found by summing the actual temperatures and dividing by the number of temperatures, the value of the mean would be

29.86°F, or 29.9°F. There is a small difference between the two calculated means. This is because we are using the midpoints of the intervals rather than the actual data. However, how easy the means are to find and how many fewer mistakes are made make the small difference in values acceptable.

Let us again find the mean of the data given in Example 1, this time using an interval length of 5. Again, the range of the data is 53, which is close to 55, a number divisible by 5. Since 55 ÷ 5 = 11, we will use 11 intervals, each of length 5. We will now make a frequency distribution using 4 as the first midpoint, with 1.5 - 6.5 as the first interval. The frequency distribution then becomes:

Temperature °F	Midpoint x	Frequency f	Product xf
1.5 - 6.5	4	1	4
6.5 - 11.5	9	3	27
11.5 - 16.5	14	9	126
16.5 - 21.5	19	9	171
21.5 - 26.5	24	15	360
26.5 - 31.5	29	20	580
31.5 - 36.5	34	7	238
36.5 - 41.5	39	11	429
41.5 - 46.5	44	11	484
46.5 - 51.5	49	3	147
51.5 - 56.5	54	3	162
		92	2728

Find the mean temperature:

$$\text{mean} = \frac{\text{sum of } xf}{\text{sum of } f}$$

$$= \frac{2728}{92}$$

$$= 29.7°F$$

EXERCISES 14.7   1. From the following grouped data, find the mean.

Interval	Midpoint x	Frequency f	Product xf
41.5 - 48.5		12	
48.5 - 55.5		15	
55.5 - 62.5		20	
62.5 - 69.5		25	
69.5 - 76.5		4	
76.5 - 83.5		2	

2. The following are scores from a mathematics test. Make a frequency distribution and use it to find the mean.

85, 73, 74, 69, 87, 81, 68, 76, 78, 75, 88, 85, 67, 83, 82, 95, 63, 84, 94, 66, 84, 78, 96, 67, 63, 59, 100, 90, 100, 94, 79, 79, 74

3. A laboratory technician records the life span in months of rats treated at birth with a fertility hormone. From the frequency distribution below, find the mean.

Months	Midpoint x	Frequency f	Product xf
-0.5 - 2.5		12	
2.5 - 5.5		18	
5.5 - 8.5		22	
8.5 - 11.5		30	
11.5 - 14.5		18	

4. The life expectancy of a certain fluorescent light bulb is given by the number of hours that it will burn. From the frequency distribution below, find the mean life of the bulb.

Hours	Midpoint x	Frequency f	Product xf
-0.5 - 499.5		2	
499.5 - 999.5		12	
999.5 - 1499.5		14	
1499.5 - 1999.5		17	
1999.5 - 2499.5		28	
2499.5 - 2999.5		33	
2999.5 - 3499.5		14	
3499.5 - 3999.5		5	

5. The average shipment time in hours for a load of goods from a factory to market is tabulated in the frequency distribution below. Find the mean.

Hours	Midpoint x	Frequency f	Product xf
22.5 - 27.5		2	
27.5 - 32.5		41	
32.5 - 37.5		79	
37.5 - 42.5		28	
42.5 - 47.5		15	
47.5 - 52.5		6	

6. The cost of goods stolen from a department store during the month of December has been tabulated by dollar amounts in the frequency distribution below. Find the mean.

Dollars	Midpoint x	Frequency f	Product xf
-0.5 - 24.5		2	
24.5 - 49.5		17	
49.5 - 99.5		25	
99.5 - 149.5		51	
149.5 - 199.5		38	
199.5 - 249.5		32	

7. The number of passengers and their luggage weight in pounds on flight 2102 have been tabulated in the frequency distribution below. Find the mean luggage weight.

Weight (lb)	Midpoint x	Frequency f	Product xf
0.5 - 10.5		1	
10.5 - 20.5		3	
20.5 - 30.5		22	
30.5 - 40.5		37	
40.5 - 50.5		56	
50.5 - 60.5		19	
60.5 - 70.5		17	
70.5 - 80.5		10	
80.5 - 90.5		5	
90.5 - 100.5		2	

8. The income of a neighborhood was tabulated and the results shown in the frequency distribution below. Find the mean income.

Salary	Midpoint x	Frequency f	Product xf
$ 500 - $ 5,500		1	
5,500 - 10,500		2	
10,500 - 15,500		15	
15,500 - 20,500		25	
20,500 - 25,500		8	

9. The number of defective parts per shipment has been tabulated in the frequency distribution below. Find the mean.

Defective parts	Midpoint x	Frequency f	Product xf
0.5 - 3.5		1	
3.5 - 6.5		7	
6.5 - 9.5		20	
9.5 - 12.5		9	
12.5 - 15.5		32	
15.5 - 18.5		3	

10. The following are the traffic fines collected in one day in a large city. Make a frequency distribution and use it to find the mean.

$30, $28, $15, $14, $32, $67, $45, $30, $17, $25, $30, $19, $27, $32, $51, $45, $36, $42, $72, $50, $18, $41, $23, $32, $35, $46, $50, $61, $82, $78, $39, $42, $27, $20

14.8   Variance and Standard Deviation

The mean measurement gives the technician an average value of a group of measurements, but the mean does not give any information about how the actual data vary in values. Some type of measurement that gives the amount of variation is often helpful in analyzing a set of data.

One way of describing the variation in the data is to find the range. The *range* is the difference between the highest value and the lowest value of the data.

    <u>Example 1</u>. Find the range of the following measurements:

$$54°, \ 57°, \ 59°, \ 55°, \ 53°, \ 57°, \ 50°, \ 56°$$

    The range is the difference between the highest value, 59°, and the lowest value, 50°. The range is 59° − 50° = 9°.

While the range gives us an idea of how much the data are spread out, another measure, the standard deviation, is often more helpful. The *standard deviation* tells how the data typically vary from the mean average.

To find the standard deviation, we first need to find what is called the variance. The *variance* is the mean average of the sum of the squares of the differences between each measurement (piece of data) and the mean average of the measurements.

$$\text{variance} = \frac{\text{sum of (measurement - mean)}^2}{\text{number of measurements}}$$

    The unit of measurement for the variance is actually the square of the original unit of measurement of the data. For this reason, the standard deviation is found. The standard deviation is the square root of the variance and so has the same unit of measurement as the original measurements.

$$\text{standard deviation} = \sqrt{\text{variance}}$$

Example 2. Find the standard deviation for the data given in Example 1.

Step 1: Find the mean.

$$\text{mean} = \frac{\text{sum of measurements}}{\text{number of measurements}}$$

$$= \frac{54° + 57° + 59° + 55° + 53° + 57° + 50° + 56°}{8}$$

$$= \frac{441°}{8} = 55.1°$$

Step 2: Find the difference between each piece of data and the mean:

$$54 - 55.1 = -1.1$$
$$57 - 55.1 = 1.9$$
$$59 - 55.1 = 3.9$$
$$55 - 55.1 = -0.1$$
$$53 - 55.1 = -2.1$$
$$57 - 55.1 = 1.9$$
$$50 - 55.1 = -5.1$$
$$56 - 55.1 = 0.9$$

Step 3: Square each difference and find the sum:

$$(-1.1)^2 = 1.21$$
$$(1.9)^2 = 3.61$$
$$(3.9)^2 = 15.21$$
$$(-0.1)^2 = 0.01$$
$$(-2.1)^2 = 4.41$$
$$(1.9)^2 = 3.61$$
$$(-5.1)^2 = 26.01$$
$$(0.9)^2 = \underline{0.81}$$
$$54.88$$

Step 4: Divide the sum in Step 3 by the number of measurements to find the variance:

$$\text{variance} = \frac{54.88}{8} = 6.86 \quad \text{or} \quad 6.9$$

Step 5: To find the standard deviation, find the square root of the variance from Step 4:

$$\text{standard deviation} = \sqrt{6.9}$$

$$= 2.6°$$

For grouped data, the variance and standard deviation are found in a way similar to the way the mean average is found. A frequency table is used. Columns are added to show the difference, D, between the midpoints and the mean ($D = x - $ mean); the square of D, D^2; and the frequency times D^2, D^2f. The following formulas give the variance and standard deviation for grouped data:

$$\text{variance} = \frac{\text{sum of } D^2f}{n}, \text{ where } n \text{ is the number of pieces of data}$$

$$\text{standard deviation} = \sqrt{\text{variance}}$$

<u>Example 3.</u> Given the following grouped data, find (a) the mean and (b) the standard deviation.

Interval (cm)	Frequency f
3.5 - 12.5	3
12.5 - 21.5	3
21.5 - 30.5	7
30.5 - 39.5	6
39.5 - 48.5	4
48.5 - 57.5	1

As noted above, we then add the following columns to the frequency distribution.

Interval (cm)	Midpoint x	Frequency f	Product xf	x − mean D	D^2	D^2f
3.5 - 12.5	8	3	24	−21	441	1323
12.5 - 21.5	17	3	51	−12	144	432
21.5 - 30.5	26	7	182	−3	9	63
30.5 - 39.5	35	6	210	6	36	216
39.5 - 48.5	44	4	176	15	225	900
48.5 - 57.5	53	1	53	24	576	576
		$n = 24$	696			3510

(a) The mean $= \dfrac{\text{sum of } xf}{n} = \dfrac{696}{24} = 29.0$ cm

(b) The variance $= \dfrac{\text{sum of } D^2 f}{n} = \dfrac{3510}{24} = 146.25$ cm^2

And the standard deviation $= \sqrt{\text{variance}} = \sqrt{146.25 \text{ cm}^2}$
$= 12.1$ cm. So, the average measurement is 29.0 cm, and the data tend to vary typically from the mean by 12.1 cm.

EXERCISES 14.8   1-12. Find the standard deviation for each set of measure-
                         ments in Exercises 1-12 on pages 535—536 (Sec. 14.5).
                   13-22. Find the standard deviation for each set of data in
                         Exercises 1-10 on pages 543—545 (Sec. 14.7).

Chapter 14 Review

1. Find 35% of 360°.   2. Find 56.1% of 360°.
3. Draw a circle graph using the following data. In 1983-84,
   30,780,000 students attended primary school; 13,495,000
   attended grades 9 to 12; and 12,400,000 attended
   college.
4. Draw a broken-line graph using the data in Exercise 3.
5. In the figure on page 532, what was the temperature
   at 10:00 p.m.?

For Exercises 6, 7, and 8, use the data below:

A technician, using a very precise tool, measured a piece
of metal to be used in a staellite. He recorded the following measurements: 7.0036 mm; 7.0035 mm; 7.0038 mm;
7.0035 mm; 7.0036 mm.

6. What is the mean measurement?
7. What is the median?
8. What is the standard deviation?
9. Given the frequency chart below, find (a) the mean;
   (b) the standard deviation.

Interval	Frequency f
10.5 – 21.5	4
21.5 – 32.5	17
32.5 – 43.5	10
43.5 – 54.5	28
54.5 – 65.5	13
65.5 – 76.5	12
76.5 – 87.5	9

APPENDIX A:
CALCULATORS

A.1 Using a Hand Calculator

Many brands and types of calculators are available. Some are simple and do only the basic operations of addition, subtraction, multiplication, and division. Others do a wide variety and a large number of operations. All calculators operate in one of two ways, using either *algebraic logic* or *Reverse Polish notation*. We show the most common operations for technical students using only algebraic logic because it follows the steps commonly found in mathematics. If your calculator uses Reverse Polish notation, consult your instruction manual.

To demonstrate how to use a calculator, we will (a) use a flowchart to give the directions for each step, (b) show what buttons are pushed and the order in which they are pushed, and (c) show the display at each step.

We have chosen to illustrate the most common types of algebraic logic calculators. Yours may differ in the number of digits displayed. If yours has a Function (F) button, consult your manual.

Note: We will always assume that your calculator and its memory are cleared before you begin any calculation.

A.2 Using a Calculator to Add and Subtract

Use a calculator to add as follows:

Example 1. Add: 9463
                     125
                        9
                      80

Flowchart	Buttons Pushed	Display
Enter 9463	9 → 4 → 6 → 3	9463
Push plus	+	9463
Enter 125	1 → 2 → 5	125
Push plus	+	9588
Enter 9	9	9
Push plus	+	9597
Enter 80	8 → 0	80
Push equals	=	9677

The sum is 9677.

Example 2. Add: $14.62 + $0.78 + $1.40 + $0.05

Flowchart	Buttons Pushed	Display
Enter 14.62	1 → 4 → . → 6 → 2	14.62
Push plus	+	14.62
Enter 0.78	. → 7 → 8	0.78
Push plus	+	15.4

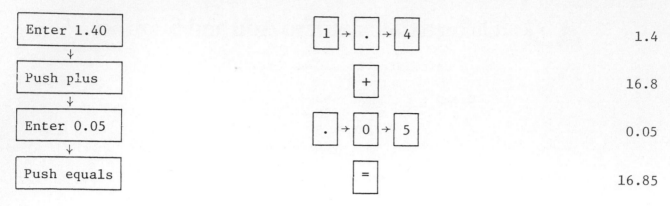

	Display
Enter 1.40 → Push plus → Enter 0.05 → Push equals	
$1 \rightarrow . \rightarrow 4$	1.4
$+$	16.8
$. \rightarrow 0 \rightarrow 5$	0.05
$=$	16.85

The sum is \$16.85.

Use a calculator to subtract as follows:

Example 3. Subtract: 3500
                       1628

Flowchart	Buttons Pushed	Display
Enter 3500	$3 \rightarrow 5 \rightarrow 0 \rightarrow 0$	3500
Push minus	$-$	3500
Enter 1628	$1 \rightarrow 6 \rightarrow 2 \rightarrow 8$	1628
Push equals	$=$	1872

The result is 1872.

Combinations of addition and subtraction may be found on a calculator as follows:

Example 4. Do as indicated: 74.6 − 8.57 + 5 − 0.0031

Flowchart	Buttons Pushed	Display
Enter 74.6	$7 \rightarrow 4 \rightarrow . \rightarrow 6$	74.6
Push minus	$-$	74.6
Enter 8.57	$8 \rightarrow . \rightarrow 5 \rightarrow 7$	8.57
Push plus	$+$	66.03

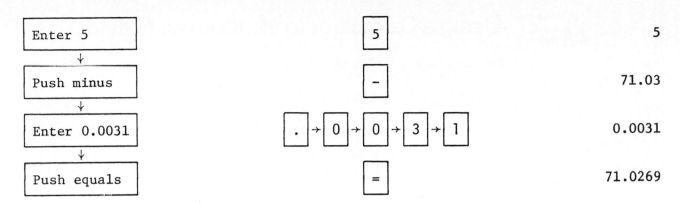

The result is 71.0269.

EXERCISES A.2   *Add*:

1. 967   2.   15   3. $38.65   4.   0.87   5. 8.9
   42   892 $92.77   63   17.2
   180   41 $10.05   5.72   152
   634   1063 $63.42   0.9   3.35
                        $10.50   120   0.016

6. 183 + 645 + 1821   7. 160 + 42.5 + 1.9

8. 59.63 + 1.875 + 4500   9. 8.76 + 0.163 + 0.0068

10. 82,000 + 45,000 + 275,000 + 118,000 + 865,000

Subtract:

11. 2562   12. 49.76   13. 24,500   14. 184.5
    1590   38.9   11,639   17.97

15. 63,000   16. 4825 – 1641   17. 1821 – 706
    21,608

18. 44 – 18.63   19. 21.5 – 0.053   20. 0.683 – 0.0045

Do as indicated:

21. 18 + 7 – 4 + 11 – 21   22. 18.5 – 4.2 + 1.7 – 6.2

23. 21 – 18.2 + 9.35 – 16.25

24. 15.372 – 3.073 – 4.63 + 16.2

25. 36 – 0.48 – 9.7 + 14.06

26. 8 – 1.605 + 18.75 – 6.23

27. 45.63 + 19.72 – 4.61 – 9.25

28. 19.301 – 4.9 + 46.973 – 40.085 – 18.1

A.3 Using a Calculator to Multiply and Divide

You may use a calculator to multiply as follows:

<u>Example 1</u>. Multiply: 125 × 68

Flowchart	Buttons Pushed	Display
Enter 125	1 → 2 → 5	125
↓		
Push times	×	125
↓		
Enter 68	6 → 8	68
↓		
Push equals	=	8500

The product is 8500.

<u>Example 2</u>. Multiply: 8.23 × 65 × 0.4

Flowchart	Buttons Pushed	Display
Enter 8.23	8 → . → 2 → 3	8.23
↓		
Push times	×	8.23
↓		
Enter 65	6 → 5	65
↓		
Push times	×	534.95
↓		
Enter 0.4	. → 4	0.4
↓		
Push equals	=	213.98

The product is 213.98.

To divide numbers using a calculator, follow the steps in the examples that follow.

Example 3. Divide $3.6 \div 8.64$ and round the quotient to three significant digits.

Flowchart	Buttons Pushed	Display
Enter 3.6	$3 \to \cdot \to 6$	3.6
↓		
Push divide	$\div$	3.6
↓		
Enter 8.64	$8 \to \cdot \to 6 \to 4$	8.64
↓		
Push equals	$=$	0.416667*

The quotient rounded to three significant digits is 0.417.

Example 4. Evaluate $\dfrac{(18.3)(46)}{(0.352)(240)}$ and round the result to three significant digits.

Flowchart	Buttons Pushed	Display
Enter 18.3	$1 \to 8 \to \cdot \to 3$	18.3
↓		
Push times	$\times$	18.3
↓		
Enter 46	$4 \to 6$	46
↓		
Push divide	$\div$	841.8
↓		
Enter 0.352	$\cdot \to 3 \to 5 \to 2$	0.352
↓		
Push divide	$\div$	2391.48
↓		
Enter 240	$2 \to 4 \to 0$	240
↓		
Push equals	$=$	9.96449

The result is 9.96 rounded to three significant digits.

*In order to allow for variation in the accuracy (number of significant digits displayed), we have consistently rounded to six (6) significant digits in each display, when necessary.

EXERCISES A.3   *Multiply:*

1. 423×618 2. 54.5×3600 3. 0.358×0.028

4. 374.2×180 5. 9.72×0.183 6. 250×580

7. $3.5 \times 6.8 \times 4.92$ 8. $0.65 \times 0.04 \times 280$

9. $9.3 \times 4.6 \times 0.23 \times 120$ 10. $180 \times 0.026 \times 0.42 \times 7.7$

Do as indicated and round each result to three significant digits:

11. $360 \div 4.8$ 12. $215 \div 13.5$ 13. $30.8 \div 0.942$

14. $63 \div 408$ 15. $0.18 \div 490$ 16. $72 \div 62.1$

17. $\dfrac{3.51 \times 6.32}{580 \times 4.2}$ 18. $\dfrac{0.197 \times 632}{4.12 \times 50.2}$

19. $\dfrac{31.3 \times 0.47 \times 6810}{0.193 \times 50.8}$ 20. $\dfrac{911 \times 6.4}{0.83 \times 2.75 \times 36}$

21. $\dfrac{939 \times 6.4 \times 8.7}{4.2 \times 1.91 \times 26.6}$ 22. $\dfrac{0.586 \times 610 \times 820}{3.5 \times 0.97 \times 4.6}$

23. $\dfrac{19.8}{41.2 \times 0.036 \times 15.3}$ 24. $\dfrac{5600}{180 \times 27.5 \times 31.5}$

A.4 Reciprocals, Squares, and Square Roots

The reciprocal of a number may be found by using the $1/x$ button.

> Example 1. Find the reciprocal of 12 rounded to three significant digits.

Flowchart	Buttons Pushed	Display
Enter 12		12
Push reciprocal	$1/x$	0.0833333

Thus $\dfrac{1}{12} = 0.0833$ rounded to three significant digits.

> Example 2. Find $\dfrac{1}{41.2}$ rounded to three significant digits.

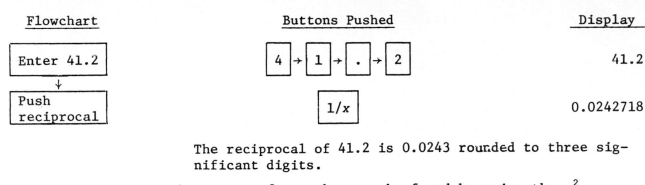

The reciprocal of 41.2 is 0.0243 rounded to three significant digits.

The square of a number may be found by using the x^2 button.

Example 3. Find 73.6^2 rounded to three significant digits.

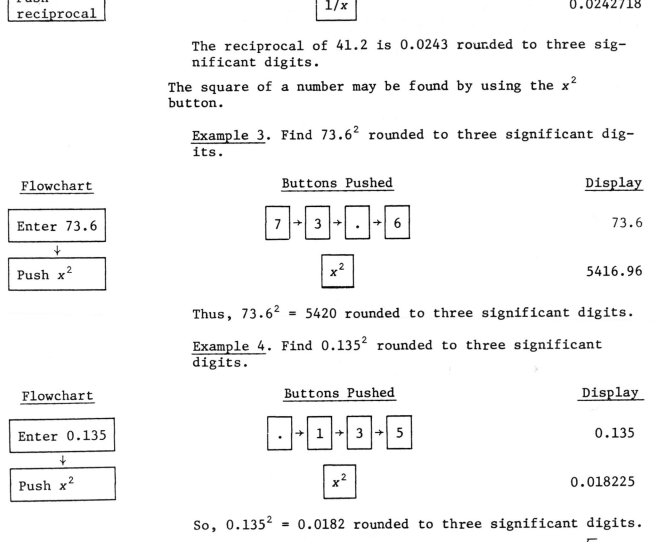

Thus, $73.6^2 = 5420$ rounded to three significant digits.

Example 4. Find 0.135^2 rounded to three significant digits.

So, $0.135^2 = 0.0182$ rounded to three significant digits.

The square root of a number may be found using the $\sqrt{}$ button.

Example 5. Find $\sqrt{21.4}$ rounded to three significant digits.

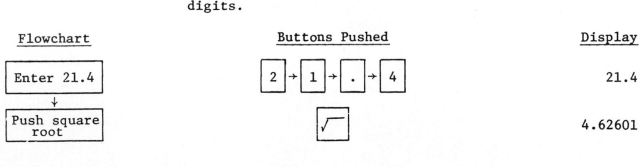

Then, $\sqrt{21.4} = 4.63$ rounded to three significant digits.

Example 6. Find $\sqrt{0.000594}$ rounded to three significant digits.

Flowchart	Buttons Pushed	Display

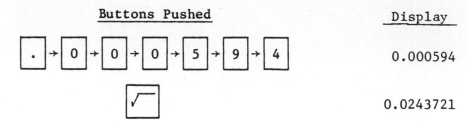

Enter 0.000594	$. \rightarrow 0 \rightarrow 0 \rightarrow 0 \rightarrow 5 \rightarrow 9 \rightarrow 4$	0.000594
Push square root	$\sqrt{}$	0.0243721

Thus, $\sqrt{0.000594} = 0.0244$ rounded to three significant digits.

EXERCISES A.4   *Do as indicated and round the result to three significant digits:*

1. $\dfrac{1}{8}$ 2. $\dfrac{1}{6}$ 3. $\dfrac{1}{27}$ 4. $\dfrac{1}{54}$

5. $\dfrac{1}{0.25}$ 6. $\dfrac{1}{0.375}$ 7. $\dfrac{1}{18.3}$ 8. $\dfrac{1}{240}$

9. $\dfrac{1}{0.361}$ 10. $\dfrac{1}{0.0257}$ 11. 18^2 12. 25^2

13. 31.2^2 14. 83.5^2 15. 120^2 16. 615^2

17. 0.086^2 18. 0.0217^2 19. 0.405^2 20. 0.061^2

21. $\sqrt{3136}$ 22. $\sqrt{256}$ 23. $\sqrt{38.9}$ 24. $\sqrt{60.7}$

25. $\sqrt{0.0412}$ 26. $\sqrt{0.00925}$ 27. $\sqrt{0.29}$ 28. $\sqrt{38,500}$

29. $\sqrt{197,000}$ 30. $\sqrt{0.0000515}$

A.5 Using a Calculator with Memory

For some combinations of operations, a calculator with a memory key or keys is very helpful. A memory stores numbers for future use. Operations can then be done with numbers and the number in the memory. <u>Note</u>: If your calculator has more than one memory, consult your instruction manual.

Example 1. Find the value of $\dfrac{14}{\sqrt{5}} - \sqrt{\dfrac{15}{7}}$ and round the
result to three significant digits.

Flowchart	Buttons Pushed	Display
Enter 14	$\boxed{1} \rightarrow \boxed{4}$	14
Push divide	$\boxed{\div}$	14
Enter 5	$\boxed{5}$	5
Push square root	$\boxed{\sqrt{}}$	2.23607
Push equals	$\boxed{=}$	6.26099
Push store or add to memory	$\boxed{\text{STO}}$ or $\boxed{\text{M+}}$	6.26099
Enter 15	$\boxed{1} \rightarrow \boxed{5}$	15
Push divide	$\boxed{\div}$	15
Enter 7	$\boxed{7}$	7
Push equals	$\boxed{=}$	2.14286
Push square root	$\boxed{\sqrt{}}$	1.46385
Push subtract from memory	$\boxed{\text{+/-}} \rightarrow \boxed{\text{SUM}}$ or $\boxed{\text{M-}}$	1.46385
Push memory recall	$\boxed{\text{RCL}}$ or $\boxed{\text{MR}}$	4.79714

The result is 4.80 rounded to three significant
digits.

Example 2. Find the value of

$$\sqrt{\left(\frac{16}{1.35}\right)^2 + \left(\frac{1}{2\pi(60)(4 \times 10^{-5})}\right)^2}$$

rounded to three significant digits.

Flowchart	Buttons Pushed	Display
Enter 16	$\boxed{1} \rightarrow \boxed{6}$	16
Push divide	$\boxed{\div}$	16
Enter 1.35	$\boxed{1} \rightarrow \boxed{.} \rightarrow \boxed{3} \rightarrow \boxed{5}$	1.35
Push equals	$\boxed{=}$	11.8519
Push square	$\boxed{x^2}$	140.466
Push store or add to memory	$\boxed{STO}$ or $\boxed{M+}$	140.466
Enter 2	$\boxed{2}$	2
Push times	$\boxed{\times}$	2
Enter π	$\boxed{\pi}$	3.14159
Push times	$\boxed{\times}$	6.28319
Enter 60	$\boxed{6} \rightarrow \boxed{0}$	60
Push times	$\boxed{\times}$	376.991
Enter 4×10^{-5}	$\boxed{4} \rightarrow \boxed{EE} \rightarrow \boxed{5} \rightarrow \boxed{+/-}$	4 −05
Push equals	$\boxed{=}$	1.508 −02

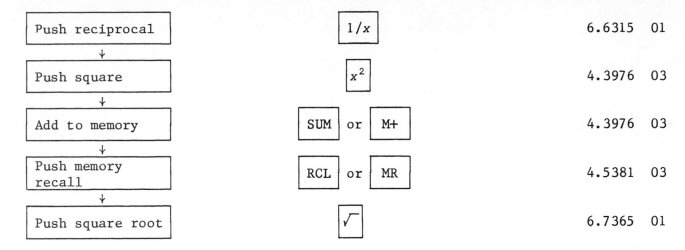

Push reciprocal	$1/x$	6.6315 01
Push square	x^2	4.3976 03
Add to memory	SUM or M+	4.3976 03
Push memory recall	RCL or MR	4.5381 03
Push square root	$\sqrt{}$	6.7365 01

The result is 67.4 rounded to three significant digits.

EXERCISES A.5   *Do as indicated and round each final result to three significant digits:*

1. $16\sqrt{3} + \left(\dfrac{18.1}{14}\right)^2$

2. $\dfrac{5}{\sqrt{2}} + \sqrt{\dfrac{48}{13.5}}$

3. $36^2 - \sqrt{\dfrac{19.8}{0.0246}}$

4. $\dfrac{18.3}{6\sqrt{2}} - \left(\dfrac{245}{167}\right)^2$

5. $\dfrac{49.2 + 6.8}{21.5 - 14.2}$

6. $\dfrac{29.2 - 14.2}{29.2 - 21.3}$

7. $\sqrt{\dfrac{16.8}{4.23}} + \left(\dfrac{1.45}{0.94}\right)^2$

8. $\dfrac{25.6^2}{5.2} + \left(\dfrac{41.9}{8.25}\right)^2$

9. $(13.5^2 + 18.6^2)^2 + (25.6^2 + 31.7^2)^2$

10. $(46.5^2 + 19.8^2)^2 - (44.6^2 - 23.7^2)^2$

11. $\sqrt{25.5^2 + 29.5^2} + \sqrt{36.5^2 - 27.3^2}$

12. $\sqrt{41.5^2 + 36.7^2} - \sqrt{51.7^2 - 48.7^2}$

13. $\sqrt{\dfrac{196}{3.91}} + \left(\dfrac{21.7}{0.361}\right)^2 - \dfrac{23.6}{\sqrt{2\pi}}$

14. $\dfrac{35.1}{(1.97)^2} + \left(\dfrac{25.1}{3.75}\right)^2 - \sqrt{\dfrac{14.2}{3.64}}$

15. $\sqrt{\left(\dfrac{35.6}{2.93}\right)^2 + \left(\dfrac{1}{2\pi(60)(8 \times 10^{-6})}\right)^2}$

16. $\sqrt{\left(\dfrac{855}{16.2}\right)^2 + \left(\dfrac{1}{2\pi(60)(9 \times 10^{-8})}\right)^2}$

17. $\sqrt{\left(\dfrac{90.5}{21.6}\right)^2 + \left(\dfrac{1}{2\pi(60)(1.8 \times 10^{-5})}\right)^2}$

18. $\sqrt{\left(\dfrac{185}{31.5}\right)^2 + \left(\dfrac{1}{2\pi(60)(6.5 \times 10^{-10})}\right)^2}$

19. $\dfrac{(19.2)(10.6) + (9)(15.6) - (6)(15)}{(5)(12) + (5.5)(15) + (10)(16.5)}$

20. $\dfrac{(18.6)(5.5) + (6.75)(18.5) - (9.25)(12.55)}{(5.25)(10.5) + (6.75)(18.5) + (10.5)(16.1)}$

A.6 The Fraction Key

Calculators are more commonly having a fraction key, which makes operations with fractions much easier.

The fraction key often looks similar to $\boxed{a\ ^{b}\!/_{c}}$.

Example 1. Reduce $\dfrac{108}{144}$ to lowest terms.

Flowchart	Buttons Pushed	Display
$\boxed{\text{Enter } \dfrac{108}{144}}$	$\boxed{1} \rightarrow \boxed{0} \rightarrow \boxed{8} \rightarrow \boxed{a\ ^{b}\!/_{c}} \rightarrow \boxed{1} \rightarrow \boxed{4} \rightarrow \boxed{4}$	108 ⌐ 144
$\downarrow$		
$\boxed{\text{Push equals}}$	$\boxed{=}$	3 ⌐ 4

Then, $\dfrac{108}{144} = \dfrac{3}{4}$ in lowest terms.

Example 2. Add $\dfrac{2}{3} + \dfrac{7}{12}$.

Flowchart	Buttons Pushed	Display
$\boxed{\text{Enter } \dfrac{2}{3}}$	$\boxed{2} \rightarrow \boxed{a\ ^{b}\!/_{c}} \rightarrow \boxed{3}$	2 ⌐ 3
$\downarrow$		
$\boxed{\text{Push plus}}$	$\boxed{+}$	2 ⌐ 3
$\downarrow$		
$\boxed{\text{Enter } \dfrac{7}{12}}$	$\boxed{7} \rightarrow \boxed{a\ ^{b}\!/_{c}} \rightarrow \boxed{1} \rightarrow \boxed{2}$	7 ⌐ 12
$\downarrow$		
$\boxed{\text{Push equals}}$	$\boxed{=}$	1 ⌐ 1 ⌐ 4

Thus, $\dfrac{2}{3} + \dfrac{7}{12} = 1\dfrac{1}{4}$.

Note. The symbol ⌐ separates the whole number, the numerator, and the denominator in the display.

Example 3. Subtract: $8\frac{1}{4} - 3\frac{5}{12}$.

Flowchart	Buttons Pushed	Display
Enter $8\frac{1}{4}$	$8 \rightarrow \boxed{a\ b/c} \rightarrow \boxed{1} \rightarrow \boxed{a\ b/c} \rightarrow \boxed{4}$	8 ⌐ 1 ⌐ 4
Push minus	$\boxed{-}$	8 ⌐ 1 ⌐ 4
Enter $3\frac{5}{12}$	$\boxed{3} \rightarrow \boxed{a\ b/c} \rightarrow \boxed{5} \rightarrow \boxed{a\ b/c} \rightarrow \boxed{1} \rightarrow \boxed{2}$	3 ⌐ 5 ⌐ 12
Push equals	$\boxed{=}$	4 ⌐ 5 ⌐ 6

So, $8\frac{1}{4} - 3\frac{5}{12} = 4\frac{5}{6}$.

Example 4. Divide: $5\frac{5}{7} \div 8\frac{1}{3}$.

Flowchart	Buttons Pushed	Display
Enter $5\frac{5}{7}$	$\boxed{5} \rightarrow \boxed{a\ b/c} \rightarrow \boxed{5} \rightarrow \boxed{a\ b/c} \rightarrow \boxed{7}$	5 ⌐ 5 ⌐ 7
Push divide	$\boxed{\div}$	5 ⌐ 5 7
Enter $8\frac{1}{3}$	$\boxed{8} \rightarrow \boxed{a\ b/c} \rightarrow \boxed{1} \rightarrow \boxed{a\ b/c} \rightarrow \boxed{3}$	8 ⌐ 1 ⌐ 3
Push equals	$\boxed{=}$	24 ⌐ 35

Thus, $5\frac{5}{7} \div 8\frac{1}{3} = \frac{24}{35}$.

For practice using the fraction key, do Exercises 2.1, 2.2, and 2.4 using a calculator.

APPENDIX B:
TABLES

Table 1. English Weights and Measures

LENGTH

Standard unit: inch (in. or ")

$$12 \text{ inches} = 1 \text{ foot (ft or ')}$$
$$3 \text{ feet} = 1 \text{ yard (yd)}$$
$$5\frac{1}{2} \text{ yards or } 16\frac{1}{2} \text{ feet} = 1 \text{ rod (rd)}$$
$$5280 \text{ feet} = 1 \text{ mile (mi)}$$

WEIGHT

Standard unit: pound (lb)

$$16 \text{ ounces (oz)} = 1 \text{ pound}$$
$$2000 \text{ pounds} = 1 \text{ ton}$$

VOLUME

Liquid

$$16 \text{ fluid ounces (fl oz)} = 1 \text{ pint (pt)}$$
$$2 \text{ pints} = 1 \text{ quart (qt)}$$
$$4 \text{ quarts} = 1 \text{ gallon (gal)}$$

Dry

$$2 \text{ pints (pt)} = 1 \text{ quart (qt)}$$
$$8 \text{ quarts} = 1 \text{ peck (pk)}$$
$$4 \text{ pecks} = 1 \text{ bushel (bu)}$$

Table 2. Metric System Prefixes

Multiple or submultiple* decimal form	Power of 10	Prefix**	Prefix symbol	Pronun- ciation	Meaning
1,000,000,000,000	10^{12}	tera	T	tĕr´ă	one trillion times
1,000,000,000	10^{9}	giga	G	jĭg´ă	one billion times
1,000,000	10^{6}	mega	M	mĕg´ă	one million times
1,000	10^{3}	kilo	k	kĭl´ō	one thousand times
100	10^{2}	hecto	h	hĕk´tō	one hundred times
10	10^{1}	deka	da	dĕk´ă	ten times
0.1	10^{-1}	deci	d	dĕs´ĭ	one tenth of
0.01	10^{-2}	centi	c	sĕnt´ĭ	one hundredth of
0.001	10^{-3}	milli	m	mĭl´ĭ	one thousandth of
0.000001	10^{-6}	micro	μ	mī´krō	one millionth of
0.000000001	10^{-9}	nano	n	nă n´ō	one billionth of
0.000000000001	10^{-12}	pico	p	pē´kō	one trillionth of

*Factor by which the unit is multiplied.
**The same prefixes are used with all SI metric units.

As an example, the prefixes are used below with the metric standard unit of length, metre (m).

1 *tera*metre (Tm)	= 1,000,000,000,000 m	1 m = 0.000000000001 Tm
1 *giga*metre (Gm)	= 1,000,000,000 m	1 m = 0.000000001 Gm
1 *mega*metre (Mm)	= 1,000,000 m	1 m = 0.000001 Mm
1 *kilo*metre (km)	= 1,000 m	1 m = 0.001 km
1 *hecto*metre (hm)	= 100 m	1 m = 0.01 hm
1 *deka*metre (dam)	= 10 m	1 m = 0.1 dam
1 *deci*metre (dm)	= 0.1 m	1 m = 10 dm
1 *centi*metre (cm)	= 0.01 m	1 m = 100 cm
1 *milli*metre (mm)	= 0.001 m	1 m = 1,000 mm
1 *micro*metre (μm)	= 0.000001 m	1 m = 1,000,000 μm
1 *nano*metre (nm)	= 0.000000001 m	1 m = 1,000,000,000 nm
1 *pico*metre (pm)	= 0.000000000001 m	1 m = 1,000,000,000,000 pm

Table 3. Metric and English Conversion

LENGTH

	English	*Metric*

```
    1 inch (in.) = 2.54  cm
     1 foot (ft) = 30.5  cm
     1 yard (yd) = 91.4  cm
     1 mile (mi) = 1610  m
            1 mi = 1.61  km
      0.0394 in. = 1 mm
       0.394 in. = 1 cm
        39.4 in. = 1 m
         3.28 ft = 1 m
         1.09 yd = 1 m
        0.621 mi = 1 km
```

WEIGHT

	English	*Metric*

```
     1 ounce (oz) = 28.3 g
     1 pound (lb) = 454 g
             1 lb = 0.454 kg
       0.0353 oz = 1 g
      0.00220 lb = 1 g
         2.20 lb = 1 kg
```

CAPACITY

	English	*Metric*

```
    1 gallon (gal) = 3.79 L
      1 quart (qt) = 0.946 L
        0.264 gal = 1 L
         1.06 qt = 1 L
```

1 metric ton = 1000 kg

Table 3.   (Continued)

AREA

<table>
<tr><td colspan="2">English</td><td colspan="2">Metric</td></tr>
</table>

English:

$1 \text{ ft}^2 = 144 \text{ in}^2$

$1 \text{ yd}^2 = 9 \text{ ft}^2$

$1 \text{ rd}^2 = 30.25 \text{ yd}^2$

$1 \text{ acre} = 160 \text{ rd}^2$

$\qquad = 4840 \text{ yd}^2$

$\qquad = 43,560 \text{ ft}^2$

$1 \text{ mi}^2 = 640 \text{ acres}$

$\qquad = 1 \text{ section}$

Metric:

$1 \text{ m}^2 = 10,000 \text{ cm}^2 \text{ or } 10^4 \text{ cm}^2$

$\qquad = 1,000,000 \text{ mm}^2 \text{ or } 10^6 \text{ mm}^2$

$1 \text{ cm}^2 = 100 \text{ mm}^2$

$\qquad = 0.0001 \text{ m}^2$

$1 \text{ km}^2 = 1,000,000 \text{ m}^2$

$1 \text{ hectare (ha)} = 10,000 \text{ m}^2$

$\qquad = 1 \text{ hm}^2$

English-Metric

$1 \text{ in}^2 = 6.45 \text{ cm}^2$

$\qquad = 645 \text{ mm}^2$

$1 \text{ ft}^2 = 929 \text{ cm}^2$

$\qquad = 0.0929 \text{ m}^2$

$1 \text{ yd}^2 = 8361 \text{ cm}^2$

$\qquad = 0.8361 \text{ m}^2$

$1 \text{ rd}^2 = 25.3 \text{ m}^2$

$1 \text{ acre} = 4047 \text{ m}^2$

$\qquad = 0.004047 \text{ km}^2$

$\qquad = 0.4047 \text{ ha}$

$1 \text{ mi}^2 = 2.59 \text{ km}^2$

Metric-English

$1 \text{ m}^2 = 10.76 \text{ ft}^2$

$\qquad = 1550 \text{ in}^2$

$\qquad = 0.0395 \text{ rd}^2$

$\qquad = 1.196 \text{ yd}^2$

$1 \text{ cm}^2 = 0.155 \text{ in}^2$

$1 \text{ km}^2 = 247 \text{ acres}$

$\qquad = 1.08 \times 10^7 \text{ ft}^2$

$\qquad = 0.386 \text{ mi}^2$

$1 \text{ ha} = 2.47 \text{ acres}$

VOLUME

English

$1 \text{ ft}^3 = 1728 \text{ in}^3$

$1 \text{ yd}^3 = 27 \text{ ft}^3$

Metric

$1 \text{ m}^3 = 10^6 \text{ cm}^3$

$1 \text{ cm}^3 = 10^{-6} \text{ m}^3$

$\qquad = 10^3 \text{ mm}^3$

English-Metric

$1 \text{ in}^3 = 16.39 \text{ cm}^3$

$1 \text{ ft}^3 = 28,317 \text{ cm}^3$

$\qquad = 0.028317 \text{ m}^3$

$1 \text{ yd}^3 = 0.7646 \text{ m}^3$

Metric-English

$1 \text{ cm}^3 = 0.06102 \text{ in}^3$

$1 \text{ m}^3 = 35.3 \text{ ft}^3$

$\qquad = 1.31 \text{ yd}^3$

Table 4. Formulas from Geometry

PLANE FIGURES

In the following figures, a, b, c, and d are lengths of sides and h, altitude.

		Perimeter	Area

Rectangle 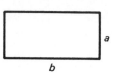 $P = 2(a + b)$ $A = ab$

Square $P = 4b$ $A = b^2$

Parallelogram 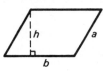 $P = 2(a + b)$ $A = bh$

Rhombus $P = 4b$ $A = bh$

Trapezoid 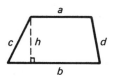 $P = a + b + c + d$ $A = \left(\dfrac{a + b}{2}\right)h$

Triangle 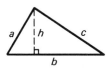 $P = a + b + c$ $A = \dfrac{1}{2}bh$, or

$$A = \sqrt{s(s - a)(s - b)(s - c)},$$

$$\text{where } s = \frac{1}{2}(a + b + c).$$

The sum of the measures of the angles of a triangle $= 180°$.

In a right triangle:

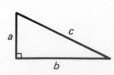

$$c^2 = a^2 + b^2 \text{ or } c = \sqrt{a^2 + b^2}.$$

Table 4. (Continued)

	Circumference	Area

Circle

$$C = 2\pi r$$
$$C = \pi d$$
$$(d = 2r)$$

$$A = \pi r^2$$

The sum of the measures of the central angles of a circle = 360°.

GEOMETRIC SOLIDS

In the following figures, B, r, and h are the area of base, length of radius, and height, respectively.

	Volume	Lateral surface area

Prism

$$V = Bh$$

Cylinder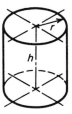

$$V = \pi r^2 h$$

$$A = 2\pi rh$$

Pyramid

$$V = \frac{1}{3}Bh$$

Cone

$$V = \frac{1}{3}\pi r^2 h$$

$A = \pi rs$, where s is the slant height.

Sphere

$$V = \frac{4}{3}\pi r^3$$

$$A = 4\pi r^2$$

Table 5. Electrical Symbols

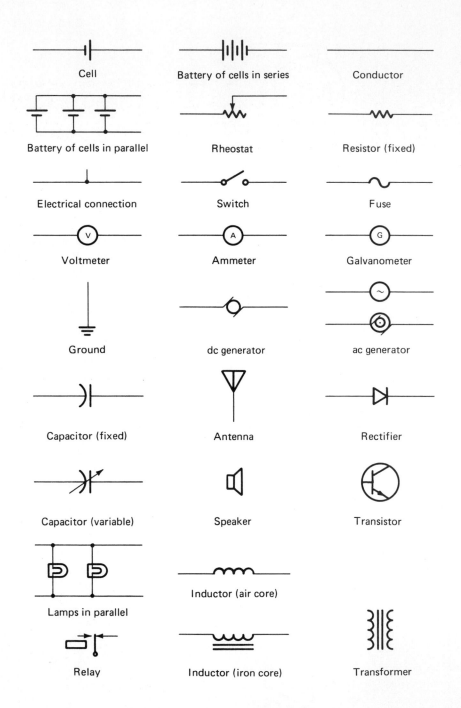

Cell	Battery of cells in series	Conductor
Battery of cells in parallel	Rheostat	Resistor (fixed)
Electrical connection	Switch	Fuse
Voltmeter	Ammeter	Galvanometer
Ground	dc generator	ac generator
Capacitor (fixed)	Antenna	Rectifier
Capacitor (variable)	Speaker	Transistor
Lamps in parallel	Inductor (air core)	
Relay	Inductor (iron core)	Transformer

ANSWERS TO ODD-NUMBERED EXERCISES AND CHAPTER REVIEWS

Chapter 1

1.1 Pages 5—8

1. 3255	3. 2466	5. 1298
7. 1454	9. 579	11. 1925
13. 1162	15. 97,596	17. 131
19. 5164	21. 144,973	23. 1241
25. 26,008	27. 18,036	29. 2119
31. 2820	33. 198 Ω	35. 4195 Ω
37. 224	39. $2217; $4891; $7108	
41. 225 mg	43. I : 1925 cm^3 ; O : 1425 cm^3	

1.2 Pages 10—13

1. 1792	3. 27,216	5. 818,802
7. 18,172,065		9. 26,527,800
11. 35,360,000		13. 1809
15. 576	17. 1125 r4	19. 389
21. 1980 r2	23. 844 r40	25. 528 mi
27. 3400 cm^3	29. 325 cm^3	31. 8 km/L
33. 4820 ft	35. 1554 lb	
37. 67 in. from either corner		39. 44 bu/acre
41. 220	43. (a) 881 lb	(b) 15 lb
45. 85 bu/acre		47. $4200
49. 5 A	51. 3 A	53. 24 V
55. 120 V	57. 880 oz	59. 4
61. 4		

1.3 Pages 14—15

1. 2	3. 10	5. 18	7. 131
9. 73	11. 23	13. 16	15. 34
17. 102	19. 137	21. 230	23. 55
25. 0	27. 4	29. 0	31. 13
33. 19			

1.4 Pages 16—17

1. 96 yd^2	3. 84 cm^2	5. 48 in^2
7. 52 in^2	9. 108	11. 32 gal
13. (a) $41,040	(b) $89,072	

1.5 Pages 20—21

1. 96 m^3	3. 720 cm^3	5. 3900 in^3
7. 600 cm^3	9. 69,480 lb	

1.6 Page 22

1. 103 r1	3. 28 r4	5. 1784
7. 9210	9. 5081	11. 106,136 r1
13. Yes	15. Yes	17. Yes
19. No	21. No	23. Yes
25. No	27. Yes	

1.7 Pages 25—26

1. Prime	3. Not prime
5. Not prime	7. Not prime
9. Divisible by 2	11. Not divisible by 2
13. Not divisible by 2	15. Divisible by 3
17. Not divisible by 3	19. Divisible by 3
21. Divisible by 5	23. Not divisible by 5
25. Not divisible by 5	27. Yes 29. No
31. No 33. Yes	35. Yes 37. Yes
39. $2 \cdot 2 \cdot 5$	41. $2 \cdot 3 \cdot 11$
43. $2 \cdot 2 \cdot 3 \cdot 3$	45. $3 \cdot 3 \cdot 3$ 47. $3 \cdot 17$
49. $2 \cdot 3 \cdot 7$	51. $2 \cdot 2 \cdot 2 \cdot 3 \cdot 5$
53. $3 \cdot 3 \cdot 19$	55. $3 \cdot 5 \cdot 7$
57. $2 \cdot 2 \cdot 3 \cdot 3 \cdot 7$	59. $2 \cdot 3 \cdot 3 \cdot 3 \cdot 5$

1.8 Pages 29—31

1. 3	3. 6	5. 4	7. 17
9. 15	11. 10	13. 7	15. −2
17. −12	19. 6	21. 12	23. 4
25. −14	27. −5	29. −12	31. −9
33. 4	35. −3	37. −1	39. −4
41. −6	43. −2	45. 4	47. 9
49. 3	51. 4	53. 7	55. −7
57. 5	59. −20	61. −2	63. 0
65. −5	67. −19	69. −12	71. 7
73. 11	75. 4	77. 4	79. −14

1.9 Pages 32—33

1. −2	3. 11	5. 12	7. 6
9. 18	11. −6	13. −4	15. 0
17. 35	19. 14	21. 1	23. −10
25. −7	27. 15	29. −16	31. −10
33. −2	35. 14	37. 23	39. −15
41. 8	43. −4	45. 8	47. 2
49. −15	51. −3	53. −23	55. −4
57. −1	59. 2	61. −8	63. −10

1.10 Page 35

1. 24	3. −18	5. −35	7. 27
9. −72	11. 27	13. 0	15. 49
17. −300	19. −13	21. −6	23. 48
25. 21	27. −16	29. 54	31. −6
33. 24	35. 27	37. −63	39. −9
41. −6	43. 36	45. 30	47. −168
49. −162	51. 5	53. −9	55. 8
57. 2	59. 9	61. −4	63. 3
65. 17	67. 40	69. 15	71. 7
73. 4	75. −3	77. 10	79. 8

1.11 Pages 36—37

1. 600	3. 11,250	5. 57,376	7. 38,400
9. 8	11. 5017	13. 56 m^2	15. 540 ft^2
17. 131	19. 16		

Chapter 1 Review, Pages 37—39

1. 8243	2. 55,197	3. 9,178,000	
4. 226 r240		5. 3	6. 43
7. 37	8. 31	9. 340 cm^2	10. 30 cm^3
11. 54 r4	12. No	13. 2 · 3 · 3 · 3	
14. 2 · 3 · 5 · 11		15. 5	16. 16
17. 13	18. 3	19. −8	20. −3
21. −13	22. −3	23. −11	24. 19
25. −2	26. 0	27. −19	28. −24
29. 36	30. 72	31. −84	32. 6
33. −6	34. 5	35. 3000	

Chapter 2

2.1 Page 46

1. $\frac{3}{7}$	3. $\frac{6}{7}$	5. $\frac{3}{16}$	7. $\frac{1}{3}$
9. $\frac{4}{5}$	11. 1	13. 0	15. Undefined
17. $\frac{7}{8}$	19. $\frac{3}{4}$	21. $\frac{3}{4}$	23. $\frac{4}{5}$
25. $\frac{3}{10}$	27. $\frac{2}{3}$	29. $\frac{7}{8}$	31. $\frac{6}{7}$
33. $\frac{3}{5}$	35. $\frac{19}{37}$	37. $\frac{2}{3}$	39. $\frac{7}{9}$
41. $15\frac{3}{5}$	43. $9\frac{1}{3}$	45. $1\frac{1}{4}$	47. $9\frac{1}{2}$
49. $20\frac{4}{5}$	51. $24\frac{5}{9}$	53. $6\frac{1}{4}$	55. $\frac{23}{6}$
57. $\frac{17}{8}$	59. $\frac{23}{16}$	61. $\frac{55}{8}$	63. $\frac{53}{5}$

2.2 Pages 52—53

1. $\frac{11}{16}$	3. $\frac{15}{32}$	5. 16	7. 210
9. 48	11. 210	13. 78	15. $\frac{5}{6}$
17. $\frac{5}{32}$	19. $\frac{11}{28}$	21. $\frac{29}{64}$	23. $\frac{7}{20}$
25. $\frac{13}{10}$ or $1\frac{3}{10}$		27. $\frac{13}{10}$ or $1\frac{3}{10}$	
29. $\frac{19}{16}$ or $1\frac{3}{16}$		31. $\frac{31}{32}$	33. $\frac{3}{23}$
35. $\frac{37}{48}$	37. $\frac{13}{120}$	39. $\frac{67}{105}$	
41. $\frac{43}{27}$ or $1\frac{16}{27}$		43. $1\frac{51}{242}$	45. $\frac{1}{8}$
47. $\frac{1}{8}$	49. $\frac{1}{2}$	51. $\frac{4}{7}$	53. $\frac{1}{32}$
55. $7\frac{1}{4}$	57. $10\frac{1}{16}$	59. $2\frac{5}{8}$	61. $4\frac{3}{4}$
63. $2\frac{5}{16}$	65. $13\frac{31}{45}$	67. $\frac{79}{84}$	69. $11\frac{5}{16}$
71. $9\frac{13}{24}$			

2.3 Pages 55—60

1. $2237\frac{1}{4}$ ft   3. (a) $1\frac{1}{2}$ in.   (b) $22\frac{23}{64}$ in.

5. (a) $\frac{1}{4}$ in.   (b) $12\frac{5}{16}$ in.

7. $3\frac{1}{4}$ A   9. $1\frac{43}{48}$ A   11. $9\frac{5}{8}$ in.

13. (a) $10\frac{1}{2}$ in.   (b) $\frac{3}{4}$ in.

15. (a) $21\frac{13}{16}$ in.   (b) $\frac{7}{8}$ in.

17. $\frac{5}{64}$ in.   19. $1\frac{25}{32}$ in.; $1\frac{19}{32}$ in.

21. $3\frac{3}{4}$ in.; $10\frac{13}{16}$ in.   23. $5\frac{11}{16}$ in.

25. $52\frac{5}{32}$ in.

2.4 Pages 63—64

1. $\frac{7}{3}$ or $2\frac{1}{3}$	3. 9	5. $\frac{35}{64}$	
7. $\frac{2}{3}$	9. 10	11. $\frac{1}{8}$	
13. $\frac{5}{4}$ or $1\frac{1}{4}$	15. $\frac{2}{5}$	17. $\frac{18}{25}$	
19. 18	21. 40	23. $\frac{35}{33}$ or $1\frac{2}{33}$	
25. $\frac{88}{45}$ or $1\frac{43}{45}$	27. $\frac{9}{4}$ or $2\frac{1}{4}$	29. $\frac{27}{32}$	
31. $\frac{1}{126}$	33. $\frac{7}{12}$	35. $\frac{5}{4}$ or $1\frac{1}{4}$	

37. $\frac{1}{12}$ 39. $\frac{3}{8}$ 41. $\frac{5}{8}$

43. $\frac{27}{8}$ or $3\frac{3}{8}$ 45. $\frac{21}{8}$ or $2\frac{5}{8}$ 47. $\frac{37}{8}$ or $4\frac{5}{8}$

49. $\frac{145}{16}$ or $9\frac{1}{16}$ 51. $\frac{177}{8}$ or $22\frac{1}{8}$ 53. $\frac{141}{16}$ or $8\frac{13}{16}$

55. $\frac{101}{32}$ or $3\frac{5}{32}$ 57. $\frac{14}{9}$ or $1\frac{5}{9}$ 59. $\frac{3}{10}$

2.5 Pages 66—71

1. $31\frac{1}{2}$ gal 3. $20\frac{1}{6}$ ft 5. $4\frac{15}{32}$ in.

7. 80 bd ft 9. $1633\frac{1}{3}$ bd ft 11. $3\frac{27}{32}$ in.

13. $1\frac{7}{8}$ in. 15. $8\frac{5}{8}$ in. 17. $8\frac{2}{3}$ min

19. $1\frac{5}{11}$ in. 21. 2695 W 23. $203\frac{29}{32}$ W

25. $2\frac{2}{7}$ A 27. $7\frac{1}{4}$ ft or 7 ft 3 in.

29. 63 lb 31. 40 lb 33. 160 bu/acre

35. $\frac{1}{2}$ oz 37. $\frac{1}{2}$ 39. $1\frac{1}{4}$ gr

41. 2 43. 3 45. $7\frac{1}{2}$

2.6 Pages 74—75

1. 43 3. 83 5. 9 7. 10
9. 96 11. 6 13. 8 15. 5
17. $5\frac{1}{2}$ 19. $3\frac{1}{2}$ 21. $3\frac{1}{2}$ 23. $30\frac{2}{3}$

25. 3520 27. $15\frac{5}{8}$ 29. 384 31. 153 in.

33. 7 qt 35. $\frac{66}{125}$ Ω

2.7 Pages 75—77

1. $6\frac{3}{32}$ in^2 3. $17\frac{13}{16}$ in^2 5. $39\frac{7}{8}$ in^2

7. 175 mi 9. 52 mi/h 11. $52\frac{4}{7}$ mi/h

13. $4\frac{22}{35}$ 15. 25 17. $27\frac{1}{2}$ in^3;
110 in^3

19. $18\frac{6}{13}$ Ω

2.8 Page 81

1. $-\frac{3}{16}$ 3. $\frac{1}{16}$ 5. $-12\frac{3}{20}$ 7. $-\frac{13}{18}$

9. $\frac{15}{16}$ 11. $-\frac{1}{20}$ 13. $\frac{1}{3}$ 15. $\frac{14}{3}$ or $4\frac{2}{3}$

17. $-\frac{25}{12}$ or $-2\frac{1}{12}$ 19. -48

21. $-\frac{23}{10}$ or $-2\frac{3}{10}$ 23. $-\frac{10}{3}$ or $-3\frac{1}{3}$

25. $\frac{13}{4}$ or $3\frac{1}{4}$ 27. $-\frac{9}{8}$ or $-1\frac{1}{8}$

29. $\frac{1}{4}$ 31. $-\frac{1}{4}$

Chapter 2 Review, Pages 81—82

1. $\frac{9}{14}$ 2. $\frac{5}{6}$ 3. $4\frac{1}{6}$ 4. $6\frac{3}{5}$

5. $\frac{21}{8}$ 6. $\frac{55}{16}$ 7. 2 8. $1\frac{1}{2}$

9. $\frac{103}{180}$ 10. $14\frac{53}{84}$ 11. $1\frac{19}{24}$ 12. $11\frac{3}{5}$

13. 5 14. $\frac{1}{4}$ 15. 18 16. $\frac{1}{16}$

17. $\frac{3}{8}$ 18. $\frac{336}{25}$ or $13\frac{11}{25}$

19. (a) $3\frac{3}{8}$ in. (b) $5\frac{7}{16}$ in.

20. 105 21. 2016 22. 24 23. 63,360

24. $11\frac{1}{6}$ in. 25. $-\frac{1}{42}$ 26. $\frac{1}{12}$

27. $-3\frac{1}{8}$ 28. $-\frac{25}{32}$

Chapter 3

3.1 Pages 87—88

1. Four thousandths 3. Five ten-thousandths
5. One and four hundred twenty-one hundred-thousandths
7. Sixty-two and three hundred eighty-four thousandths
9. Six and ninety-two thousandths
11. Two hundred forty-six and one hundred twenty-four thousandths

13. 5.02; $5\frac{2}{100}$ or $5\frac{1}{50}$ 15. 71.0021; $71\frac{21}{10,000}$

17. 65,000; 65,000 19. 43.0101; $43\frac{101}{10,000}$

21. 0.375 23. $0.7\overline{3}$ 25. 0.34 27. $1.\overline{27}$

29. $18.\overline{285714}$ 31. $34.\overline{2}$ 33. $\frac{7}{10}$

35. $\frac{11}{100}$ 37. $\frac{337}{400}$ 39. $14\frac{613}{1000}$ 41. $10\frac{19}{25}$

43. $\frac{1}{16}$, $\frac{5}{16}$, $\frac{1}{8}$, $\frac{7}{8}$, $\frac{1}{32}$

3.2 Pages 91—94

1. 150.888 3. 163.204 5. 239.161
7. 1.58 9. 123.588 11. 8.68
13. −4.122 15. −10.0507
17. $a = 4.56$ cm; $b = 4.87$ cm 19. 7.94 in.
21. 70.8 mm 23. 2.605 A 25. 1396.8 Ω
27. 0.532 in. 29. 0.22 in. 31. 1.727 in.
33. 0.0032 in.

3.3 Page 96

1. (a) 1700 (b) 1650
3. (a) 3100 (b) 3130
5. (a) 18,700 (b) 18,680
7. (a) 3.1 (b) 3.142
9. (a) 0.1 (b) 0.057

	Hundred	Ten	Unit	Tenth	Hundredth	Thousandth
11.	700	660	659	659.2	659.17	659.173
13.	17,200	17,160	17,159	17,159.2	17,159.17	17,159.167
15.	1,543,700	1,543,680	1,543,679	—	—	—
17.	10,600	10,650	10,650	10,649.8	10,649.83	—

3.4 Pages 100—103

1. 0.555   3. 10.5126   5. 1507.2
7. 15,372   9. 9,280,000   11. 30
13. 15   15. 148.12   17. 5.8
19. 248.23   21. 3676.47   23. 37.76 m
25. 38 mm   27. 2.95 in.   29. 3000 sheets
31. $812   33. 51.20 in.   35. 450 in^3
37. 2.1 L   39. (a) 0.056 in.   (b) 4.535 in.
41. 240 gal   43. $132.50   45. 94.2 Ω
47. 0.2835 W   49. 6.20 A   51. 136.9 Ω
53. 6.39 A   55. 1.5 gr   57. 5

3.5 Pages 106—107

1. 2.36×10^2 3. 8.624×10^2
5. 9.218×10^0 7. 1.18×10^{-3}
9. 1.88×10^{-4} 11. 8.23×10^{17}
13. 2.68   15. 0.000136
17. 0.867   19. 888,000,000
21. 39.6   23. 16,900
25. 0.0048   27. 0.00091
29. 0.00037   31. 0.0613
33. 296,888   35. 0.0009
37. 1.0009   39. 0.00000000998
41. 0.000271   43. 18,215

3.6 Pages 113—114

1. 10^{13} 3. 10^{12} 5. $\frac{1}{10^{10}}$ 7. 10^3
9. $\frac{1}{10^{12}}$ 11. $\frac{1}{10^5}$ 13. $\frac{1}{10^4}$ 15. 10^{17}
17. 10^6 19. 10^{14} 21. 2.4×10^{-15}
23. 3×10^{24} 25. 1×10^{-6}
27. 1.728×10^{18} 29. 1.1284×10^{-1}
31. 1.11×10^{-25} 33. 9×10^{-1}
35. 4.6575×10^{13} 37. 7.455×10^5
39. 1.1664×10^{10} 41. 9.06×10^{-11}
43. 2.66×10^{24}

3.7 Pages 123—124

1. 0.27   3. 0.06   5. 0.56   7. 1.56
9. 0.292   11. 0.087   13. 1.287   15. 9.478
17. 0.0028   19. 0.00068   21. 0.000086
23. 0.0425   25. 0.00375   27. $0.50\overline{3}$ 29. $0.101\overline{6}$
31. 54%   33. 8%   35. 62%   37. 27%
39. 217%   41. 435%   43. 18.5%   45. 29.7%
47. 519%   49. 1.87%   51. 9.16%   53. 0.825%
55. 80%   57. 12.5%
63. 60%   65. 32.5%   59. $16\frac{2}{3}\%$ 61. $44\frac{4}{9}\%$
67. 36%   69. 43.75%
71. $66\frac{2}{3}\%$ 73. 240%   75. 1060%   77. 175%
79. $241\frac{2}{3}\%$ 81. $466\frac{2}{3}\%$ 83. 650%   85. $\frac{3}{4}$

87. $\frac{4}{25}$ 89. $\frac{3}{5}$ 91. $\frac{93}{100}$ 93. $\frac{11}{4}$ or $2\frac{3}{4}$
95. $\frac{5}{4}$ or $1\frac{1}{4}$ 97. $\frac{54}{25}$ or $2\frac{4}{25}$
99. $\frac{6}{5}$ or $1\frac{1}{5}$ 101. $\frac{43}{400}$
103. $\frac{107}{1000}$ 105. $\frac{69}{400}$ 107. $\frac{97}{600}$ 109. $\frac{67}{900}$

3.8 Page 125

1. P = 60; R = 25%; B = 240
3. P = 250; R = 10%; B = 2500
5. P = 108; R = 40%; B = 270
7. P = 252; R = 9%; B = 2800
9. P = 129; R = 20%; B = 645
11. P = Unknown; R = 4%; B = 28,000
13. P = 21; R = 60%; B = Unknown
15. P = 2050; R = 6%; B = Unknown
17. P = Unknown; R = 1%; B = 12 feet

3.9 Pages 127—128

1. 13   3. 42   5. 5.1   7. 22.95
9. 174.6   11. 62.14   13. 0.16   15. 0.0256
17. 5.92   19. 114.24   21. 8.96 V
23. 46.8 hp   25. $30,520   27. $38,665.20
29. $160.64   31. 176 acres   33. 35 mg
35. 3.375 grains

3.10 Pages 129—130

1. 500   3. 90   5. $133.\overline{3}$ or $133\frac{1}{3}$
7. 2000   9. 27,200   11. $1600
13. $24,731.71   15. $95,619.05
17. $125,000   19. $321,625
21. 30 lb   23. 54,000 lb
25. $110   27. 60,000 W
29. 120 V   31. 1725 lb
33. 150 mL   35. 8.75 grains

3.11 Pages 132—133

1. 50%   3. 60%   5. 28%   7. 49.0%
9. 50%   11. 33.3%   13. 25%   15. 20%
17. 10%   19. 20.5%   21. 24%   23. 3.7%
25. 0.9%   27. 75.6%   29. 56.4%   31. 55%
33. 23.4%

3.12 Pages 135—142

1. 3.3%   3. 10.35 A   5. 0.66%   7. $425
9. 37.5%; 62.5%
11. 54 lb antimony, 22.5 lb tin, 373.5 lb lead
13. 54 workers had accidents, 1746 workers had perfect safety records
15. 25%   17. 150%   19. 8.75 lb
21. 35.7%, 27,000 miles more

23. 127,977,800 head
25. 440 lb, 352 lb active ingredients, 88 lb inert
    ingredients
27. 875 gal, 36.75 gal butterfat
29. 0.3 gr   31. 25%   33. 120 mL   35. 53.3%
37. 14.2%   39. 11.9%   41. $2247.78   43. $567.19

Chapter 3 Review, Pages 143—144

1. 0.5625   2. 0.41$\overline{6}$ 3. $\frac{9}{20}$ 4. 19$\frac{5}{8}$

5. 168.278   6. -24.75   7. 68.665   8. 33.72
9. 3206.5   10. 1.96745   11. 3.18   12. 20.6
13. (a) 200   (b) 248.2   (c) 250
14. (a) 5.6   (b) 5.65   (c) 5.6491
15. 4.76 × 10^5 16. 1.4 × 10^{-3}
17. 0.0000535   18. 61,000,000
19. 0.00105   20. 0.06   21. 0.000075   22. 0.00183
23. $\frac{1}{10^5}$ 24. 10^9 25. $\frac{1}{10^{12}}$ 26. 1
27. 4.37 × 10^{-2} 28. 2.8 × 10^{14}
29. 2 × 10^{20} 30. 2.025 × 10^{-15}
31. 1.6 × 10^{37} 32. 0.15   33. 0.0825
34. 6.5%   35. 120%   36. $1050   37. 38.1%
38. 2250   39. 42.3%   40. 48 tons

Chapter 4

4.1 Page 148

1. kilo   3. centi   5. milli   7. mega
9. h   11. d   13. c   15. μ
17. 65 mg   19. 82 cm   21. 36 μA   23. 19 hL
25. 18 metres   27. 36 kilograms
29. 24 picoseconds   31. 135 millilitres
33. 45 milliamperes   35. metre
37. ampere   39. litre and cubic metre

4.2 Pages 151—153

1. 1 metre   3. 1 kilometre
5. 1 centimetre   7. 1000   9. 0.01
11. 0.001   13. 10   15. cm   17. mm
19. m   21. km   23. mm   25. mm
27. km   29. m   31. cm   33. mm; mm
35. (a) 52 mm; 5.2 cm; 0.052 m
    (b) 11 mm; 1.1 cm; 0.011 m
    (c) 137 mm; 13.7 cm; 0.137 m
    (d) 95 mm; 9.5 cm; 0.095 m
    (e) 38 mm; 3.8 cm; 0.038 m
    (f) 113 mm; 11.3 cm; 0.113 m
37. 52 mm; 5.2 cm; 0.052 m
39. 79 mm; 7.9 cm; 0.079 m
41. 77 mm; 7.7 cm; 0.077 m
43. 102 mm; 10.2 cm; 0.102 m
45. 0.675 km   47. 1.54 m   49. 0.65 m
51. 730 cm   53. 12.5 cm
55. Answers will vary.   57. Answers will vary.

4.3 Pages 154—155

1. 1 gram   3. 1 kilogram
5. 1 milligram   7. 1000   9. 0.01
11. 1000   13. 1000   15. g   17. kg
19. g   21. metric ton   23. mg
25. g   27. mg   29. g   31. kg
33. metric tons   35. 0.875 kg
37. 85,000 mg   39. 3600 g
41. 0.27 g   43. 2500 kg   45. Answers will vary.

4.4 Pages 158—159

1. 1 litre   3. 1 cubic centimetre
5. 1 square kilometre   7. 1000
9. 1,000,000   11. 100   13. 1000
15. L   17. m^2 19. cm^2 21. m^3
23. mL   25. ha   27. m^3 29. L
31. m^2 33. ha   35. cm^2 37. cm^2
39. 1.5 L   41. 1,500,000 cm^3 43. 85 mL
45. 0.085 km^2 47. 8.5 ha
49. 500 g   51. 0.675 ha

4.5 Page 162

1. 1 amp   3. 1 second   5. 1 megavolt
7. 43 kW   9. 17 ps   11. 3.2 MW   13. 450 Ω
15. 1000   17. 0.000000001
19. 1,000,000   21. 0.000001
23. 350 mA   25. 0.35 s   27. 3 h 52 min 30 s
29. 0.175 mF   31. 1.5 MHz

4.6 Pages 164—165

1. (b)   3. (c)   5. (b)   7. (c)
9. (d)   11. 25   13. 617   15. 3.2
17. -26.7   19. -108.4

4.7 Pages 167—169

1. 3.63   3. 15.0   5. 366   7. 66
9. 81.3   11. 46.6 mi   13. 10.8 mm   15. 30.3 L
17. (a) 300 ft   (b) 91.4 m
19. (a) 12.7 cm   (b) 127 mm
21. 2.51 m^2 23. 117 m^2 25. 116 cm^2 27. 1170 ft^2
29. 11.5 m^3 31. 279,000 mm^3
33. 2,380,000 cm^3
35. $2.25/ft^2; $405/frontage ft
37. 34.4 acres   39. 13.9 ha
41. 0.619 acre   43. 80 acres
45. 9397 lb/acre; 168 bu/acre   47. 0.606 acre

Chapter 4 Review, Page 170

1. milli   2. kilo   3. M   4. μ
5. 42 mL   6. 8.3 ns   7. 18 kilometres
8. 350 milliamperes   9. 50 microseconds
10. 1 L   11. 1 MW   12. 1 km^2 13. 1 m^3
14. 0.65   15. 0.75   16. 6100
17. 4.2 × 10^6 18. 1.8 × 10^7
19. 25,000   20. 25,000   21. 2.5   22. 6 × 10^5
23. 250   24. 22.2   25. -13   26. 0
27. 100   28. 81.7   29. 38.4   30. 142
31. 1770   32. 162   33. 177   34. (c)
35. (a)   36. (d)   37. (d)   38. (b)
39. (b)   40. (a)

Chapter 5

5.1 Page 175

1. 3   3. 4   5. 4   7. 4
9. 3   11. 4   13. 3   15. 3
17. 3   19. 4   21. 4   23. 5
25. 2   27. 6   29. 4   31. 6
33. 4   35. 2

5.2 Page 179

1. (a) 0.01 A   (b) 0.005 A
3. (a) 0.01 cm   (b) 0.005 cm
5. (a) 1 km   (b) 0.5 km
7. (a) 0.01 mi   (b) 0.005 mi
9. (a) 0.001 A   (b) 0.0005 A
11. (a) 0.0001 W   (b) 0.00005 W
13. (a) 10 Ω (b) 5 Ω
15. (a) 1000 L   (b) 500 L
17. (a) 0.1 cm   (b) 0.05 cm
19. (a) 10 V   (b) 5 V
21. (a) 0.001 m   (b) 0.0005 m
23. (a) $\frac{1}{3}$ yd   (b) $\frac{1}{6}$ yd
25. (a) $\frac{1}{32}$ in.   (b) $\frac{1}{64}$ in.
27. (a) $\frac{1}{16}$ mi   (b) $\frac{1}{32}$ mi
29. (a) $\frac{1}{9}$ in^2 (b) $\frac{1}{18}$ in^2

5.3 Pages 183—185

1. 27.20 mm   3. 63.50 mm   5. 8.00 mm
7. 115.90 mm   9. 71.00 mm   11. 34.60 mm
13. 68.60 mm   15. 3.40 mm   17. 43.50 mm
19. 76.15 mm

5.3A Pages 187—188

1. 1.362 in.   3. 2.695 in.   5. 0.133 in.
7. 1.715 in.   9. 2.997 in.   11. 1.073 in.
13. 2.503 in.   15. 0.317 in.   17. 4.565 in.
19. 2.799 in.

5.4 Pages 191—192

1. 4.25 mm   3. 3.90 mm   5. 1.65 mm
7. 7.77 mm   9. 5.59 mm   11. 10.28 mm
13. 7.13 mm   15. 8.75 mm   17. 6.07 mm
19. 5.42 mm

5.4A Pages 194—196

1. 0.237 in.   3. 0.314 in.   5. 0.147 in.
7. 0.820 in.   9. 0.502 in.   11. 0.200 in.
13. 0.321 in.   15. 0.170 in.   17. 0.658 in.
19. 0.245 in.

5.5 Page 200

1. (a) 14.7 in.   (b) 0.017 in.
3. (a) 16.01 mm   (b) 0.737 mm
5. (a) 0.0350 A   (b) 0.00050 A
7. (a) All have same accuracy.   (b) 0.391 cm
9. (a) 205,000 Ω (b) 205,000 Ω and 45,000 Ω
11. (a) 0.04 in.   (b) 15.5 in.
13. (a) 0.48 cm   (b) 43.4 cm
15. (a) 0.00008 A   (b) 0.91 A
17. (a) 0.6 m   (b) All have same precision.
19. (a) 500,000 Ω (b) 500,000 Ω

5.6 Pages 202—203

1. 18.1 m   3. 94.8 cm   5. 97,000 W
7. 840,000 V   9. 19 V
11. 25.27 cm or 252.7 mm   13. 126.4 cm
15. 8600 mi   17. 35 mm or 3.5 cm
19. 65.4 g   21. 0.330 in.   23. 26.1 mm

5.7 Pages 205—206

1. 4400 m^2 3. 1,230,000 cm^2
5. 901 m^2 7. 0.13 ΩA   9. 7,360 cm^3
11. 4.7 × 10^9 m^3 13. 35 A^2 Ω
15. 2500 in^2 17. 40 m   19. 340 V/A
21. 2.1 km/s   23. 3$\overline{0}$0 V^2/Ω 25. 4.0 g/cm^3
27. 9$\overline{0}$0 ft^3 29. 28 m   31. 28.8 hp
33. 1$\overline{0}$00 m^3

5.8 Page 208

1. 100 lb; 50 lb; 0.0357; 3.57%
3. 10,000 Ω; 5000 Ω; 0.0208; 2.08%
5. 1 rpm; 0.5 rpm; 0.000571; 0.06%
7. 0.1 g; 0.05 g; 0.0227; 2.27%
9. 10 W; 5 W; 0.000278; 0.03%
11. 0.001 A; 0.0005 A; 0.0122; 1.22%
13. $\frac{1}{4}$ in.; $\frac{1}{8}$ in.; 0.0714; 7.14%
15. 1 oz; 0.5 oz; 0.00649; 0.65%
17. Same   19. 3450 Ω

5.9 Page 211

	Upper limit	Lower limit	Tolerance interval
3.	$6\frac{21}{32}$ in.	$6\frac{19}{32}$ in.	$\frac{1}{16}$ in.
5.	$3\frac{29}{64}$ in.	$3\frac{27}{64}$ in.	$\frac{1}{32}$ in.
7.	$3\frac{25}{128}$ in.	$3\frac{23}{128}$ in.	$\frac{1}{64}$ in.
9.	1.24 cm	1.14 cm	0.10 cm
11.	0.0185 A	0.0175 A	0.0010 A
13.	26,000 V	22,000 V	4000 V
15.	10.36 km	10.26 km	0.10 km

17. $53,075   19. $1,567,020.60

5.10 Pages 216—217

1. 360 Ω; ±10%   3. 830,000 Ω; ±20%
5. 1,400,000 Ω; ±20%   7. 70 Ω; ±20%
9. 500,000 Ω; ±20%   11. 10,000,0C0 Ω; ±20%
13. Yellow, gray, red   15. Violet, red, orange
17. Blue, green, yellow
19. Red, green, silver
21. Yellow, green, green
23. Violet, blue, gold
25. (a) 30 Ω (b) 330 Ω
     (c) 270 Ω (d) 60 Ω
27. (a) 166,000 Ω (b) 996,000 Ω
     (c) 664,000 Ω (d) 332,000 Ω
29. (a) 14 Ω (b) 84 Ω
     (c) 56 Ω (d) 28 Ω

5.11 Pages 222—226

1. 314,830 ft^3 3. 571,081 ft^3
5. 616,284 ft^3 7. $34.13
9. $153.81   11. 3471 kWh
13. 7568 kWh   15. +1.16 mm

17. -0.67 mm   19. -3.40 mm
21. +1.74 mm   23. -4.08 mm
25. +0.113 in.   27. +0.110 in.
29. -0.006 in.   31. -0.159 in.
33. -0.878 in.

5.12 Page 229

1. 6 V   3. 6.4 V   5. 0.4 V   7. 1.4 V
9. 40 V   11. 230 V

5.13 Pages 231—232

1. 7 Ω 3. 12 Ω 5. 11 Ω 7. 35 Ω
9. 85 Ω 11. 300 Ω

Chapter 5 Review, Pages 232—234

1. 3   2. 2   3. 3   4. 3
5. 2   6. 3   7. 4   8. 5
9. (a) 0.01 m   (b) 0.005 m
10. (a) 0.1 mi   (b) 0.05 mi
11. (a) 100 L   (b) 50 L
12. (a) 100 V   (b) 50 V
13. (a) 0.01 cm   (b) 0.005 cm
14. (a) 10,000 V   (b) 5000 V
15. (a) $\frac{1}{8}$ in.   (b) $\frac{1}{16}$ in.
16. (a) $\frac{1}{16}$ mi   (b) $\frac{1}{32}$ mi
17. 42.35 mm   18. 1.670 in.
19. 11.84 mm   20. 0.438 in.
21. (a) 36,500 V   (b) 9.6 V
22. (a) 0.0005 A   (b) 0.425 A
23. 730,000 W   24. 4$\overline{0}$0 m
25. 400,000 V   26. 1900 cm^3
27. 5.88 m^2 28. 1.4 N/m^2
29. 130 V^2/Ω 30. 0.0057; 0.57%
31. 0.00032; 0.03%   32. 2200 Ω; 1800 Ω
33. 120,000 Ω; ±20%   34. 0.85 Ω; ±5%
35. 947,602 ft^3 36. -0.563 in.
37. 8.4 V   38. 2.6 Ω

Chapter 6

6.1 Page 237

1. 83   3. 8   5. 42   7. -6
9. -128   11. 1   13. -5   15. 3
17. -24   19. $-9\frac{3}{5}$ or $-\frac{48}{5}$ 21. -13
23. $-1\frac{3}{5}$ or $-\frac{8}{5}$ 25. 189   27. 331,776
29. -72   31. 60   33. 8   35. -78
37. -1

6.2 Pages 240—242

1. $a + b + c$ 3. $a + b + c$
5. $a - b - c$ 7. $x + y - z + 3$
9. $x - y - z - 3$ 11. $2x + 4 + 3y + 4r$
13. $3x - 5y + 6z - 2w + 11$
15. $-5x - 3y - 6z + 3w + 3$
17. $2x + 3y - z - w + 3r - 2s - 10$
19. $-2x + 3y - z - 4w - 4r + s$
21. $5b$ 23. $10h$ 25. $3m$ 27. $a + 12b$
29. $6a^2 - a + 1$ 31. $3x^2 + 3x$
33. $-1.8x$ 35. $15a^3 + 2a^2 - 15$
37. $2x^2y - 2xy + 2y^2 - 3x^2$
39. $5x^3 + 3x^2y - 5y^3 + y$

41. 1   43. $3x + 4$ 45. $5 - x$
47. $3y - 7$ 49. $4y + 5$
51. 5   53. 6   55. 28   57. $\frac{5}{4}x - \frac{8}{3}$
59. $2x + 19$ 61. $-8.5y - 4$
63. -42   65. $4n + 8$ 67. $1.8x - 7$
69. $-x + 33$ 71. $-6n - 2$
73. $-1.05x - 8.4$

6.3 Pages 245—247

1. Binomial   3. Binomial
5. Monomial   7. Trinomial
9. Binomial
11. $x^2 - x + 1$; 2nd   13. $x^2 + x - 1$; 2nd
15. $5x^3 - 4x^2 - 2$; 3rd
17. $4y^3 - 6y^2 - 3y + 7$; 3rd
19. $-x^5 + 1$; 5th   21. $2a$
23. $2x^2 - 2x$ 25. $8y^2 - 5y + 6$
27. $4a^3 - 3a^2 - 5a$ 29. $7a^2 - 10a + 1$
31. $9x^2 - 5x$ 33. $9a^3 + 4a^2 + 5a - 5$
35. $4x + 4$ 37. $5x^2 + 18x - 22$
39. $5x^3 + 13x^2 - 8x + 7$
41. $a + b$ 43. $16m - 58n + 45$
45. $-14a^2 - 11a - 3$ 47. $6y^2 + 48y - 15$
49. $2x^2 + 6x + 2$ 51. $-3x^2 + 2x + 2$
53. $x^3 + 3x - 3$
55. $8x^5 - 18x^4 + 5x^2 + 1$
57. $a + 3b$ 59. $7a - 4b - 3x + 4y$
61. $x^2 - 3x + 5$ 63. $7w^2 - 24w - 6$
65. $x^2 - 4x - 10$ 67. $4x^2 - 4x + 12$
69. $-6x^2 + 2x - 10$

6.4 Page 250

1. $-15a$ 3. $28a^3$ 5. $28m^4$ 7. $32a^8$
9. $-26p^2q$ 11. $30n^3m$ 13. $21a^4b$ 15. $\frac{3}{8}x^3y^4$
17. $24a^3b^4c^3$ 19. $\frac{3}{16}x^3y^5z^3$
21. $-371.64m^3n^2p^2$ 23. $-255a^6b$
25. x^6 27. x^{20} 29. $9x^8$ 31. $-x^9$
33. x^{10} 35. x^{30} 37. $25x^6y^4$ 39. $225m^4$
41. $15,625n^{12}$ 43. $9x^{12}$
45. $8x^9y^{12}z^3$ 47. $-32h^{15}k^{30}m^{10}$
49. -408   51. -144   53. 17,712   55. 16
57. -1728   59. 291,600   61. -1080
63. 324   65. -324   67. 16

6.5 Pages 252—253

1. $4a + 24$ 3. $-18x^2 - 12y$
5. $4ax^2 - 6ay + a$ 7. $3x^3 - x^2 + 5x$
9. $6a^3 + 12a^2 - 20a$ 11. $-12x^3 + 21x^2 + 6x$
13. $-28x^3 - 12xy + 8x^2y$ 15. $3x^3y^2 - 3x^2y^3 + 12x^2y^2$
17. $-6x^3 + 36x^5 - 54x^7$ 19. $5a^4b^2 - 5ab^5 - 5a^2b^3$
21. $\frac{28}{3}mn - 8m^2$ 23. $16y^2z^3 - \frac{8}{35}yz^4$
25. $-5.2a^6 - 10a^3 - 4a$ 27. $1334.4a^3 + 1668a^2$
29. $24x^4y - 16x^3y^2 + 20x^2y^3$
31. $\frac{1}{2}a^3b^3 - \frac{1}{3}a^2b^5 + \frac{5}{9}ab^6$
33. $-19x^2 + 8x$
35. $3x^2y - 2x^2y^2 + 2x^2y^3 - 7xy^3$
37. $x^2 + 7x + 6$ 39. $x^2 + 5x - 14$
41. $x^2 - 13x + 40$ 43. $3a^2 - 17a + 20$
45. $12a^2 - 10a - 12$ 47. $24a^2 + 84a + 72$
49. $15x^2 - 4xy - 4y^2$ 51. $4x^2 - 12x + 9$
53. $4c^2 - 25d^2$ 55. $91m^2 + 32m - 3$

57. $x^8 - 2x^5 + x^2$ 59. $10y^3 - 24y^2 - 32y + 16$
61. $24x^2 - 78x - 6y^2 - 39y$
63. $g^2 - 3g - h^2 + 9h - 18$
65. $10x^7 - 3x^6 - x^5 + 44x^4 + x^3 - x^2 + 16x - 2$

6.6 Pages 254—255

1. $3x^2$ 3. $\dfrac{3x^8}{2}$ 5. $\dfrac{6}{x^2}$ 7. $\dfrac{2x}{3}$

9. x 11. $\dfrac{17}{3}$ 13. $\dfrac{5a^2}{b}$ 15. $\dfrac{8}{mn}$

17. 0 19. $23p^2$ 21. -2 23. $\dfrac{36}{r^2}$

25. $\dfrac{-92x^2}{7y^2}$ 27. $\dfrac{8}{7a^3b^2}$ 29. $\dfrac{4x^2z^4}{9y}$

31. $2x^2 - 4x + 3$ 33. $x^2 + x + 1$
35. $x - y - z$ 37. $3a^4 - 2a^2 - a$
39. $b^9 - b^6 - b^3$ 41. $-x^3 + x^2 - x + 1$
43. $12x + 6x^2y - 3$ 45. $8x^3z - 6x^2yz^2 - 4y^2$

47. $4y^3 - 3y - \dfrac{2}{y}$ 49. $\dfrac{1}{2x^2} - 3 - 2x^2$

6.7 Pages 257—258

1. $x + 2$ 3. $3a + 3$ r 11
5. $4x - 3$ r -3 7. $y + 2$ r -3
9. $3b - 4$ 11. $6x^2 + x - 1$
13. $4x^2 + 7x - 15$ r 5 15. $2x^2 - 2x - 12$
17. $2x^2 - 16x + 32$ 19. $3x^2 + 10x + 20$ r 34
21. $2x^2 + 6x + 30$ r 170
23. $2x^3 + 4x + 6$ 25. $4x^2 - 2x + 1$ r -2
27. $x^3 - 2x^2 + 4x - 8$
29. $3x^2 - 4x + 1$

Chapter 6 Review, Pages 258—259

1. a 2. 0 3. 1 4. 1

5. 9 6. -6 7. $-\dfrac{9}{2}$ 8. -8

9. 5 10. $6y - 5$ 11. $6 - 8x$
12. $10x + 27$ 13. $3x^3 + x^2y - 3y^3 - y$
14. Binomial   15. 4
16. $8a^2 + 5a + 2$ 17. $9x^3 + 4x^2 + x + 2$
18. $10x^2 - 7x + 4$ 19. $24x^5$
20. $-56x^5y^3$ 21. $27x^6$ 22. $15a^2 + 20ab$
23. $-32x^2 + 8x^3 - 12x^4$ 24. $15x^2 - 11x - 12$

25. $6x^3 - 24x^2 + 26x - 4$ 26. $\dfrac{7}{x}$

27. $5x^2$ 28. $4a^2 - 3a + 1$
29. $3x - 4$ 30. $3x^2 - 4x + 2$

Chapter 7

7.1 Pages 266—267

1. 6 3. 17 5. $10\dfrac{1}{2}$ 7. 19.5
9. 5.2 11. 301 13. 7 15. 20

17. $4\dfrac{1}{14}$ or $\dfrac{57}{14}$ 19. 0 21. 0

23. 16 25. 392 27. -4 29. -2
31. 2 33. -2 35. 32 37. -4

39. 18 41. $5\dfrac{2}{3}$ or $\dfrac{17}{3}$ 43. 42

45. $10\dfrac{1}{2}$ or $\dfrac{21}{2}$ 47. $-\dfrac{1}{3}$ 49. -1

51. $\dfrac{77}{135}$ 53. $1\dfrac{5}{7}$ or $\dfrac{12}{7}$

55. $2\dfrac{12}{29}$ or $\dfrac{70}{29}$ 57. $-3\dfrac{1}{3}$ or $-\dfrac{10}{3}$

59. 25

7.2 Pages 270—271

1. 270 3. 20 5. 3 in./h   7. 10 in.
9. 130 bu/acre   11. 1750 lb
13. $19.50/h   15. 0.2 A
17. 0.4 mL/stroke   19. 66 plugs/wk
21. 26 syringes/day   23. 17 vials/day

7.3 Page 274

1. 8 3. 4 5. 4 7. 3
9. 6 11. 6 13. -5 15. 7

17. $2\dfrac{1}{5}$ 19. -9 21. 3 23. 13

25. $\dfrac{3}{4}$ 27. -1 29. 4

7.4 Pages 276—277

1. 5 3. $\dfrac{2}{5}$ 5. $-\dfrac{4}{3}$ or $-1\dfrac{1}{3}$ 7. 6

9. -5 11. $\dfrac{36}{5}$ or $7\dfrac{1}{5}$ 13. -5

15. 0 17. $\dfrac{8}{7}$ or $1\dfrac{1}{7}$ 19. 9

21. 14 23. 45 25. 4 27. 22
29. -1 31. 10 33. 18 35. 2
37. -8 39. -2 41. -1 43. -7

45. $\dfrac{33}{4}$ or $8\dfrac{1}{4}$ 47. $-\dfrac{4}{3}$ or $-1\dfrac{1}{3}$

49. -3 51. 6 53. -2 55. 0

57. $\dfrac{19}{24}$ 59. $\dfrac{273}{44}$ or $6\dfrac{9}{44}$

7.5 Page 282

1. 8 3. 5 5. 3 7. 16
9. 1 11. 50 13. 6 15. 1
17. 24 19. 4 21. -56 23. 6

25. -3 27. $\dfrac{2}{3}$ 29. $-\dfrac{1}{2}$ 31. 1

33. $\dfrac{1}{5}$ 35. 3 37. $\dfrac{2}{3}$ 39. 5

41. $\dfrac{1}{7}$

7.6 Pages 285—286

1. $\dfrac{E}{I}$ 3. $\dfrac{F}{m}$ 5. $\dfrac{C}{\pi}$ 7. $\dfrac{V}{\ell h}$ 9. $\dfrac{A}{2\pi r}$

11. $\dfrac{v^2}{2g}$ 13. $\dfrac{Q}{I}$ 15. vt 17. $\dfrac{V}{I}$

19. $4E\pi r^2$ 21. $\dfrac{1}{2\pi X_c C}$ 23. $\dfrac{2A}{h}$ 25. $\dfrac{QJ}{I^2 t}$

27. $\dfrac{5}{9}(F - 32)$ or $\dfrac{5F - 160}{9}$

29. $C_T - C_1 - C_3 - C_4$ 31. $\dfrac{-By - C}{A}$

33. $\dfrac{Q_1 + PQ_1}{P}$ or $\dfrac{Q_1}{P} + Q_1$ 35. $\dfrac{2A}{a + b}$

37. $\dfrac{\ell - a}{n - 1}$ 39. $\dfrac{Ft}{V_2 - V_1}$

41. $\dfrac{Q}{w(T_1 - T_2)}$ 43. $\dfrac{PV}{2\pi} - 3960$

7.7 Pages 288—290

1. 23   3. 20   5. 2000   7. 9.9
9. 576   11. 3   13. 1936   15. 48.9
17. 3   19. 16.6   21. 324 W
23. 6.72 ft   25. 12.1 m
27. 15.0 Ω   29. 0.43 in.

Chapter 7 Review, Pages 290—291

1. $\frac{3}{2}$ or $1\frac{1}{2}$ 2. -4   3. 57

4. 24   5. -7   6. -9   7. 22 ft
8. Bill $960; Tony $480   9. 8

10. 3   11. 6   12. $\frac{3}{5}$ 13. 1

14. 6   15. 4   16. -2   17. 6

18. $\frac{1}{2}$ 19. $\frac{8}{3}$ or $2\frac{2}{3}$

20. $-\frac{15}{2}$ or $-7\frac{1}{2}$ 21. 26   22. 2

23. 5   24. $\frac{F}{W}$ 25. $\frac{W}{P}$

26. $2L - 2A - 2B$ 27. $\frac{2k}{v^2}$

28. $\frac{P_1 T_2}{P_2}$ 29. $2v - v_t$ 30. -148

31. 19.35   32. 36.8

Chapter 8

8.1 Pages 294—296

1. $\frac{1}{5}$ 3. $\frac{1}{3}$ 5. $\frac{5}{3}$ 7. $\frac{1}{5}$

9. $\frac{2}{1}$ or 2   11. $\frac{1}{32}$ 13. $\frac{9}{14}$ 15. $\frac{11}{16}$

17. $\frac{1}{4}$ 19. $\frac{4}{1}$ or 4   21. $\frac{7}{5}$

23. 32 to 5 or 6.4 to 1   25. 3.4 gal/min
27. 1 to 275   29. 12 to 1
31. 36 lb/bu   33. 25 gal/acre
35. 7 to 5 or 1.4 to 1   37. 35 to 18
39. 85¢/ft   41. $42/ft^2
43. 50 mg/cm^3 45. 4 mg/cm^3
47. 30 mi/gal   49. 46 gal/h
51. 50 mi/h   53. $\frac{1}{4}$ lb/gal
55. $36/ft^2

8.2 Pages 299—301

1. (a) 2, 3   (b) 1, 6   (c) 6   (d) 6
3. (a) 9, 28   (b) 7, 36   (c) 252   (d) 252
5. (a) 7, w   (b) x, z   (c) 7w   (d) xz
7. Yes; 30 = 30   9. No; 60 ≠ 90
11. Yes; 12 = 12   13. No; 324 ≠ 328

15. 3   17. $5\frac{3}{5}$ 19. 14   21. $3\frac{1}{3}$

23. 35   25. -7.5   27. 0.28   29. 0.5

31. 126   33. 20.6   35. 37.3   37. 38.2

39. 818   41. 9050   43. 21.3

45. 44 ft^3 47. $98,000   49. $144   51. $8500
53. 52.5 lb   55. 108 lb
57. 1250 ft   59. 595 turns
61. 177 hp   63. 21 lb
65. 14 cm^3 67. 6 cm^3

8.3 Page 302

1. 20%   3. 53.1%   5. 39.2%   7. 2.2%
9. 23.3%   11. 5.9%   13. (a) 17% (b) 110.5 lb
15. 18.8%   17. 20%

19. (a) 13.3%   (b) 33.3%   (c) 53.3%

8.4 Pages 308—310

All answers allowable error ±8 mi or ±8 km.

1. 48 mi   3. 60 mi   5. 86 mi   7. 104 mi
9. 96 mi   11. 527 km   13. 298 km   15. 323 km
17. 12.5 cm   19. 100 cm^2
21. 1 cm × 1 cm   23. 6.5 cm   25. 6.5 cm
27.   29.

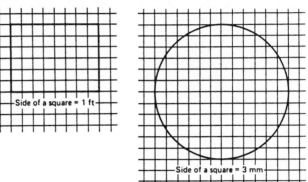

Side of a square = 1 ft   Side of a square = 3 mm

31. Yes; no   33. 20 : 1   35. 9 : 1
37. 68 lb   39. 238,500 lb   41. 81 : 1
43. 121 : 1   45. 175 : 1   47. 400 lb
49. 400 lb   51. 640 lb

8.5 Pages 314—316

1. 36 rpm   3. 80 rpm   5. 22 cm
7. $357\frac{1}{7}$ rpm   9. 520 rpm   11. 160 in.
13. 896 rpm   15. 10 cm; 17 cm
17. 160 teeth   19. 1008 rpm   21. 576 teeth
23. 96 rpm   25. 144 rpm   27. 40 teeth
29. $d_2 = 10$ 31. $F_2 = 120$ 33. 135 lb
35. $\frac{1}{2}$ ton or 1000 lb   37. 64 cm   39. $25\frac{1}{3}$ in.

Chapter 8 Review Pages 316—318

1. $\frac{1}{4}$ 2. $\frac{3}{2}$ 3. $\frac{2}{1}$ 4. $\frac{11}{18}$

5. Yes   6. No   7. 1   8. 30
9. 8   10. 40   11. 106   12. 41.3
13. 788   14. 529   15. $187.50
16. 1,250 ft   17. 216 h   18. 300 lb
19. Jones 58%; Smith 42%   20. 8.8%
21. direct   22. direct   23. inverse
24. 362.5 mi   25. 52.8 in.
26. 500 rpm   27. 75 rpm
28. 50 kg   29. 180 lb   30. 2 A

Chapter 9

9.1 Pages 324—326

1. $A(-2, 2)$ 3. $C(5, -1)$ 5. $E(-4, -5)$
7. $G(0, 5)$ 9. $I(4, 2)$

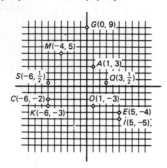

31. $(3, 2)$ $(8, -3)$ $(-2, 7)$
33. $(2, -1)$ $\left(0, \frac{7}{3}\right)$ $\left(-2, \frac{17}{3}\right)$
35. $(0, -2)$ $\left(2, -\frac{1}{2}\right)$ $(-4, -5)$
37. $(5, 4)$ $(0, 2)$ $\left(-3, \frac{4}{5}\right)$
39. $(2, 4)$ $(0, -5)$ $(-4, -23)$
41. $(2, 10)$ $(0, 4)$ $(-3, -5)$
43. $(2, -3)$ $(0, 7)$ $(-4, 27)$
45. $(3, 10)$ $(0, 4)$ $(-1, 2)$
47. $(3, 10)$ $(0, 4)$ $(-2, 0)$
49. $(1, 2)$ $\left(0, -\frac{5}{2}\right)$ $(-3, -16)$
51. $(2, 3)$ $(0, 3)$ $(-4, 3)$
53. $(5, 4)$ $(5, 0)$ $(5, -2)$
55. $y = \frac{6 - 2x}{3}$ 57. $y = \frac{7 - x}{2}$
59. $y = \frac{x - 6}{2}$ 61. $y = \frac{2x - 9}{3}$
63. $y = \frac{2x + 6}{3}$ 65. $y = \frac{-2x + 15}{3}$

9.2 Page 329

1.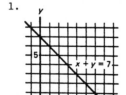

3.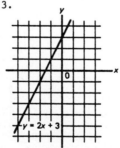

5.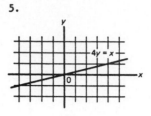

7.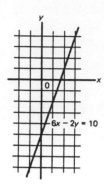

9.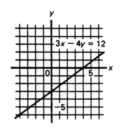

11.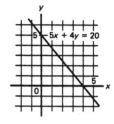

13.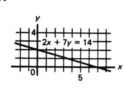

15.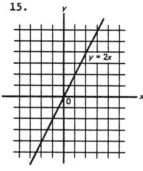

17.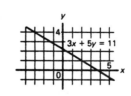

19.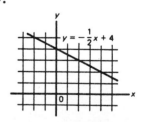

21.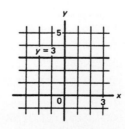

23.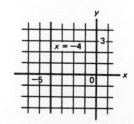

9.3 Page 336

1.

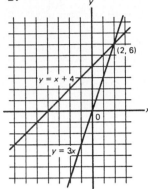

3.

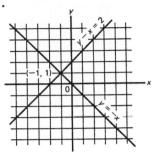

5.

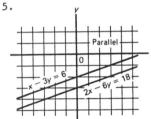

7.

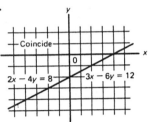

9.

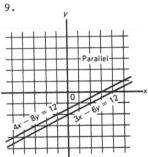

11.

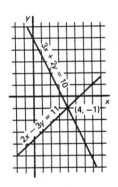

13.

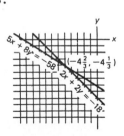

15.

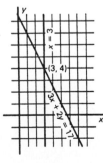

17.

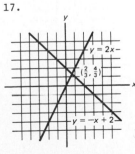

19.

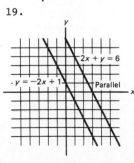

21.

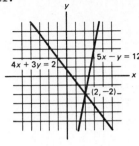

23.

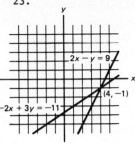

25.

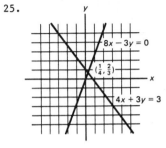

9.4 Pages 340—341

1. (2, 1) 3. (4, 2) 5. (-2, 7) 7. (3, 1)
9. (-5, 5) 11. (5, 1) 13. (1, -2) 15. (-1, 4)
17. (7, -5) 19. (4, -1) 21. (-2, 5)
23. $\left(2, \frac{5}{4}\right)$ or $\left(2, 1\frac{1}{4}\right)$ 25. (-2, 4) 27. Coincide
29. (2, -1)   31. Parallel
33. $\left(\frac{1}{2}, 5\right)$ 35. Coincide   37. Parallel
39. $\left(\frac{1}{4}, \frac{2}{5}\right)$ 41. (72, 30)   43. (9, -2)

9.5 Pages 345—347

1. 42 cm, 54 cm
3. $2\frac{1}{2}$ h @ 180 gal/h; $3\frac{1}{2}$ h @ 250 gal/h
5. 18 ft^3 gravel; 4.5 ft^3 cement
7. 30 lb @ 5%; 70 lb @ 15%
9. 2700 bu corn; 450 bu beans
11. 200 gal of 6%; 100 gal of 12%
13. 5 @ 3 V; 4 @ 4.5 V
15. 105 mL @ 8%; 35 mL @ 12%
17. 5 min @ 850 rpm; 9 min @ 1250 rpm
19. 2 h at setting 1; 3 h at setting 2
21. 40 yd^2 @ $14.50; 35 yd^2 @ $18
23. 21 @ $14.95; 2 @ $21.75
25. 80 mL solution; 60 mL water
27. 750 cm^3 @ 4%; 250 cm^3 @ 8%

9.6 Pages 349—350

1. $\left(\frac{12}{5}, \frac{36}{5}\right)$ or (2.4, 7.2)   3. (10, 2)
5. (6, 6)   7. (4, 2)   9. (-3, -1)
11. (1, 4)   13. (-2, 2)   15. $\left(2\frac{1}{2}, -12\frac{1}{2}\right)$
17. (1, 3)   19. (9, -2)   21. 100 Ω, 450 Ω
23. 30 cm, 90 cm
25. width 40 cm; length 80 cm   27. 150 mA

Chapter 9 Review, Pages 350—351

1. $A: (3, 5)$
3. $C: (2, -1)$
2. $B: (-2, -6)$
4. $D: (4, 0)$

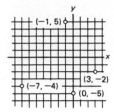

9. $\left(3, \dfrac{5}{2}\right)$ $(0, 4)$ $(-4, 6)$
10. $(3, -2)$ $(0, -4)$ $(-3, -6)$
11. $y = -6x + 15$
12. $y = \dfrac{3x + 10}{5}$ or $y = \dfrac{3}{5}x + 2$

13.
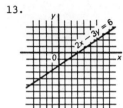

14.
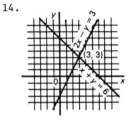

15. $(3, 4)$ 16. $(3, 1)$ 17. $(4, 3)$
18. No common solution—parallel lines.
19. Infinitely many solutions—lines coincide.
20. $(4, 8)$ 21. $(-1, -2)$ 22. $\left(1\dfrac{9}{25}, 10\dfrac{11}{25}\right)$

23. Resistor 10¢; capacitor 20¢
24. 132.5 ft by 57.5 ft

Chapter 10

10.1 Pages 353—354

1. $4(a + 1)$
5. $5(3b - 4)$
9. $a(a - 4)$
13. $5x(2x + 5)$
17. $4x(x^3 + 2x^2 + 3x - 6)$
21. $10(x + y - z)$
25. $7xy(2 - xy)$
29. $12a(5x - 1)$
33. $2(26m^2 - 7m + 1)$
37. $3m(2m^3 - 4m + 1)$

3. $b(x + y)$
7. $x(x - 1)$
11. $4n(n - 2)$
15. $3r(r - 2)$
19. $9a(a - x^2)$
23. $3(y - 2)$
27. $m(12x^2 - 7)$
31. $13mn(4mn - 1)$
35. $18y^2(2 - y + 3y^2)$
39. $-2x^2y^3(2 + 3y + 5y^2)$

10.2 Page 357

1. $x^2 + 7x + 10$
5. $x^2 - 11x + 30$
9. $2x^2 + 19x + 24$
13. $x^2 - 19x + 90$
17. $8x^2 - 18x - 35$
21. $14x^2 + 41x + 15$
25. $6x^2 + 47x + 35$
29. $120x^2 + 54x - 21$
33. $4x^2 - 16x + 15$
37. $6x^2 + 5x - 56$
41. $16x^2 + 14x - 15$
45. $6n^2 + 3ny - 30y^2$
49. $\dfrac{1}{8}x^2 - 5x + 48$

3. $6x^2 + 17x + 12$
7. $x^2 - 14x + 24$
11. $x^2 + 4x - 12$
15. $x^2 - 6x - 72$
19. $8x^2 + 6x - 35$
23. $3x^2 - 19x - 72$
27. $169x^2 - 104x + 16$
31. $100x^2 - 100x + 21$
35. $4x^2 + 4x - 15$
39. $6x^2 + 37x + 56$
43. $2y^2 - 11y - 21$
47. $8x^2 + 26xy - 7y^2$

10.3 Pages 361—362

1. $(x + 2)(x + 4)$
5. $3(r + 5)(r + 5)$
9. $(x + 8)(x + 9)$
13. $(x - 4)(x - 3)$
17. $3(x - 7)(x - 3)$
21. $(x - 9)(x - 10)$
25. $(x + 4)(x - 2)$
29. $(a + 8)(a - 3)$
33. $3(x - 4)(x + 3)$
37. $(y + 14)(y + 3)$
41. $(m - 20)(m - 2)$
45. $(a + 23)(a + 4)$
49. $(a + 25)(a + 4)$
53. $(y - 16)(y - 2)$
57. $6(x^2 + 2x - 1)$
61. $(a + 9)(a - 7)$
65. $2(y - 15)(y - 3)$
69. $(x + 15)(x + 15)$
73. $(x + 12)(x + 16)$
77. $2b(a + 6)(a - 4)$

3. $(y + 4)(y + 5)$
7. $(b + 5)(b + 6)$
11. $5(a + 4)(a + 3)$
15. $2(a - 7)(a - 2)$
19. $(w - 6)(w - 7)$
23. $(t - 10)(t - 2)$
27. $(y + 5)(y - 4)$
31. $(c - 18)(c + 3)$
35. $(c + 6)(c - 3)$
39. $(r - 7)(r + 5)$
43. $(x - 15)(x + 6)$
47. $2(a - 11)(a + 5)$
51. $(y - 19)(y + 5)$
55. $7(x + 2)(x - 1)$
59. $(y - 7)(y - 5)$
63. $(x + 4)(x + 14)$
67. $3x(y - 3)(y - 3)$
71. $(x - 9)(x - 17)$
75. $(x + 22)(x - 8)$
79. $(y - 9)(y + 8)$

10.4 Page 364

1. $x^2 - 9$ 3. $a^2 - 25$ 5. $4b^2 - 121$
7. 9991 or $(10,000 - 9)$ 9. $9y^4 - 196$
11. $r^2 - 24r + 144$ 13. $16y^2 - 25$
15. $x^2y^2 - 8xy + 16$ 17. $a^2b^2 + 2abd + d^2$
19. $z^2 - 22z + 121$ 21. $s^2t^2 - 14st + 49$
23. $x^2 - y^4$ 25. $x^2 + 10x + 25$
27. $x^2 - 49$ 29. $x^2 - 6x + 9$
31. $a^2b^2 - 4$ 33. $x^4 - 4$
35. $r^2 - 30r + 225$ 37. $y^6 - 10y^3 + 25$
39. $100 - x^2$

10.5 Pages 367—368

1. $(a + 4)(a + 4)$
5. $(x - 2)(x - 2)$
9. $(y + 6)(y - 6)$
13. $(1 + 9y)(1 - 9y)$
17. $(7x + 8y)(7x - 8y)$
21. $(2x - 3)(2x - 3)$
25. $(7x + 5)(7x - 5)$
29. $(b + 3)(b - 3)$
33. $(2m + 3)(2m - 3)$
37. $3(3x + 1)(3x - 1)$

3. $(b + c)(b - c)$
7. $(2 + x)(2 - x)$
11. $5(a + 1)(a + 1)$
15. $(7 + a^2)(7 - a^2)$
19. $(1 + xy)(1 - xy)$
23. $(R + r)(R - r)$
27. $(y - 5)(y - 5)$
31. $(m + 11)(m + 11)$
35. $4(x + 3)(x + 3)$
39. $a(m - 7)(m - 7)$

10.6 Page 370

1. $(5x + 2)(x - 6)$
5. $(6x - 5)(2x - 3)$
9. $(16x + 5)(x - 1)$
13. $(5y + 3)(3y - 2)$
17. $(7a + 1)(5a - 1)$
21. $(3x - 7)(x + 9)$
25. $(5y + 2)(3y - 4)$
29. $(3x - 5)(2x - 1)$
33. $(2b + 13)(2b + 13)$
37. $x(14x + 35)(2x + 5)$

3. $(2x - 3)(5x - 7)$
7. $(2x + 9)(4x - 5)$
11. $4(3x + 2)(x - 1)$
15. $(4m + 1)(2m - 3)$
19. $(4y - 1)(4y - 1)$
23. $(4b - 1)(3b + 2)$
27. $(10 + 3c)(9 - c)$
31. $(2y^2 - 5)(y^2 + 7)$
35. $(7x - 8)(2x - 5)$
39. $5a(2b + 7)(b - 5)$

Chapter 10 Review, Pages 370—371

1. $c^2 - d^2$ 2. $x^2 - 36$ 3. $y^2 + 3y - 28$
4. $4x^2 - 8x - 45$ 5. $x^2 + 5x - 24$
6. $x^2 - 13x + 36$ 7. $x^2 - 6x + 9$
8. $4x^2 - 24x + 36$ 9. $1 - 10x^2 + 25x^4$
10. $6(a + 1)$ 11. $5(x - 3)$
12. $x(y + 2z)$ 13. $y^2(y + 18)(y - 1)$

14. $(y - 7)(y + 1)$
15. $(z + 9)(z + 9)$
16. $(x + 8)(x + 2)$
17. $4(a^2 + x^2)$
18. $(x - 9)(x - 8)$
19. $(x - 9)(x - 9)$
20. $(x + 15)(x + 4)$
21. $(y - 1)(y - 1)$
22. $(x - 7)(x + 4)$
23. $(x - 12)(x + 8)$
24. $(x + 11)(x - 10)$
25. $(x + 7)(x - 7)$
26. $(4y + 3x)(4y - 3x)$
27. $(x + 12)(x - 12)$
28. $(5x + 9y)(5x - 9y)$
29. $4(x - 13)(x + 7)$
30. $5(x + 12)(x - 13)$
31. $(2x + 7)(x + 2)$
32. $(4x - 1)(3x - 4)$
33. $(6x + 5)(5x - 3)$
34. $(12x - 1)(x + 12)$
35. $2(2x - 1)(x - 1)$
36. $(6x + 7y)(6x - 7y)$
37. $2(7x + 3)(2x + 5)$
38. $3(5x - 7)(2x + 1)$
39. $4x(x + 1)(x - 1)$
40. $25(y + 2)(y - 2)$

Chapter 11

11.1 Pages 375—376

1. $-4, 3$
3. $-5, 4$
5. $2, -1$
7. $1, -1$
9. $2, -2$
11. $-3, -2$
13. $7, -3$
15. $10, -4$
17. $0, 9$
19. $12, -9$
21. $-8, 2$
23. $-\frac{2}{5}, -\frac{5}{2}$
25. $\frac{5}{2}, -\frac{5}{2}$
27. $\frac{4}{3}$
29. $0, -3$
31. $\frac{1}{2}x(x + 1) = 66$; base is 12 m; height is 11 m

11.2 Pages 379—381

1. $a = 1, b = -7, c = 4$
3. $a = 3, b = 4, c = 9$
5. $a = -3, b = 4, c = 7$
7. $a = 3, b = 0, c = -14$
9. $-3, 2$
11. $-9, 1$
13. $0, -\frac{2}{5}$
15. $\frac{5}{4}, -\frac{7}{12}$
17. $1.35, -1.85$
19. $0, \frac{5}{3}$ or $0, 1\frac{2}{3}$
21. $-1, \frac{3}{2}$ or $-1, 1\frac{1}{2}$
23. $-1.38, -0.121$
25. $-\frac{1}{4}, -1$
27. $2.38, -2.38$
29. $15.5, -0.453$
31. $4.15, -2.49$
33. $0.206, -2.40$
35. (a) 4 s; 8 s   (b) 1.42 s; 10.6 s   (c) 16 s
37. Length 7 ft, width 3 ft
39. 18 inches on a side   41. 2 ft

11.3 Page 385

1.
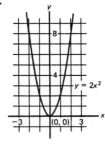

3.
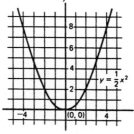

5.
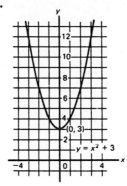

7.
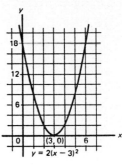

9.
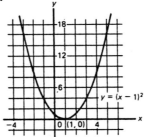

11.
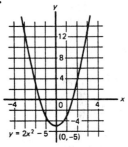

13.
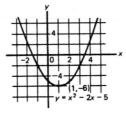

15.
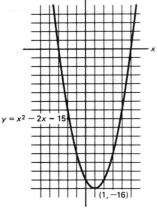

17.
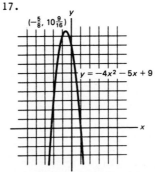

19.
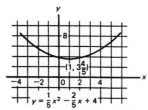

<u>11.4 Pages 388—389</u>

1. 7*j*   3. 3.74*j*   5. 1.41*j*   7. 7.48*j*
9. 13*j*   11. 5.20*j*   13. -*j*   15. *j*
17. -*j*   19. 1   21. -1   23. -*j*
25. Two rational roots   27. Two imaginary roots
29. Two irrational roots   31. Two imaginary roots
33. Two imaginary roots   35. 3 + *j*, 3 - *j*
37. 7 + 2*j*, 7 - 2*j*   39. -4 + 5*j*, -4 - 5*j*
41. -0.417 + 1.08*j*, -0.417 - 1.08*j*
43. 1 + 1.15*j*, 1 - 1.15*j*
45. -0.8 + 0.4*j*, -0.8 - 0.4*j*
47. 0.2, -3   49. -0.5 + 0.866*j*, -0.5 - 0.866*j*

<u>Chapter 11 Review, Pages 389—390</u>

1. *a* = 0 or *b* = 0   2. 0, 2   3. 2, -2

4. 3, -2   5. 0, $\frac{6}{5}$ 6. 7, -4   7. 9, 5

8. 6, -3   9. -$\frac{8}{3}$, -4   10. 6, -$\frac{2}{3}$

11. 1.79, -2.79   12. $\frac{5}{2}$, -3

13. 4.45, -0.449   14. 1.47, -1.14
15. Length 9 ft, width 4 ft
16. (a) 4 μs, 8 μs   (b) 6 μs   (c) 2.84 μs, 9.16 μs
17.   18.

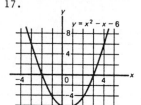
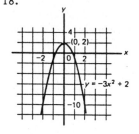

19. 6*j*   20. 8.54*j*   21. 1   22. -*j*
23. One rational root   24. Two imaginary roots
25. 2 + *j*, 2 - *j*   26. 0.6 + 0.663*j*, 0.6 - 0.663*j*

<u>Chapter 12</u>

<u>12.1 Pages 398—402</u>

1. 47°, acute   3. 90°, right
5. 41°, acute   7. 156°, obtuse
9. Right, perpendicular
11. (a) 1, 2; 2, 4; 3, 4; 1, 3;
         5, 6; 6, 8; 7, 8; 5, 7
     (b) 1, 4; 2, 3; 5, 8; 6, 7
13. ∠2 = 57°; ∠3 = 57°; ∠4 = 123°
15. ∠COB = 61°   17. 52°
19. (a) Yes; (b) Yes   21. 90°   23. 15°
25. 108°
27. ∠1 = 58°; ∠2 = 75°; ∠3 = 105°
29. Triangle   31. Hexagon
33. Quadrilateral   35. Heptagon

<u>12.2 Pages 405—408</u>

1. *P* = 60.0 cm; *A* = 225 cm²
3. *P* = 32.0 m; *A* = 48 m²   5. *P* = 45.0 m; *A* = 78 m²
7. *P* = 89.4 in.; *A* = 411 in²
9. *P* = 36.8 cm; *A* = 85 cm²
11. 24.0 cm   13. 11.6 mm   15. 5.87 ft   17. 13.4 ft
19. 8.14 m   21. 38.27 cm
23. 45.556 km   25. 9   27. 36 ft

29. 6 pieces   31. $178,500
33. (a) 208 in²   (b) 590 in²
35. 400,000 acres   37. 672 ft²
39. $927.35   41. 250 ft²   43. 568.6 ft

<u>12.3 Pages 416—422</u>

1. 10 m   3. 25 m   5. 17 m   7. 7.0 m
9. 91.8 m   11. 13.5 cm   13. 1460 km   15. 350 ft
17. 19.7 in.   19. 12 in.   21. 2.83 in.
23. 16.6 in.   25. 116 V   27. 21 A
29. 127 Ω   31. 2.6 Ω   33. 27 cm²   35. 3070 m²
37. 48.6 dm²   39. 18.0 cm; 12 cm²
41. 24.0 cm; 24 cm²   43. 6.8 m; 2.0 m²
45. 63.0 m; 170 m²   47. 52 ft²   49. 36 in²
51. 8.8 ft²   53. 28°   55. 60°
57. *x* = 36°, *y* = 72°

<u>12.4 Pages 424—426</u>

1. 5
3. Yes, all angles of ΔABO are equal to corresponding
     angles of ΔDCO.

5. 5$\frac{1}{3}$ 7. 120 ft   9. *DE* = 12.0 m

11. *A* = 1.33 ft; *B* = 2.67 ft; *C* = 4.00 ft;
     D = 5.33 ft; *E* = 6.67 ft; cross-piece is 10.0 ft

<u>12.5 Pages 431—436</u>

1. (a) 31.4 in.; (b) 78.5 in²
3. (a) 28.9 mm; (b) 66.6 mm²
5. (a) 352 mi; (b) 9890 mi²
7. 171°   9. 43.4°   11. 3.00 cm   13. 2.60 in.
15. 9.99 m   17. 32.1 ft   19. 90°   21. 144°
23. 25%   25. 10.4%   27. 226°   29. 2670 ft
31. (a) 12.6 ft²   (b) 563.4 ft²
33. 11 pipes   35. 1.33 in²
37. (a) 2.50 ft; (b) 3.54 ft
39. 72°   41. 29°   43. 30°
45. 18,600 mi   47. 171°   49. 14°

<u>12.6 Pages 439—440</u>

1. 180   3. 0.37   5. 60°

7. $\frac{3\pi}{4}$ rad; 2.36 rad   9. $\frac{4\pi}{3}$ rad; 4.19 rad

11. $\frac{4\pi}{5}$ rad; 2.51 rad   13. 31.4 cm   15. 4.7 cm

17. 104 cm   19. 2.80   21. 98.5   23. 1.59 rps
25. 92.2 cm   27. 25.8 cm

<u>12.7 Pages 444—445</u>

1. (a) 440 in²; (b) 662 in²; (c) 912 in³
3. (a) 2490 m²; (b) 1230 m²; (c) 188 m³; (d) 4900 m²
5. 324 ft³   7. 30 in.

<u>12.8 Pages 448—451</u>

1. 13,600 mm³   3. 8,280,000 L
5. 42.6 ft   7. 14 : 1
9. 1180 ft³   11. 1.79 in.
13. 2050 mm²   15. 6370 cm²
17. 101,000 cm²   19. 18.0 in³
21. 1.38 in²   23. 6.1 gal   25. 25 kg

12.9 Pages 454—457

1. 150 in³   3. 11̄00 m³   5. 69,700 mm³
7. 180 m̄m³   9. 312 ft³   11. 1010 cm³
13. 1380 cm²   15. 1520 bu
17. 92,300 lb   19. 1690 cm³
21. 128 lb

12.10 Page 458

1. (a) 804 m²   (b) 2140 m³
3. (a) 824 in²   (b) 2230 in³
5. 114,000 m³   7. 933,000 gal

Chapter 12 Review, Pages 459—462

1. 72°   2. 121°
3. ∠1 = 59°; ∠2 = 121°; ∠3 = 59°; ∠4 = 59°
4. Adjacent   5. 20
6. (a) Quadrilateral   (b) Pentagon
   (c) Hexagon   (d) Triangle
   (e) Octagon
7. P = 36.0 cm; A = 60.0 cm²
8. P = 360.4 m; A = 80̄00 m²
9. P = 43.11 cm; A = 102 cm²
10. 7.73 m   11. 1.44 cm
12. A = 159 m²; P = 58.7 m
13. A = 796 m²; P = 159.5 m
14. A = 4.73 cm²; P = 10.93 cm
15. 41.9 m   16. 3.85 cm   17. 13.8 in. 18. 86°
19. 8.0 m   20. A = 423 cm²; C = 72.9 cm
21. 12.1 cm 22. 144°   23. 42%   24. 76°
25. $\frac{2\pi}{15}$ rad   26. 10°   27. 42.2 cm 28. 1.55 rad
29. 291 cm   30. 1792 m²   31. 2152 m² 32. 5760 m³
33. 1260 cm³   34. 70.4 cm³
35. 70.4 cm²   36. 95.5 cm²
37. 131 m³   38. 57,300 m³
39. 6370 m² 40. 869 m³ 41. 44̄0 m²

Chapter 13

13.1 Pages 468—470

1. a   3. c   5. a   7. B
9 B   11. 60.0 m   13. 55.2 mi 15. 17.7 cm
17. 265 ft 19. 35̄0 m   21. 1530 km 23. 59,900 m
25. 0.5000 27. 0.5741   29. 0.5000   31. 0.7615
33. 2.174   35. 0.8686   37. 0.5246   39. 0.2536
41. 0.05328 43. 0.6104   45. 0.3929   47. 0.6800
49. 52.6°   51. 62.6°   53. 54.2°   55. 9.2°
57. 30.8°   59. 40.8°   61. 10.98°   63. 69.07°
65. 8.66°   67. 75.28°   69. 43.02°   71. 20.13°
73. sin A and cos B; cos A and sin B

13.2 Page 473

1. A = 35.3°; B = 54.7°   3. A = 42.3°; B = 47.7°
5. A = 43.3° B = 46.7°   7. A = 45.0°; B = 45.0°
9. A = 64.05°; B = 25.95°
11. A = 52.60°; B = 37.40°
13. A = 39.84°; B = 50.16°
15. A = 41.85°; B = 48.15°
17. A = 11.8°; B = 78.2°
19. A = 31.11°; B = 58.89°
21. A = 15.9°; B = 74.1°
23. A = 64.65°; B = 25.35

13.3 Page 476

1. b = 40.6 m; c = 54.7 m
3. b = 278 km; c = 365 km
5. a = 29.3 cm; b = 39.4 cm
7. a = 10.4 cm; c = 25.9 cm
9. a = 6690 km; c = 30,̄000 km
11. b = 54.92 m; c = 58.36 m
13. a = 161.7 mi; b = 197.9 mi
15. a = 4564 m; c = 13,170 m
17. c = 6270 m; b = 5950 m
19. a = 288.3 km; c = 889.6 km
21. a = 88.70 m; b = 163.0 m
23. a = 111 cm; b = 231 cm

13.4 Page 479

1. B = 39.4°; a = 37.9 m; b = 31.1 m
3. A = 48.8°; b = 234 ft; c = 355 ft
5. A = 22.6°; B = 67.4°; a = 30.0 mi
7. B = 37.9°; b = 56.1 mm; c = 91.2 mm
9. B = 21.2°; a = 36.7 m; b = 14.2 m
11. A = 26.05°; B = 63.95°; c = 27.33 m
13. B = 60.81°; a = 1451 ft; b = 2597 ft
15. A = 67.60°; B = 22.40°; c = 50.53 m
17. B = 48.9°; a = 319 m; b = 365 m
19. A = 80.55°; b = 263.8 ft; c = 1607 ft
21. A = 44.90°; B = 45.10°; a = 268.6 m
23. A = 8.5°; a = 14̄00 ft; c = 9470 ft

13.5 Pages 481—487

1. 18.8 m   3. 5.2 m   5. A = 58.2°; 32.3 ft
7. 5.5°   9. 92.1 ft   11. 5.70 cm 13. 95.3 m
15. (a) 430 Ω; (b) 38°; (c) 37°; 3̄00 Ω
17. (a) 53.7 V; (b) 267 V; (c) 236 V
19. 9.5°   21. (a) 3.46 in.; (b) 2.04 in.
23. x = 22.8 ft, A = 55.2°

13.6 Pages 493—494

1. 0.6820   3. -0.4067   5. -0.4352   7. -0.9724
9. 0.7701   11. -0.4390

13.
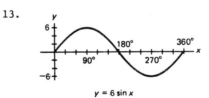
y = 6 sin x

15.
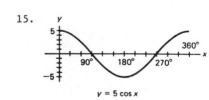
y = 5 cos x

17.
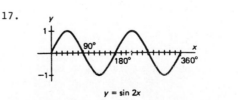
y = sin 2x

19.

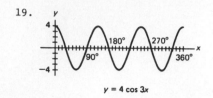

$y = 4 \cos 3x$

21.

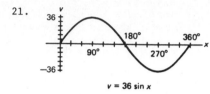

$v = 36 \sin x$

23.

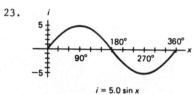

$i = 5.0 \sin x$

25. 25 V, -33 V   27. 3.5 A, -3.5 A
29. 8.8×10^9 Hz or 8800 MHz

13.7 Pages 497—499

1. $c = 24.9$ m; $B = 41.5°$; $C = 70.5°$
3. $a = 231$ ft; $c = 179$ ft; $C = 42.9°$
5. $C = 130.3°$; $a = 14.4$ cm; $b = 32.9$ cm
7. $B = 82.9°$; $b = 32.4$ m; $c = 22.4$ m
9. $b = 79.3$ mi; $B = 145°$; $C = 14.9°$
11. $A = 74.9°$; $a = 6910$ m; $c = 5380$ m
13. $a = 26.4$ km; $A = 54.2°$; $C = 20.3°$
15. $c = 38\overline{0}$ ft; $B = 15.1°$; $C = 148.4°$
17. 63.5 m   19. 7.29 km   21. 214 ft

13.8 Pages 506—507

1. (a) One solution
   (b) $B = 28.2°$; $C = 113.8°$; $c = 62.9$ m
3. (a) Two solutions
   (b) $B = 28.7°$; $C = 125.7°$; $c = 517$ m
      $B = 151.3°$; $C = 3.1°$; $c = 34.4$ m
5. No solution
7. (a) Two solutions
   (b) $A = 77.0°$; $B = 31.8°$; $b = 132$ cm
      $A = 103.0°$; $B = 5.8°$; $b = 25.4$ cm
9. (a) One solution
   (b) $A = 44.6°$; $C = 30.4°$; $c = 17.3$ mi
11. (a) Two solutions
   (b) $C = 34.3°$; $B = 114.2°$; $b = 656$ m
      $C = 145.7°$; $B = 2.8°$; $b = 35.2$ m
13. No solution
15. (a) Two solutions
   (b) $C = 15.3°$; $A = 156.7°$; $a = 1280$ m
      $C = 164.7°$; $A = 7.3°$; $a = 412$ m
17. 496 ft or 78.5 ft

13.9 Pages 512—514

1. $a = 21.0$ m; $C = 69.2°$; $B = 55.8°$
3. $c = 476$ ft; $B = 37.0°$; $A = 28.0°$
5. $A = 49.6°$; $B = 60.2°$; $C = 70.2°$
7. $c = 6.72$ km; $B = 58.4°$; $A = 50.0°$
9. $A = 33.7°$; $B = 103.5°$; $C = 42.8°$
11. $A = 148.7°$; $C = 12.0°$; $b = 3070$ ft
13. $A = 30.7°$; $B = 107.3°$; $C = 42.0°$
15. 65.3 m   17. $A = 40.9°$; $C = 30.5°$
19. (a) 59.5°   (b) 70.5°   (c) 4.92 m   (d) 6.33 m
21. 114.3°

13.10 Pages 516—517

1. 1.155	3. 1.000	5. 2.790	7. 1.453
9. 1.140	11. 2.519	13. 0.2263	15. 1.881
17. 1.299	19. 35.1°	21. 52.5°	23. 24.6°
25. 16.2°	27. 82.8°	29. 58.3°	31. 60.50°
33. 31.59°	35. 21.76°		

Chapter 13 Review, Pages 517—520

1. 29.7 m   2. B 3. Hypotenuse
4. 43.8 m   5. sin A
6. Length of side adjacent to angle A
7. $\dfrac{\text{length of side opposite angle } B}{\text{length of side adjacent to angle } B} = \dfrac{32.2 \text{ m}}{29.7 \text{ m}}$
8. 0.8070   9. 1.138   10. 0.4023   11. 45.5°
12. 10.4°   13. 65.8°   14. 39.2°   15. 50.8°
16. 7.17 km   17. 8.25 km
18. $b = 21.9$ m; $a = 18.6$ m; $A = 40.4°$
19. $b = 102$ ft; $c = 119$ ft; $B = 58.8°$
20. $A = 30.0°$; $B = 60.0°$; $b = 118$ mi
21. 4950 m   22. -0.7536   23. -0.4970   24. 0.7361
25.

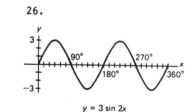

$y = 10 \cos x$

26.

$y = 3 \sin 2x$

27. $c = 258$ m; $A = 42.5°$; $C = 84.8°$
28. $B = 43.2°$; $b = 62.2$ cm; $a = 79.6$ cm
29. $b = 129$ m; $C = 109.0°$; $A = 53.5°$
30. $A = 66.8°$; $B = 36.2°$; $C = 77.0°$
31. $c = 33.3$ m; $B = 33.1°$; $C = 117.4°$
    or $B = 146.9°$; $C = 3.6°$; $c = 2.36$ m
32. $c = 3010$ m; $A = 20.5°$; $C = 141.0°$
    or $c = 167$ m; $A = 159.5°$; $C = 2.0°$
33. $A = 27.3°$; $B = 59.7°$; $C = 93.0°$
34. No solution   35. 36.6°
36. 52.9 m   37. 9.40 in.
38. (a) 23.5 m   (b) 16.7 m   (c) 15.1 m   (d) 7.06 m
39. 1.227   40. 0.2272   41. 1.104
42. 60.6°   43. 61.4°   44. 41.6°

Chapter 14

14.1 Pages 522—524

1. $1800   3. $780   5. $850   7. $1650
9. $1500   11. 1.8 million
13. United States   15. 4.5 million
17. 15 million   19. 2.3 million
21. 310   23. 25   25. 15   27. 190
29. 10

14.2 Pages 526—528

1. 94°   3. 55°   5. 270°   7. 25°
9. 15°   11. 328°

13.

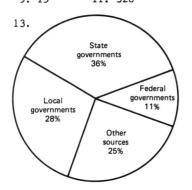

Expenditures for education, 1980

15.

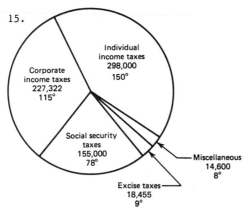

Federal budget receipts, 1980
(millions of dollars)

17.

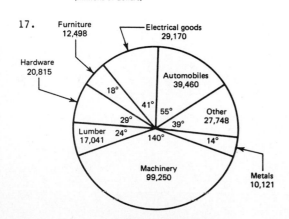

Wholesale sales, durable goods, 1982
(millions of dollars)

14.3 Pages 529—532

1.

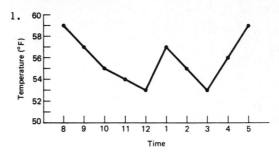

3.

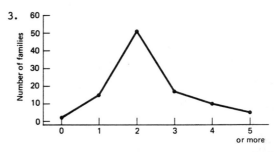

5.

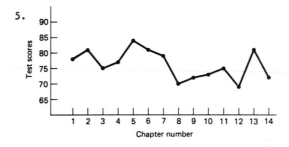

7.

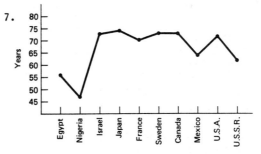

9. 29.2 in.   11. 29.09 in.
13. 29.18 in.   15. 66°
17. 66° at 4:00 **a.m.**   19. 68% at 2:30 a.m.

14.4 Pages 533—534

1. 900 mW; 14 db   3. 80 mW
5. 30 db and 32 db   7. 16 db
9. 14 db and 16 db

14.5 Pages 535—536

1. 47.89 cm   3. 0.2619 in.
5. 25,770 mi   7. 2036 km
9. 709 lb   11. 6911

14.6 Page 538

1. 47.87 cm   3. 0.2618 in.
5. 25,770 mi   7. 2025 km
9. 781 lb   11. 6185.5
13. 163   15. 107   17. 23

14.7 Pages 543—545

1. 59   3. 7.7 months
5. 35.9 hours   7. 46.6 lb
9. 11 defective parts

14.8 Page 549

1. 0.23 cm   3. 0.0003 in.
5. 30 mi   7. 33 km 9. 240 lb   11. 1848
13. 8.75   15. 3.82 months
17. 5.2 hours   19. 16.6 lb
21. 3.6 defective parts

Chapter 14 Review, Page 549

1. 126°   2. 202°

3.   4.

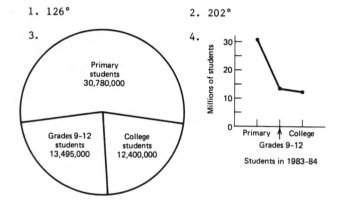

Students in 1983-84

5. 74°F   6. 7.0036 mm
7. 7.0036 mm   8. 0.0001 mm
9. (a) 49.9 (b) 18.2

Appendix A

A.2 Page 553

1. 1823   3. $215.39   5. 181.466   7. 204.4
9. 8.9298   11. 972   13. 12,861   15. 41,392
17. 1115   19. 21.447   21. 11   23. -4.1
25. 39.88   27. 51.49

A.3 Page 556

1. 261,414   3. 0.010024   5. 1.77876   7. 117.096
9. 1180.728   11. 75   13. 32.7
15. 0.000367   17. 0.00911   19. 10,200
21. 245   23. 0.873

A.4 Page 558

1. 0.125   3. 0.0370   5. 4   7. 0.0546
9. 2.77   11. 324   13. 973   15. 14,400
17. 0.00740 19. 0.164   21. 56   23. 6.24
25. 0.203   27. 0.539   29. 444

A.5 Pages 561—562

1. 29.4   3. 1270   5. 7.67   7. 4.37
9. 3,040,000   11. 63.2   13. 3610
15. 332   17. 147   19. 0.826

INDEX